OSPF and IS-IS

From Link State Routing
Principles to Technologies

OSPF and IS-IS

From Link State Routing Principles to Technologies

Rui Valadas

CRC Press
Taylor & Francis Group
Boca Raton London New York

CRC Press is an imprint of the
Taylor & Francis Group, an **informa** business

CRC Press
Taylor & Francis Group
6000 Broken Sound Parkway NW, Suite 300
Boca Raton, FL 33487-2742

© 2019 by Taylor & Francis Group, LLC
CRC Press is an imprint of Taylor & Francis Group, an Informa business

No claim to original U.S. Government works

Printed on acid-free paper
Version Date: 20181122

International Standard Book Number-13: 978-1-138-50455-4 (Hardback)

Visit the Taylor & Francis Web site at
http://www.taylorandfrancis.com

and the CRC Press Web site at
http://www.crcpress.com

to Matilde and Teresa

Contents

Preface

Link State Routing (LSR) is a networking topic that has deserved much attention in scientific papers, standards, and textbooks. There are voluminous textbooks dedicated to specific LSR technologies, such as OSPF (Open Shortest Path First) and IS-IS (Intermediate System to Intermediate System) [11, 15, 31, 35, 37, 50]. We have learned a lot from these books, but they are mostly targeted to the explanation of the many technological details of OSPF and IS-IS—certainly something that network managers need to be aware of.

The subject of LSR is often considered complex. However, we believe that it is much simpler than it seems and that the key to its comprehension is the correct segregation of the fundamental mechanisms and data structures of LSR protocols, i.e. the *principles* that guide their operation. Once these principles are mastered, it is much easier to grasp the specific technologies and navigate in their otherwise dense specifications, as well as be critical about some of the options taken.

The book aims at being a *complete* reference on OSPF and IS-IS, from the methodological point of view. We ground the learning process on the principles of LSR, explaining the mechanisms and data structures that are fundamental to its operation and common to the various technologies. Then, we describe OSPF and IS-IS in great detail, covering both their IPv4 and IPv6 versions. The level of detail is enough for specialists needing to configure, manage, and plan networks supported on these technologies. Finally, we complement the book with the discussion of a large set of *experiments*, illustrating the many features of OSPF and IS-IS. Our experiments cover not only the static view of networks—while they remain stable—, but also their dynamics—while converging following some perturbation (e.g. a network configuration or a failure). However, we are not exhaustive in covering the many extensions of OSPF and IS-IS, since new ones keep appearing at a fast rate and are not changing the fundamental nature of the technologies.

Understanding the principles behind the technologies makes the learning process easier and more solid. Moreover, it helps uncovering the dissimilarities and commonalities of OSPF and IS-IS, and exposing their stronger and weaker features. The experiments addressed in the book help in acquiring the ability to *troubleshoot* networks, an important skill required from computer networking professionals. This skill is also more effectively achieved when the understanding of the LSR subject goes beyond the specificities of particular technologies.

Interestingly, the Routing Area of the Internet Engineering Task Force

(IETF), which defines Internet standards for routing protocols, decided to merge the OSPF and IS-IS work groups into a single group called Link State Routing. This is very much in line with the spirit of this book.

Overview of Link State Routing

In LSR protocols, routers exchange routing information so that each router individually builds and maintains a *global view* of the network. This global view includes two aspects: (i) the topological information (or network map), i.e. a description of the routers and links between routers, and (ii) the addressing information, i.e. the address prefixes (IPv4 or IPv6) assigned to the routers and links. The global view is obtained in a *distributed* way: each router disseminates to all others information on its *local view* of the network, describing itself and its links (the local topological information), and the address prefixes assigned to itself and its links (the local addressing information). Then, each router autonomously joins together the pieces of routing information received from all the other routers, as well as its own, to construct the global network view, which it stores at a local database called the *Link State Database* (LSDB). The forwarding tables are obtained from this global view by running, at each router, a centralized shortest path algorithm, usually Dijkstra's algorithm. Much of the effort in LSR protocols is to make sure that the global view remains the same across all routers, to avoid inconsistencies in the forwarding tables. This is the reason why these protocols are sometimes called *replicated distributed database* protocols [35], a designation that is certainly insightful. The mechanisms required to maintain the consistent global view of the network are usually called *synchronization* mechanisms.

The need to store the complete network view raises *scalability* concerns, since routers may lack memory resources when networks become large. To address this issue, LSR protocols include the possibility of structuring a network in multiple smaller subnetworks called *areas*, such that a router only needs to keep the complete network view of its own area. However, this possibility requires several enhancements to the basic LSR protocol, e.g. the need to introduce an *inter-area* routing protocol.

OSPF and IS-IS are presently the main technologies implementing LSR. They are supported by virtually all vendors of networking equipment and are widely deployed by network operators worldwide. They form an important part of the *skeleton* of the current Internet and are here to stay. Their specifications keep constantly being updated and enhanced by the IETF.

Both OSPF and IS-IS include a restriction on the way multi-area networks can be structured. Specifically, the network topology is restricted to a two-level hierarchy, with only one area in the upper level, and where the communication between lower level areas can only be through the upper one. Due to its hierarchical structure, the multi-area networks OSPF and IS-IS are usually referred to as *hierarchical* networks.

Structure of the book

The book is structured in four parts: Introduction (Part I), Principles (Part II), Technologies (Part III), and Case Studies (Part IV).

- **Introduction** - The Introduction includes Chapter 1, which reviews the basic notions of routing and addressing. This part can be skipped by readers familiar with TCP/IP networking. Nevertheless, you may find here less traditional approaches to some subjects. Give it a look!

- **Principles** - The Principles part includes two chapters addressing the principles of LSR. Chapter 2 concentrates on single-area networks and Chapter 3 on multi-area networks. These chapters are, probably, the main added value of the book.

- **Technologies** - The Technologies part describes the IPv4 and IPv6 versions of OSPF and IS-IS. Chapter 4 provides a brief overview of the technologies. The next chapters discuss the LSDB structure of single-area networks (Chapter 5), the distributed database synchronization mechanisms (Chapter 6), and the extensions required for the support of hierarchical networks (Chapter 7).

- **Case studies** - The Case Studies part presents and discusses several experiments designed to illustrate the various features of OSPF and IS-IS. Chapter 8 introduces the tools used in the experiments and explains their basic configurations. The next chapters discuss experiments related to the forwarding tables and LSDB structure (Chapter 9), the synchronization mechanisms (Chapter 10), and the hierarchical networks (Chapter 11).

We include at the end of the book a glossary explaining the main concepts and the abbreviations.

How to read the book

This book can be read in several different ways. As mentioned above, you can skip Chapter 1 if you are familiar with the basics of routing and addressing. You can also read first the chapters concerned with single-area networks, and only subsequently the ones related with multi-area networks. This means reading first Chapter 2, on principles, and Chapters 4, 5, and 6, on the LSDB structure and synchronization mechanisms. Finally, if you have already had experience with OSPF and IS-IS, you can go directly to the Case Studies part, where you may find some interesting experiments. In particular, the experiments related to the synchronization mechanisms are not usually found in the literature.

A word on nomenclature

A word is due on nomenclature. One of the most difficult barriers when trying to compare OSPF and IS-IS is the names given to the various protocol elements, which rarely coincide. We could have adopted one of the terminologies when discussing the principles of LSR. However, we decided to create a terminology of our own, to avoid favoring one technology over the other and to make the nomenclature as expressive as possible. We provide the correspondences between the various nomenclatures throughout the book. When needed, our nomenclature will be referred to as the *generic* one.

Audience

The book should be useful to teachers, students, and computer networking professionals. It provides a path to learn and teach efficiently the subject of LSR, which raises attention to its key aspects. In addition, we provide a comprehensive description of OSPF and IS-IS, the main LSR technologies, with sufficient detail to support professionals who need to configure, manage, and plan these type of networks. The book also includes a large set of experiments that (i) can be easily reproduced using a standard personal computer and (ii) can support the learning process either in the classroom or by individual students trying to master OSPF and IS-IS.

Supplements

Files containing the configurations of the experiments and the Wireshark captures are available at the CRC book website. Please visit https://www.crcpress.com/9781138504554.

We would like to hear from you

Your comments, suggestions, and corrections are very welcome. Please send any you may have to rui.valadas@tecnico.ulisboa.pt.

Acknowledgments

First, I would like to thank my friend and colleague José Brázio, from Instituto de Telecomunicações. Your contribution to this book was invaluable. Thanks for the care in reviewing the many versions of the manuscript, for the acuteness of your comments and suggestions, and for never letting me freeze on a bad idea. This book is certainly much better due to your help! I would also like to thank João Sobrinho from Instituto de Telecomunicações for reviewing early versions of the manuscript.

I also acknowledge the support of Instituto Superior Técnico from the

University of Lisbon (my University), and Instituto de Telecomunicações (my Research Institute).

Finally, I would like to thank my wife Rosário for her love and support: without you I could not have completed this dream.

Part I

Introduction

1

Routing and Addressing

1.1 The physical elements of computer networks

Computer networks include today a myriad of different physical elements, performing different functions. Figure 1.1 shows a relatively simple, yet rich, example. From the point of view of routing, we need to consider three types of physical elements: end devices, switching equipment, and communication media.

The end devices and switching equipment attach to the communication media through *interfaces*. Interfaces have distinct hardware to transmit and receive traffic; the *ingoing interface* is the part of the interface that receives traffic and the *outgoing interface* is the one that transmits traffic.

End devices, usually called *hosts*, are the sources and sinks of information. In today's Internet, they range from large servers to tiny sensors and actuators. The figure shows four types of hosts: servers, desktop computers, laptops, and tablets.

The information exchanged between hosts is carried in chunks of bits called *packets*. Packets are *routed* from source to destination through the networking infrastructure, which is formed by a mesh of switching equipment and physical communication media. The *switching equipment* transfers packets from ingoing to outgoing interfaces according to *forwarding tables*, and using the *store-and-forward* mode of operation. Store-and-forward operation means that a packet only starts being transmitted by the outgoing interface, after being completely received by the ingoing interface. Switching equipment are usually differentiated based on the type of addressing information used to make forwarding decisions. The figure shows two types of switching equipment: IP routers and Ethernet switches.

The *physical communication media* provide the physical connections between equipment (hosts or switching equipment), and include optical fibers, coaxial and twisted pair cables, as well as various types of wireless media.

The technologies covered in this book address the problem of "routing between IP routers". As will be discussed later, in this case, the connections between neighboring routers can be hidden from their details, i.e. the specific structure of these connections, whether being single physical links or networks of Ethernet switches, is of little concern.

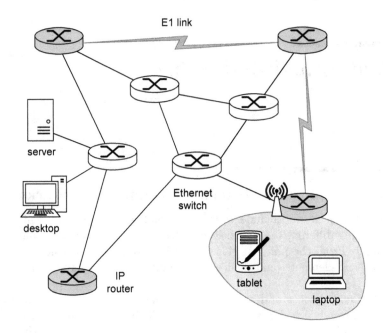

FIGURE 1.1: The physical elements of computer networks.

> *From the point of view of routing, computer networks comprise three types of physical elements: end devices (hosts), switching equipment, and communication media.*

1.2   Names, addresses, and paths

Suppose you want to obtain a service from the network, e.g. a search for some Web page. What network entities need to be named for this purpose, and what types of associations need to be established between these names, is an important issue and, by no means, a simple one. We will present here Saltzer's view on this problem [47], which is based on previous work by Shoch [49]; Day's work is also a significant contribution to the understanding of this problem [10]. We highlight that this is a conceptual model, not tied to the current Internet architecture.

Names, name spaces, and name scopes   A network entity is referenced through a *name*. Names can be characterized by their space and scope. The *space* is the set of names from which all the names for a given collection of entities are taken. The *scope* is the set of entities to which the name applies.

Obtaining a network service   The entities involved in obtaining a service

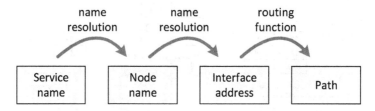

FIGURE 1.2: Saltzer's view on how to obtain a network service.

from the network are (i) services (or applications), (ii) nodes (hosts or switching equipment), (iii) network interfaces (called attachment points in [47]), and paths. The names given to network interfaces are usually called *addresses*, and we will keep this nomenclature. This process will be explained with the help of Figure 1.2.

- A service may run at one or more nodes and may need to move between nodes without losing its identity as a service. There must be some way of finding the node that runs the service. This can be provided through a *name resolution service*, binding the service name to the node name.

- A node may connect to one or more network interfaces (or attachment points) and may need to move from one interface to another without losing its identity as a node. For example, a user with a laptop computer may roam across a network and attach to different points of the network. There must be some way of finding the network interface where the node is connected and, again, this can be provided through a name resolution service, now binding the node name to the interface address.

- Finally, a pair of network interfaces may be connected by one or more paths, and these paths may need to change over time. Finding paths within the network is the role of the *routing function*.

Day later noted that the node should be thought of as a logical entity, actually the endpoint of all paths of all application processes, and that interface addresses have only local significance [10].

Routing function The goal of the routing function is then to find paths within the network. A path is a sequence of node interfaces (interfaces of hosts or switching equipment) leading to some node. Note that, due to the possibility of parallel links between nodes, a path is not fully described by the node names; the interface addresses are also required. However, the interface addresses have only local significance, i.e. they need only to be unique within a node. Routing is implemented locally at each node using forwarding tables, which indicate the next interface in the path to each destination (the next-hop interface).

Example Figure 1.3 provides an example. The network includes four routers

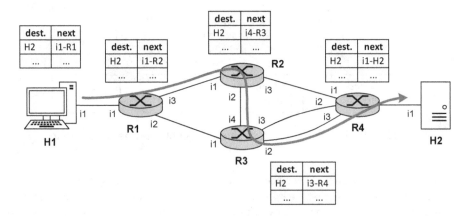

FIGURE 1.3: Paths and forwarding tables.

and two hosts. The names of hosts, routers, and interfaces are taken from separate name spaces, prefixed by the letters "H", "R", and "i", respectively. There are two parallel links between R3 and R4. Suppose that H1 needs to contact a service running on H2. Given the service name, the host name (H2) and the interface name (i1-H2) are found using a name resolution process, and the path is determined through the routing function. The path from H1 to H2 indicated in the figure is described as the sequence i1-R1 → i1-R2 → i4-R3 → i3-R4 → i1-H2. The forwarding tables indicate the next interface in the path towards H2. For example, the forwarding table of R3 indicates that the next interface in the path to H2 is i3-R4.

As simple as it may be, this model has been a subject of controversy and is not fully implemented in the current Internet architecture [10, 16, 47]. In fact, the Internet has no node names, and some authors even argue that it has only interface addresses and paths. Actually, interfaces are named using the well-known *IP addresses* (and sometimes using also MAC addresses).

> *What network entities need to be named for routing purposes, and what types of associations need to be established between these names, is an important issue subject to controversy. In principle, the services, the nodes (hosts or switching equipment), and the network interfaces should be named; moreover, the paths are sequences of interfaces leading to some node. The routing function determines the paths within the network and is implemented locally at each node using forwarding tables, which indicate the next interface in the path to each destination. In the Internet addressing architectures there are no nodes names and the interfaces are named using IP addresses.*

Distinction between forwarding and routing It is important to make a clear distinction between the terms *forwarding* and *routing*: forwarding refers to the *local* decision at a node of determining the next interface in the path to a destination, while routing refers to the *global* process of finding an end-to-end path from an origin to a destination.

Routing protocols The paths can be computed in a centralized way, i.e. with *a priori* knowledge about the network topology, or in a distributed way, i.e. through *routing protocols*. Routing protocols are distributed algorithms that run on switching equipment and cooperate through the exchange of control messages, called *routing messages* in this context. The hosts do not usually participate in routing protocols. When routing protocols are used, it is their role to build and maintain the router forwarding tables.

> *Routing protocols are distributed algorithms that run on switching equipment to determine paths within the network.*

1.3 Internet addressing architecture

The Internet uses IP addresses to identify, at a worldwide level, the *reachable* network interfaces. Moreover, the application processes running on hosts are identified through the type of transport protocol used in the communication between them and their port numbers. The port numbers have only local significance, except for the well-known port numbers, which identify specific application types running on servers.

The Internet addressing architecture also includes the DNS service for mapping host names and URLs to IP addresses. DNS defines a distributed database that stores these mappings worldwide. A comprehensive discussion of DNS can be found in [3].

IP addressing families There are currently two types of IP addresses: IPv4 addresses and IPv6 addresses. The length of IPv4 addresses is 32 bits and the length of IPv6 addresses is 128 bits. IPv6 addresses were introduced to replace IPv4 addresses, due to the scarcity of the latter. However, the transition has been a slow process, and the two address families will coexist for many years.

Address representation Figure 1.4 illustrates the address representation used in IPv4 and IPv6. IPv4 addresses are represented in *dotted-decimal notation*, i.e. using four decimal numbers separated by a dot, where each number corresponds to the decimal value of one octet, e.g. 128.10.2.30. Due to their length, IPv6 addresses are represented in *hexadecimal notation*, with 16-bit blocks separated by a colon, e.g. 2001:0db8:000d:000a:0000:0000:0000:0003. Several simplifications can be adopted to shorten the address representation, e.g. skip leading zeros in a 16-bit block and replace one group of consecutive zeros by a double colon. With these two simplifications, the previous IPv6 address would be represented as 2001:db8:d:a::3 (see Chapter 3 of [14] for additional discussion).

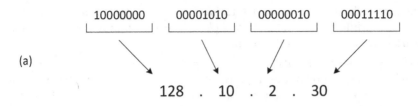

(a)

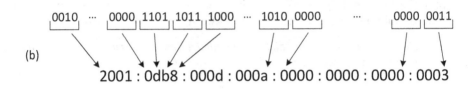

(b)

FIGURE 1.4: Address representation in (a) IPv4 and (b) IPv6.

Representing address blocks Frequently, one needs to refer to blocks of contiguous addresses that share a common prefix, rather than to individual addresses. The *prefix* corresponds to the higher-order bits of the address. The size of the block, i.e. the number of addresses it contains, is determined by the prefix length, with smaller prefixes defining larger blocks. One way to characterize a block in a compact way is to indicate (i) its lowest address (also called base address) and (ii) the prefix length. The usual notation is to follow the address by $/n$, where n is the number of bits in the prefix; this is sometimes called the *slash notation*. The block starts at the lowest address, which has the common prefix and an all-zeros suffix, and ends at the highest possible address with the same prefix and an all-ones suffix. For example, the block 123.4.8.0/24 has a prefix of 24 bits starts at 123.4.8.0, ends at 123.4.8.255, and includes a total of $2^8 = 256$ addresses. With the same lowest address, the block 123.4.8.0/21, starts at 123.4.8.0, ends at 123.4.15.255, and includes $2^{11} = 2048$ addresses. To ease the notation, the trailing decimal zeros of the address can be omitted. For example, the block 123.4.8.0/21 can also be represented as 123.4.8/21. Note that an address block can also be defined by (i) any of the addresses belonging to the block and (ii) the prefix length.

Figure 1.5 illustrates how the lowest and highest addresses of block 123.4.8/21 can be determined. This process is facilitated by using a mixed address representation, where the leading octets characterized by having all bits in the prefix are represented in dotted-decimal notation, and the remaining ones in binary notation. In this case, the first two octets (the two leftmost octets), i.e. 123.4, are represented in decimal notation, and the remaining two octets in binary notation. For example, the highest value of the third octet can be determined by noting that, due to the /21 prefix, the first five bits must be 00001, since they belong to the prefix, and the last three must be

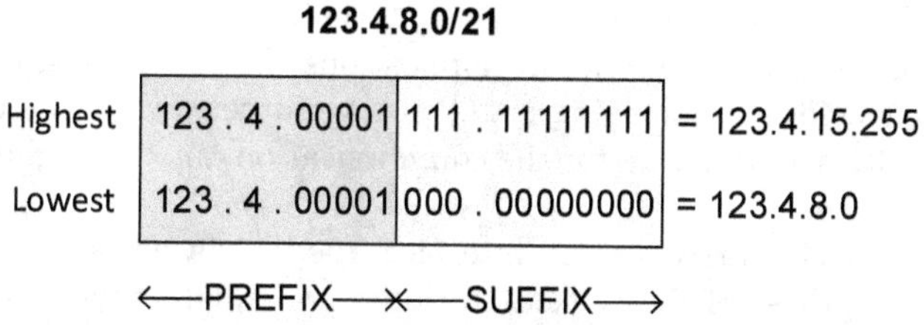

123.4.8.0/21

Highest	123 . 4 . 00001	111 . 11111111	= 123.4.15.255
Lowest	123 . 4 . 00001	000 . 00000000	= 123.4.8.0

←——PREFIX——×——SUFFIX——→

FIGURE 1.5: Determining the lowest and highest addresses of an address block.

111, since they belong to the suffix of the highest address in a block. This leads to octet 00001111, which corresponds to decimal 15.

In IPv4, the prefix length is sometimes represented by a 32-bit word expressed in dotted-decimal notation (as an IPv4 address), where for a prefix with n bits, the n highest-order bits of the word equal 1 and the remaining ones equal 0. For example, in IPv4 the prefix lengths /24 and /21 are also represented as 255.255.255.0 and 255.255.248.0, respectively. For reasons that will become clear soon, these binary words are called *subnet masks*. This type of representation is not used in IPv6.

Reserved address blocks Both the address spaces of IPv4 and IPv6 have address blocks reserved for specific purposes. These blocks are listed in Figure 1.6 and will be described next.

Multicast and broadcast communications An important classification of IP addresses is according to the number of interfaces they identify when used as destination addresses: *unicast* addresses identify a single interface, *multicast* addresses identify a group of interfaces, and *broadcast* addresses identify all interfaces. Multicast and broadcast addresses are used whenever a packet

Address types	IPv4	IPv6
Multicast	224/4	ff00::/8
Private	10/8, 172.16/12, 192.168/16, 169.254/16	fc00::/7
Unicast link-local	-	fe80::/10
Unicast global	all except reserved	2000::/3

FIGURE 1.6: Address types in IPv4 and IPv6.

needs to be sent to several interfaces at the same time. This is required in many situations, for example, in applications that perform group communications (e.g. videoconferencing), and in several networking functions. OSPF uses IP multicast addresses for the communication among neighboring routers on shared links.

When a packet arrives at an interface, the interface first analyzes the destination IP address of the packet. If it does not equal a unicast or a multicast address that has been assigned to the interface, or the broadcast address, the packet is discarded. This procedure alleviates the interface from having to analyze the complete contents of all arriving packets.

Broadcast addresses are constrained in scope: a broadcast address can only target interfaces within a specific routing domain, otherwise the whole Internet would be flooded with this type of packets. In IPv6 there are no broadcast addresses, only multicast. Most multicast addresses are also constrained in scope. Irrespective of scope, multicast and broadcast packets can be prohibited to cross network boundaries through administrative configuration.

The address blocks reserved for multicast are 224/4 (first 4 bits equal 1110) in IPv4 and ff00::/8 in IPv6.

Public versus private addresses IP addresses can also be classified as public or private addresses. *Public* addresses are used for Internet-wide communications; they are globally visible, globally unique, and can appear as destination addresses in any IP packet. *Private* addresses are used for communications inside specific domains and need only be unique inside them. Private addresses play an important role in IPv4 addressing due to the scarcity of public addresses in this addressing family. The address blocks reserved for private addressing in IPv4 are 10/8, 172.16/12, 192.168/16, and 169.254/16. IPv6 calls these addresses *unique local addresses* and reserves them block fc00::/7. However, unlike IPv4, IPv6 private addresses were designed to have a high-probability of being globally unique (see Chapter 3 of [14] for additional details).

IPv6 global and link-local addresses IPv6 introduced an important distinction between two types of unicast addresses: *global* and *link-local* addresses. Global addresses are defined by the prefix 2000::/3 and are used for Internet-wide communications, in the same way as public IPv4 addresses. Link-local addresses are defined by prefix fe80::/10 and are used for communications inside a link. Link-local addresses introduced an important benefit in relation to IPv4: interfaces can create these addresses on their own and use them to communicate on a link without any previous configuration. The structure of these two address types will be discussed in Section 1.4.2.

> *The Internet has currently two IP address families: IPv4 addresses with 32 bits, and IPv6 with 128 bits. Each of these families contains blocks of addresses reserved for special purposes, e.g. for multicast and broadcast communications, and for private communications. IPv6 includes link-local addresses, to be used for communications inside links.*

1.4   Dealing with size and heterogeneity

Needless to say, the Internet is a much larger and heterogeneous network than that of Figure 1.1. In fact, the Internet is, nowadays, *the* worldwide networking infrastructure, containing many dissimilar communication technologies. To deal with *size* and *heterogeneity* issues, the Internet is structured in several ways.

- The end-to-end routing is organized in two levels, one dealing with the communications over technologically homogeneous networks of limited geographical scope, and another abstracting the specificities of the various communication technologies (see Section 1.4.1).

- The Internet is organized in smaller networks, called *subnets*, that gather blocks of IP addresses sharing a common prefix, such that the destinations inside a given subnet can all be represented by a single entry at forwarding tables (see Section 1.4.2).

- The Internet is organized in administrative domains called *Autonomous Systems* (ASes), each running their own (intra-domain) routing protocols and communicating through an inter-AS routing protocol (currently BGP). In this way, the problem of determining end-to-end paths between hosts located far apart in the Internet is split into parts (see Section 1.4.3).

We will discuss these issues in the next three sections.

1.4.1   The TCP/IP layered architecture and the routing function

Besides routing, computer networks require other networking functions for their correct operation. Examples are error control, security, mobility, and congestion control. The set of required functions depends on the specific goals of the network. Routing is a fundamental networking function, which must always be present, except when hosts communicate directly through a link.

The TCP/IP layers The various networking functions are implemented through *protocols*. In the Internet, protocols are structured in a layered architecture of five layers known as the TCP/IP architecture. The layers are called physical layer (layer-1), link layer (layer-2), network layer (layer-3), transport layer (layer-4), and application layer (layer-5). In general, at both the source and destination hosts, a packet is processed by protocols of all five layers. This process is illustrated in Figure 1.7. At the source host, each layer adds to the packet the control information required to perform its specific networking function, and this information is impressed in the packet header (and sometimes in the packet trailer). One example of control information is the IP addresses required to route the packet within the Internet.

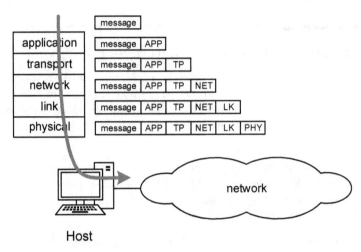

FIGURE 1.7: Addition of control information by successive TCP/IP layers.

The three lower layers of the TCP/IP stack deal with the communication between hosts, and the two higher ones with the communication between application processes running on the hosts. Examples of application layer protocols are HTTP, for Web browsing, FTP and TFTP, for file transfer, and Skype for multimedia conferencing. Regarding the transport layer protocols, the main protocols used in the Internet are TCP and UDP. They establish virtual channels between application processes. TCP provides addressing, error detection, control, and congestion control functions; UDP only provides addressing and error detection. At the other end of the stack, the physical layer deals with the problems related to the transmission of bits in the physical communication media, such as analog-to-digital and digital-to-analog conversion, modulation and coding, resilience to noise and distortion, and clock recovery. The network layer (layer-3) and the link layer (layer-2) are the most relevant to the routing problems addressed in this book and will be discussed in more detail next.

A two-level routing hierarchy The Internet is formed today by many dissimilar communications technologies, such as submarine cables, satellite links, cellular networks, and fixed and wireless local area networks, varying in speed, type of communication media, geographical span, reliability, and number of supported devices, among other properties. To deal with this heterogeneity, the TCP/IP layered architecture includes two layers, namely the network layer and the link layer, which form a two-level routing hierarchy (see Figure 1.8).

The network layer The network layer provides end-to-end communications between hosts, and abstracts the specificities of the communication technologies. The switching equipment that operates at this layer are the *routers*. Routers forward packets based on layer-3 addresses, i.e. the IP addresses. As

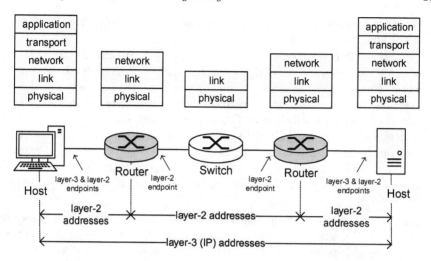

FIGURE 1.8: Relationship between layers, addresses, and devices.

shown in Figure 1.8, routers process the first three layers of the TCP/IP stack. At this layer, the network is viewed as a network of hosts and routers connected through logical links, where the links are abstract representations of actual connections. The network elements are identified at this layer by the IP addresses.

The IP protocol The network layer protocol of the Internet is the IP protocol; it is the protocol that *glues* the Internet, bringing together its many dissimilar communication technologies! The IP protocol defines the addresses used in end-to-end communications (the IP addresses) and provides the mechanisms for adapting to the heterogeneity of communication technologies. One such mechanism is the fragmentation process, whereby the end-to-end transmission of packets adjusts to communication technologies with different restrictions on the maximum packet length they can handle.

There are currently two IP protocol versions, one for IPv4 addresses and another for IPv6 addresses. IPv6 was introduced to solve the scarcity of IPv4 addresses: IPv4 addresses have 32 bits and IPv6 addresses have 128. The deployment of IPv6 started back in the 2000's but, due to its complexity, it has been a slow process. It is expected that IPv4 and IPv6 will coexist for many years.

The link layer The logical links among hosts and routers are handled by the link layer; they are sometimes referred to as *layer-2 links*. The link layer provides the basic packet-level communications between devices and delimits packets within the stream of bits arriving at a device. As highlighted above, these links can have very different characteristics and, therefore, there are many layer-2 protocols, each adapted to the link specificities. Examples of

layer-2 protocols are IEEE 802.3 (Ethernet), IEEE 802.5 (Token Ring), and
IEEE 802.11 (WiFi).

Layer-2 links can be simple point-to-point links, e.g. E1 or V.35 links,
or relatively complex networks, such as a switched Ethernet local area network. The switching equipment that operates at this layer are called *layer-2
switches*, or simply *switches*. Switches forward packets according to layer-2
addresses. As shown in Figure 1.8, unlike routers, switches only process the
first two layers of the TCP/IP stack. This actually means that a packet being transmitted over a layer-2 network, e.g. between two routers or between a
router and a host, will not have its IP address analyzed by the layer-2 switches
that it crosses. Layer-2 networks can be integrated in the global Internet or
operated in isolation.

Layer-2 addresses The prevalent layer-2 addresses are the MAC addresses
used in IEEE 802 LANs. These addresses are flat (non-hierarchical), have a
length of 48 bits and, like IP addresses, are assigned to interfaces. Thus, the
Internet addressing architecture has two address types naming the same network element. The MAC addresses are represented in hexadecimal notation,
e.g. `48:dd:a9:56:b3:47`. Like IP addresses, MAC addresses can be classified
as unicast, multicast, and broadcast, with the same meaning. The broadcast
address is the all-ones address (`ff:ff:ff:ff:ff:ff`). MAC addresses are assigned on a global basis by the IEEE, but need only have layer-2 link scope.

Layer-2 switches and forwarding tables A layer-2 network is a network
of layer-2 switches. Switches forward packets according to layer-2 addresses.
In the forwarding tables of switches, the destinations are layer-2 addresses,
and each entry indicates the outgoing interface of the switch that leads to
the destination. Unlike the forwarding tables of routers, next-hop information
makes no sense at layer-2, since it is the lowest layer to encapsulate packets.

Layer-2 and layer-3 endpoints Within the two-level routing hierarchy,
routers act as layer-2 endpoints, since they establish a frontier between layer-
2 networks, and hosts are both layer-3 and layer-2 endpoints. The layer-3
addresses have global scope and the layer-2 addresses have only link scope.
Consequently, when a packet is routed on the Internet, its layer-3 addresses
remain unchanged, but the layer-2 addresses change every time it crosses a
router; we will return to this issue in Section 1.6.

Is IP a routing protocol? The IP protocol does not solve the routing
problem in its entirety. In addition to IP, routing protocols are needed to
determine the end-to-end paths and to build and maintain the forwarding
tables of routers. Examples of routing protocols currently used in the Internet
are RIP, OSPF, IS-IS, and BGP. Besides routing, there are other issues that
the IP protocol does not handle alone and for which it needs assistance from
complementary protocols; among these issues are the provisioning of end-to-
end security and quality-of-service.

Is the routing function tied to the network layer? Routing protocols
are not tied to a specific layer of the TCP/IP architecture, since routing is

required at several layers. It is unfortunate that several textbooks still bond the routing function to the network layer, a common source of confusion among students. Indeed, routing is needed at layer-2 to find paths within networks of layer-2 switches, and at layer-3 to find paths within networks of routers, and even at layer-5 to find paths between application servers. Examples of routing protocols operating at each of these layers are the spanning tree protocol at layer-2 (see Chapter 3 of [41]), RIP, OSPF, IS-IS, and BGP at layer-3, and the routing between SIP proxy servers at layer-5 (see Chapter 7 of [27] or [22]). Interestingly, IS-IS has been recently modified to become the layer-2 routing protocol of 802.1aq shortest path bridging networks [5]; thus, even the same routing technology is used at different layers with modifications.

> *Internet protocols are organized in a layered stack known as the TCP/IP architecture. The link layer (layer-2) and the network (layer-3) form a two-level routing hierarchy that abstracts the heterogeneity of the communication technologies used in the Internet. The network layer provides the end-to-end communications between hosts; at this level the Internet can be viewed as a network of routers using IP addresses. The link layer provides communications among routers and hosts. The link layer connections can be simple point-to-point links or relatively complex networks using layer-2 addresses for internal communications. The prevalent layer-2 addresses are the MAC addresses, which have a length of 48 bits.*

1.4.2 Subnets, address structure, and forwarding tables

As it can be easily understood, it would be impossible in today's Internet to have individual destinations listed at the forwarding tables of routers. The size of the forwarding tables would be too large. Thus, IP addresses must be aggregated somehow.

Subnets To answer this problem IP networks are organized in subnets. Subnets are logical networks that gather a block of contiguous IP addresses sharing a common prefix, assigned to interfaces that can communicate among themselves without the intervention of a router. Thus, the same subnet cannot span different router interfaces, and all interfaces associated to the same subnet must be assigned IP addresses with the same prefix. Moreover, a subnet can be fully characterized by an address prefix.

Address structure To support the organization in subnets, IP addresses are structured hierarchically in two levels, where the prefix, sometimes called *netid*, identifies the subnet, and the suffix, sometimes called *hostid*, identifies the host interface on that subnet. Subnets are usually defined by (i) the lowest IP address of the address block, called *subnet address* or *prefix address*, and (ii) the prefix length or subnet mask.

In IPv4, the lowest and highest addresses of the subnet address block

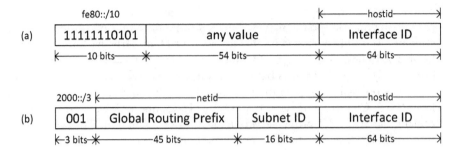

FIGURE 1.9: Structure of IPv6 (a) link-local and (b) global unicast addresses.

cannot be assigned to interfaces. The lowest address (together with the subnet mask) is reserved to identify the subnet, and the highest address is used as the broadcast address within the subnet. These kind of restrictions do not apply to IPv6.

Classful addresses In the beginnings of the Internet, there was a rigid demarcation between the netid and the hostid. By that time, there were three types of IPv4 unicast addresses, identified by their leftmost bits: class A addresses with a 24-bit hostid, class B addresses with a 16-bit hostid, and class C addresses with an 8-bit hostid. Since the address type was determined by the leftmost bits ("0" for class A, "10" for class B, and "110" for class C), there was no need for a subnet mask. Later, the introduction of subnet masks allowed variable-length subnets, bringing more flexibility into the network structure.

IPv6 global and link-local address structure Figure 1.9 shows the structure of the link-local and global addresses. Both types of addresses include the `Interface ID` field, which identifies the interface and corresponds to the hostid part of the address. Link-local addresses are to be used only within a layer-2 link and, therefore, have no netid. In global addresses, the netid part of the address is split in two parts: `Global Routing Prefix` and `Subnet ID`. The `Global Routing Prefix` is the portion of the address assigned by the ISP, and the `Subnet ID` identifies the subnet inside the ISP's network. The figure indicates typical values for the length of these fields; however, other values are possible. As in IPv4, the length of the netid part is variable and is defined by the prefix length value that accompanies the address. Irrespective of the netid subdivisions, the netid/hostid partition as defined by the prefix length is what is relevant for routing purposes at a given forwarding table.

Address configuration IP addresses can be configured at interfaces using different methods. IPv6 is much richer than IPv4 in this respect.

IPv4 addresses can be configured manually or through DHCP. When using DHCP, the interface contacts a DHCP server, which typically provides the IP address assigned to the interface, the subnet mask, and the IP addresses of the default gateway and of a DNS server (see chapter 23 of [9] for additional details).

IPv6 addresses can also be configured manually or through DHCP (called DHCPv6, in this case). However, IPv6 addresses can also be configured (i) by having the `Interface ID` assigned randomly or through the EUI-64 process or, in case of unicast global addresses, (ii) by having the address prefix obtained from a router through ICMPv6 Router Advertisement messages. The latter method is called Stateless Address Autoconfiguration (SLAAC). The former method is used in both global and link-local addresses. When the `Interface ID` is assigned randomly, a method to determine if the resulting IPv6 address is a duplicate must be involved; this method is called Duplicate Address Detection (DAD) in IPv6. The EUI-64 process builds an `Interface ID` from the MAC address of the interface. Chapters 4 and 5 of [14] provide additional details on how to configure IPv6 addresses.

Forwarding tables When the network is organized in subnets, the forwarding tables can refer to subnet prefixes (and not just to individual IP addresses), and only the router attached to a destination subnet needs to be concerned with delivering packets to individual destinations. The forwarding tables of routers have one entry per destination subnet prefix and each entry includes, in general, information on (i) the outgoing interface, (ii) the layer-3 address of the next router interface in the path to the destination, and (iii) the path cost. The path cost will be addressed in Section 1.5. In entries referring to destinations directly attached to the router, the path cost is omitted, and the next-hop address is replaced by the indication "is directly connected"; we will also use the keyword *dc* for this purpose. Note that the next-hop IP address fully determines at a router the path to be followed to each destination subnet; the remaining information is complementary.

Comparing routers and switches Routers and switches are both switching equipment that forward packets from incoming to outgoing interfaces. When a packet arrives at a switching equipment, the equipment reads the destination address contained in the packet header and searches for this address in its forwarding table, to determine how it should forward the packet. This way of operation is common to routers and switches. The fundamental distinction between them relates to the addressing information used to make forwarding decisions: routers forward packets according to IP addresses, whereas switches do it according to layer-2 addresses.

Example Subnets, as other network abstractions, are often represented by *clouds*. Figure 1.10 shows the subnet view of the network of Figure 1.1 and the forwarding table of router R1. The figure includes the identifier and cost of each router interface, placed near the interface. The interface cost is used in the path selection process, which will be explained in Section 1.5. Some interfaces also include the IP addresses assigned to them. In this network, there are five subnets, each assigned to one of the layer-2 links and characterized by a different prefix. For example, the wireless network is assigned a subnet with prefix 9.0.0.0/8, and the point-to-point link between R1 and R2 is assigned a subnet with prefix 192.168.0.0/30. The interfaces that belong to the same

destination	next hop	int	cost
9.0.0.0/8	222.0.0.4	i3	40
125.6.0.0/16	dc	i2	-
192.168.0.0/30	dc	i1	-
192.168.0.4/30	192.168.0.2	i1	30
222.0.0.0/24	dc	i3	-

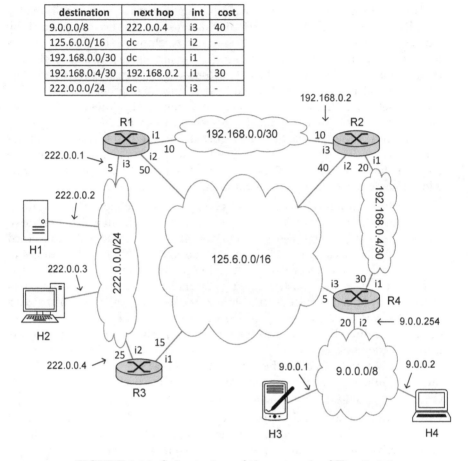

FIGURE 1.10: Subnet view of the network of Figure 1.1.

subnet are assigned IP addresses that share a common prefix. For example, all interfaces of subnet 222.0.0.0/24 are assigned IP addresses starting with 222.0.0.

The forwarding table of router R1 includes five entries, one for each subnet. Each entry includes the IP address of the next-hop interface, the identifier of the outgoing interface, and the path cost. For example, the entry relative to subnet 9.0.0.0/8 says that a packet destined to this subnet must be transmitted through interface i3 towards the interface with IP address 222.0.0.4, i.e. interface i2-R3. The forwarding table does not list the individual IP addresses of destination host interfaces, e.g. of the laptop or the tablet. A packet being transmitted from router R1 to the tablet, i.e. to IP address 9.0.0.1, is forwarded from router to router based on the subnet prefix 9.0.0.0/8, which encompasses 9.0.0.1, until it reaches router R4. It is then the responsibility of router R4 to find the tablet interface within subnet 9.0.0.0/8 and forward the packet to it. We will return to this issue in Section 1.6.

Can subnets be overlaid on layer-2 links? Since a subnet is a logical entity gathering a block of IP addresses assigned to interfaces attached to the same layer-2 link, it is possible to overlay several subnets on top of the same layer-2 link. Within a layer-2 link, there can be different interfaces belonging to different subnets; moreover, an interface can belong to more than one subnet, since it can be assigned several IP addresses. The communication between interfaces of different subnets must always be through a router, even if the interfaces are attached to the same link.

> *IP networks are organized in smaller subnetworks called subnets. Subnets are logical networks that gather a block of contiguous IP addresses sharing a common prefix, assigned to interfaces that can communicate among themselves without the intervention of a router. To support the organization in subnets, IP addresses are structured hierarchically in two levels, with the prefix identifying the subnet and the suffix identifying the host interface within the subnet. In this way, the router forwarding tables only need to refer to subnet prefixes, and not to individual IP destination addresses.*

1.4.3 Routing domains and Autonomous Systems

Routing domains and inter-domain routing End-to-end paths are not usually computed using a single routing protocol, especially if the paths are long. To ease this task, the network is organized in *routing domains*, such that each domain solves its own routing problem, i.e. determines the paths between its endpoints. The routing inside a routing domain is performed through *intra-domain* routing protocols, and the routing between routing domains is performed through *inter-domain* routing protocols. OSPF and IS-IS are two examples of intra-domain protocols. The routers located in the frontier between routing domains and running the inter-domain routing protocol are generically called *Domain Border Routers* (DBRs).

Autonomous Systems and BGP At a worldwide level, the Internet is organized in Autonomous Systems (ASes) (see Figure 1.11). ASes are domains under the responsibility of a single administration, usually an Internet Service Provider (ISP). The subnets and the routing domains are contained inside ASes. Each AS has its own routing policy and can have one or more routing domains, each running its own intra-domain routing protocol. The routers located in the frontier between ASes are called *Autonomous System Border Routers* (ASBRs). ASBRs run an inter-AS routing protocol among them, which advertises the subnets (i.e. the address prefixes) assigned to each AS, and determines paths between ASes. BGP is the current de facto standard for inter-AS routing [46]. Studying BGP is out of the scope of this book; the

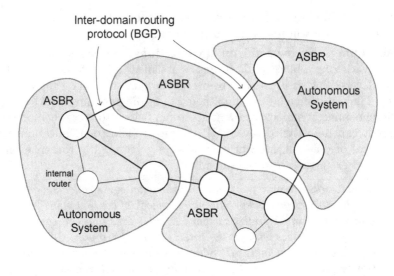

FIGURE 1.11: The Internet as a network of Autonomous Systems.

interested reader is referred to [52] and [18], two excellent textbooks on the subject.

It is important to highlight that inter-AS routing is intrinsically different from intra-domain routing. In the former, the selection of paths needs to consider political, economic, and security issues, as well as performance issues. For the latter, only the performance issues are relevant, since it applies only to routing under the responsibility of a single administration.

Multi-area routing domains Routing domains can be further structured into multiple areas, with the goal of simplifying the internal routing process (see Figure 1.12). In this case, the routers located in frontier between areas, called *Area Border Routers* (ABRs), run an *inter-area* routing protocol among them for the exchange of routing information across areas. Both OSPF and IS-IS support this possibility, and we will address it in Chapters 3 and 7. As will be seen, inter-area routing protocols share many common aspects with inter-domain routing protocols.

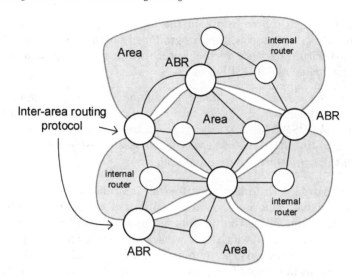

FIGURE 1.12: A routing domain structured in multiple areas.

At a worldwide level, the Internet is organized in administrative do-mains, called Autonomous Systems (ASes), each having its own rout-ing policy. The ASes communicate through the BGP protocol that runs on special routers located in the frontier between them, called Autonomous System Border Routers (ASBRs). BGP advertises the address prefixes reachable at each AS and determines paths between ASes. An AS can include one or more routing domains, each running its own intra-domain routing protocol, of which OSPF and IS-IS are two examples. Routing domains can be further structured in multiple areas, which communicate through an inter-area routing protocol run-ning on special routers located in the frontier between them, called Area Border Routers (ABRs).

Intra-domain routing approaches The two most important intra-domain routing approaches are distance vector and link state. *Distance vector* routing relies on the *distributed* and asynchronous version of the Bellman-Ford algo-rithm (see Section 5.2 of [6]). In this algorithm, routers send to their neighbors the estimates of the shortest path costs from themselves to each network des-tination. The shortest path cost to a destination, and the next-hop interface that provides it, is computed based on (i) the path cost estimates received from the neighbors and (ii) the costs of the outgoing interfaces leading to these neighbors.

Link state routing (LSR) is a two-step process: first, each router obtains information on the complete network topology and the address prefixes as-signed to network elements, using a flooding procedure; second, a *centralized* shortest-path algorithm, such as the Dijkstra or Bellman-Ford algorithms, is executed at each router to obtain the forwarding table. As will be seen in Chapter 7, LSR protocols, despite grounded on the LSR approach, may be combined with a distance vector protocol to provide inter-area routing; this is precisely the case of OSPF and IS-IS.

> *Routing inside routing domains is handled by intra-domain routing protocols, which can either take a link state or a distance vector approach. OSPF and IS-IS combine the two approaches: they are intrinsically link state routing protocols but use a distance vector approach in their inter-area routing protocol.*

Flow of routing information Routing information flows in the opposite direction of data information. In a distributed environment, routing information is originated at the destinations (where else?) when they advertise the address prefixes at which they want to be reachable. This information is then propagated towards the sources, forming a tree of paths directed to the destination, as illustrated in Figure 1.13. During this process, the original address prefix may get aggregated in successively larger address spaces, to save in the size of forwarding tables. For example, the destination subnet advertises an address prefix, its area advertises a larger prefix representing all the prefixes of the area, and its domain advertises an even larger address prefix representing all prefixes of the domain. It is through this larger address prefix that the destination will be known at the most remote sources. This process is sometimes called *supernetting* because the aggregates are described with successively shorter prefixes representing successively larger networks. In Figure 1.13, the destination D is assigned prefix 123.4.5.0/24, its area advertises 123.4.0.0/16, and its domain 123.0.0.0/8.

> *Routing information flows from destination to source, the direction opposite to the flow of data, and during this process the advertised prefixes can be aggregated in successively larger address spaces; this process is sometimes called supernetting.*

IPv4 and IPv6 coexistence and integration The IPv4 and IPv6 address families and routing protocols will have to coexist for many years. From a routing perspective, IPv4 and IPv6 run on separate routing domains and their routing protocols build separate forwarding tables. These routing domains can either be overlaid or run on separate physical networks. There are three techniques for the coexistence and integration of IPv4 and IPv6: dual-stack, tunneling, and translation. In the *dual-stack* technique the network devices (hosts and routers) implement both the IPv4 and IPv6 protocols. Thus the IPv4 and IPv6 routing domains are overlaid and hosts can communicate using one or the other. The *tunneling* technique is a method to transport IPv6 packets over IPv4-only networks. This is accomplished by encapsulating IPv6 packets inside IPv4 packets. Finally, the *translation* technique is used to interconnect IPv4-only and IPv6-only networks by converting the IP header of

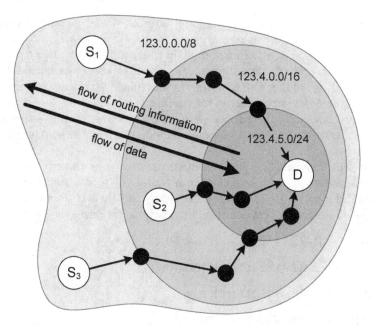

FIGURE 1.13: Flow of routing and addressing information versus flow of data.

one family into the other. Chapters 10 and 11 of [14] provide a comprehensive discussion of these techniques.

1.5 The path selection process

In computer networks, several paths are usually available to send traffic from one origin to a destination. Among these, at least one path must be selected. To this end, each candidate path is characterized by a metric, called *path cost*, expressing the preference of transmitting traffic through that path.

The path cost is a function of the attributes of routers and links. However, in fixed networks, the router attributes are usually ignored, since the resources available at routers do not constrain the communication process. The link attribute expresses the cost of transmitting packets through a link; examples of attributes are the link delay, throughput, error rate, or reliability. In many routing protocols, the link transmission costs are expressed through positive integers assigned *statically* to the outgoing interfaces of routers. These costs can be configured manually at the routers, providing network managers with a flexible tool to set up specific routing solutions. However, static costs do not reflect the real-time state of the link. We stress that the interface costs represent the cost of transmitting information, and not of receiving it; therefore, the costs are associated with outgoing interfaces, and ingoing interfaces are assumed to have a cost of zero.

LSR protocols, as other intra-domain routing protocols, select paths using the least path cost criterion. In this case, the path cost of each candidate path is obtained by adding the costs of the outgoing interfaces that are part of the path, and the path selected for routing is the one with lowest cost. Note that there can be more than one path with the lowest cost, in which case the selected path can be chosen randomly among the set of candidate paths with lowest cost.

We will illustrate the path selection process using the network of Figure 1.10. Consider the path selection from router R1 to subnet 9.0.0.0/8. There are six different candidate paths. The paths and their costs are the following (expressed in terms of transmitting interfaces to facilitate the understanding):

- i1-R1 → i1-R2 → i2-R4 with cost 50;

- i1-R1 → i2-R2 → i2-R4 with cost 70;

- i2-R1 → i1-R2 → i2-R4 with cost 90;

- i2-R1 → i2-R4 → with cost 70;

- i3-R1 → i1-R3 → i1-R2 → i2-R4 with cost 60;

- i3-R1 → i1-R3 → i2-R4 with cost 40;

To give an example of path cost computation, consider the second path above. The path cost is obtained by adding the costs of interfaces i1-R1, i2-R2, and i2-R4, which are 10, 40 and 20, respectively, giving a path cost of 70. These are the interfaces that would transmit packets if the path was selected for routing; the cost of the receiving interfaces is not accounted for in the path cost computation.

The path with the lowest cost, i.e. the one selected for routing, is the last one in the list above, with a cost of 40. This can be confirmed in the forwarding table of Figure 1.10. The first entry of the table indicates that the path to 9.0.0.0/8 is through interface i3, having 222.0.0.4 has the IP address of the next-hop interface (i.e. the address assigned to interface i2-R3), and a cost of 40.

> LSR protocols select paths according to the shortest path routing prin-
> ciple, i.e. the path selected for routing is the least cost one, and the
> cost of a path is computed by adding the costs of the outgoing interfaces
> that belong to the path.

Directly connected links There is one exception to the use of shortest paths in Internet routing, and this is when the destination is at a link *directly connected* to the router. In this case, the directly connected route takes precedence over any other path, even if the latter provides a lower cost. Consider again the example of Figure 1.10, and the problem of determining the path from router R1 to subnet 125.6.0.0/16. The shortest path is i3-R1 → i1-R3 with a cost of 20. The directly connected route has a cost of 50 (the cost of interface i2-R1) but, despite its higher cost, it gets selected for routing. The precedence of directly connected routes is implemented locally at routers through a parameter called *administrative distance*, which we discuss next.

Administrative distances The *administrative distance* provides a generic way of selecting among paths to the same destination obtained through different routing processes. Besides directly connected routes, a router may be running more than one routing protocol at the same time, e.g. OSPF and IS-IS, each providing a different path to the same destination. The administrative distance is a value assigned to each routing process that defines its precedence. Currently, the administrative distance is 0 for directly connected routes, 110 for OSPF and 115 for IS-IS, and the convention is that lower values have higher precedence. Thus, directly connected routes are preferred over any other paths and, if a router is running both OSPF and IS-IS, OSPF paths are preferred over IS-IS ones.

> *In Internet routing, directly connected routes take precedence over shortest paths. This precedence is implemented locally at routers through a parameter, called administrative distance, which also ranks among paths for the same destination obtained through different routing protocols that may be running simultaneously.*

1.6 End-to-end routing

How can a packet be routed among hosts when it needs to cross both the global layer-3 network and several layer-2 networks? As illustrated in Figure 1.14, this process can be decomposed in three steps: routing from the source host to the first-hop router (step 1), from the first-hop to the last-hop router (step 2), and from the last-hop router to the destination host (step 3). The routing protocol is only involved in the second step. All steps may require the resolution of layer-3 into layer-2 addresses.

Address resolution When an IP packet has to be transmitted over a layer-2 network, the layer-2 address associated with the next-hop IP address (i.e. the layer-3 endpoint of the layer-2 network) may be unknown. In the case of unicast packets, this problem is solved through an address resolution protocol. IPv4 uses the Address Resolution Protocol (ARP) [43] and IPv6 the Neighbor Discovery Protocol (NDP) [39]. Despite different names, these protocols operate similarly in what concerns address resolution.

Taking ARP as an example, when a device needs to determine the MAC address associated with an IPv4 address of its subnet, it questions all inter-

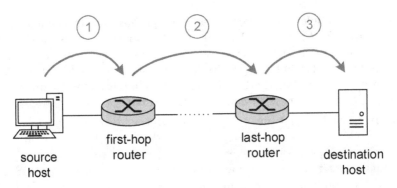

FIGURE 1.14: The three steps of end-to-end routing.

faces attached to the subnet through an ARP REQUEST message transmitted to the MAC broadcast address; the recipient that owns the IPv4 address provides the requested information through an ARP REPLY message (see Section 3.2.6 of [42]). NDP uses the same process, where the equivalent of the ARP REQUEST is the NEIGHBOR SOLICITATION message, and the equivalent of the ARP REPLY is the NEIGHBOR ADVERTISEMENT message (see Chapter 5 of [14]). To avoid exchanging these messages whenever a new packet needs to be transmitted on a layer-2 network, the devices (hosts or routers) that discover an association between IP and MAC addresses cache it in memory for some time.

The resolution of multicast IP addresses into MAC addresses is performed through a mapping process, and not through an address resolution protocol. For example, a multicast IPv4 address is converted into a multicast MAC address using the following rules: (i) bits 48 to 25 of the MAC address are $0\times01005e$; (ii) bit 24 is 0; (iii) bits 23 to 1 are the last 23 bits of the IPv4 multicast address.

From the source host to the first-hop router In the first step, there can be several possible first-hop routers to choose from, either because the source host is multi-homed and/or is connected to a shared link with multiple attached routers (see Figure 1.15). The source host could, in principle, select its first-hop router by listening passively to the routing messages transmitted by the various first-hop routers; this would allow determining the best first-hop router towards each individual destination. However, this solution has two problems: first, because hosts may roam across networks running different routing protocols, they would need to understand them all, and there are many possibilities; second, the routing information given by first-hop routers running different routing protocols could be incompatible, e.g. it could provide non-comparable path cost metrics for the same destination. Thus, instead of having a first-hop router tailored to each individual destination, the usual solution is to have a single first-hop router for each shared link interface,

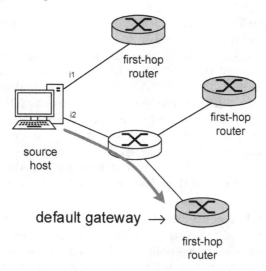

FIGURE 1.15: Multi-homed host with several possible default gateways.

which is used as the first-hop for all traffic transmitted by the interface; this router is called the *default gateway*. Note that a default gateway address is not needed for point-to-point link interfaces.

The default gateway address is the IP address of the default gateway interface attached to the shared link. In the case of IPv6, the default gateway address must be a link-local address.

Multi-homed hosts have one default gateway per shared link interface, and face the additional problem of determining which interface to use for transmitting their traffic. The traffic could be distributed among the various interfaces according to some criterion, but the common solution is to send all traffic through the highest bandwidth interface. Thus, in this setting, the source host selects first the outgoing interface and then sends all traffic to the first-hop router of that interface, which in the case of shared link interfaces is the default gateway.

The default gateway address can be superseded for specific destinations, either manually or through ICMP Redirect messages. The latter case operates in the following way: when a packet arrives at the default gateway and, based on the routing protocol, the default gateway verifies that there is a better first-hop router for the packet destination, it sends an ICMP Redirect message to the source host containing the address of that first-hop router and the corresponding destination address. The source host then installs this information in its forwarding table. This feature is optional.

In IPv4, the default gateway address is configured manually at the host or obtained from a server through the DHCP protocol (See Section 3.2.7 of [42]); in IPv6, it can be configured manually at the host, obtained from a server

through the DHCPv6 protocol, or obtained from a first-hop router through
the NDP protocol (see Chapter 5 of [14]).

When a packet gets ready for transmission at the source host, the host first
checks whether or not the destination host belongs to its subnet. It does so
by comparing (i) its subnet prefix with (ii) the prefix obtained by applying its
subnet mask to the destination address. If the prefixes match, it concludes that
the destination host belongs to its subnet and, if not, it concludes otherwise.
In the first case, the packet must be sent directly to the destination host,
and in the second case it must be sent to the default gateway. If needed,
the source host uses the address resolution protocol to determine the MAC
address associated with the IP address of the destination host or of the default
gateway.

From the first-hop router to the last-hop router The problem of routing
between routers, and therefore from the first-hop to the last-hop router, is
solved by layer-3 routing protocols, which is the main topic of this book. As
discussed above, these protocols build and maintain layer-3 forwarding tables
at each router, which indicate how routing proceeds to the next router.

When a packet arrives at a router, the router performs a forwarding table
lookup and tries to match the destination IP address contained in the packet
header with one of its forwarding table entries. The matching verification
for an incoming IP address is performed at each forwarding table entry by
comparing (i) the entry prefix with (ii) the prefix obtained by applying the
entry subnet mask to the incoming IP address. There can be more than one
forwarding table entry matching a given IP address. When this happens, the
entry with the longest prefix is selected for routing; this is known as the *longest
prefix match rule*.

The matching entry indicates the outgoing interface and, except for the
last-hop router, the IP address of the next-hop router. If needed, the router
uses the address resolution protocol to determine the MAC address associated
with the next-hop IP address.

From the last-hop router to the destination host In the last step, the
last-hop router delivers the packet to the destination host. A router knows it
is the last one for a packet when the matching forwarding table entry says
that the destination subnet is directly connected to the router. In this case,
if the destination subnet is a shared link, the router uses the address resolution protocol to obtain the layer-2 address associated with the destination IP
address, so that the packet can be transported to the destination host.

An example We give an example of end-to-end routing in Figure 1.16. The
example considers IPv4 addressing but the behavior with IPv6 addressing
is similar. The network includes three IPv4 subnets interconnected by two
routers. The subnets are assigned prefixes 223.2.3.0/24, 125.6.0.0/16, and
9.0.0.0/8; the middle subnet includes one layer-2 switch. We show, close to
the interfaces, the interface identifiers and their IP and MAC addresses. The
figure also shows the forwarding tables of all switching equipment. We il-

lustrate how a packet is sent from the server located in subnet 223.2.3.0/24 (the source host) towards a laptop located in subnet 9.0.0.0/8 (the destination host). The packet representation includes only the header fields required for forwarding purposes, i.e. the IP and MAC destination addresses. In this example, we assume that all address resolution caches are initially empty.

When the packet gets ready for transmission at the server, the server applies its /24 subnet mask to the IP destination address and concludes that the laptop is not in its subnet, since the resulting prefix, 9.0.0.0/24, is different from its own, 223.2.3.0/24. Thus, the packet must be sent to the default gateway, i.e. to IP address 223.2.3.254. Before transmitting it, the server invokes the ARP protocol to obtain the MAC address associated with 223.2.3.254, which is `00:1d:70:d7:c4:c1`. After getting this information, the IP packet is encapsulated in an Ethernet packet and transmitted to router R1.

When the packet arrives at R1, the MAC header is stripped from the packet; it was only used for transportation in the first subnet and is no longer useful. Next, R1 performs a lookup in its forwarding table, trying to match the destination IP address contained in the packet header with one of its table entries, and determines that the packet must be transmitted through interface i2 to next-hop 125.6.2.2, which is an address of router R2. Again, the router uses the ARP protocol to determine the MAC address associated with the IP address of the next-hop, which is `10:d3:51:23:d5:38`. Once R1 gets this information, the IP packet is encapsulated in an Ethernet packet and transmitted towards the switch.

When the packet arrives at the switch, the switch searches its forwarding table for an entry corresponding to the MAC destination address contained in the packet header, and concludes that the packet must be transmitted through interface i2. Note that the switch only analyzes layer-2 information to perform its forwarding decisions; the IP addresses contained in the packet header are irrelevant to its operation. Moreover, the destination MAC address is not modified by the switch, in the same way as destination IP addresses are not modified by routers.

When the packet arrives at R2, the MAC header is again stripped from the packet. While performing the forwarding table lookup, the router concludes that the destination host is located in a directly attached subnet and that the packet must be transmitted through interface i2. It then invokes the ARP protocol to determine the MAC address associated with the IP address of the laptop, 9.0.0.1, which is `d0:15:a7:5b:11:20`. Finally, the packet gets transmitted to the laptop, over the air, but this time the IP packet is encapsulated in an IEEE 802.11 packet.

Note that the destination IP address did not change during this journey. Contrarily, the destination MAC address changed every time the packet crossed a router and changed subnet.

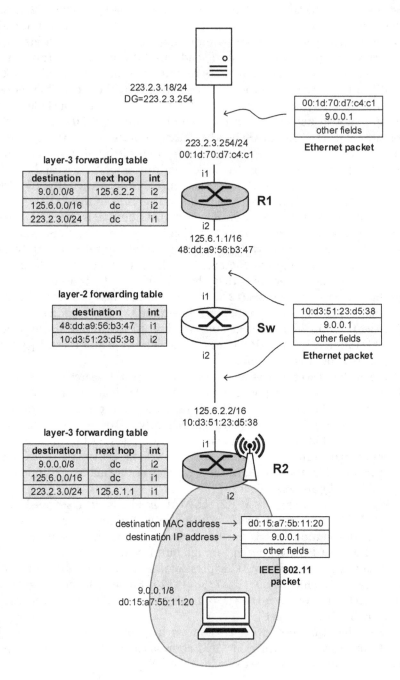

FIGURE 1.16: End-to-end routing example.

> End-to-end routing involves three steps, and each step may require the
> intervention of an address resolution protocol to map IP addresses into
> MAC addresses. In the first step, from the source host to the first-
> hop router, the source host delivers the packet to a predefined first-
> hop router, called default gateway. The default gateway address can
> be configured manually at the host or obtained from a DHCP server
> or from the default gateway itself. The second step, between the first-
> hop and the last-hop router, is handled through the routing protocol.
> At this step, the packet is routed according to the forwarding tables of
> routers. In the last step, the last-hop router delivers the packet to the
> destination host.

1.7 Representing the network topology

In many situations, it is convenient to represent networks in a simplified way,
emphasizing the key elements and hiding the irrelevant details. For example,
when studying routing protocols, the layer-2 networks that are used to connect
routers need not be represented with all its inner details, e.g. by including the
layer-2 switches and the connections between them. Having networks represented through graphs is often convenient, to better understand the network
structure and the operation of protocols. This section discusses first the typi-
fication of layer-2 connections, and then the graph representation of networks.

1.7.1 Types of layer-2 connections

Point-to-point and shared links From the point of view of layer-3 routing,
the layer-2 links between routers are just connections between routers, independently of their internal complexity. These connections can be classified in
two types: point-to-point links and shared links. A *point-to-point* link connects
two, and only two, layer-3 devices; examples are E1 or V.35 links. Typically,
point-to-point links are *bidirectional*, i.e. they provide communications in both
directions, but there are also examples of *unidirectional* point-to-point links. A
shared link is an abstraction of a layer-2 network, and can potentially connect
many layer-3 devices. Examples are Ethernet (switched and non-switched),
Token Ring, Wi-Fi, X.25, Frame Relay and ATM.

Degree of connectivity and broadcast capability Shared links can be
further classified according to two properties: degree of connectivity and
broadcast capability. Regarding the *degree of connectivity*, shared links can
either provide direct connections between all devices attached to the link (a
full mesh) or only between some devices (a partial mesh). For example, an
Ethernet network provides full connectivity among its attached devices, and

an ATM network can be configured to provide full or partial connectivity, using a full or a partial mesh of PVC (Permanent Virtual Circuits).

A shared link has *broadcast capability* if a packet transmitted on the link by one device using the broadcast destination address is received and read by all other devices attached to the link. Thus, this capability concerns the possibility of delivering information simultaneously to all devices attached to a link using just one packet. It is present in Ethernet (switched and non-switched), Token Ring and Wi-Fi networks, but not in X.25, Frame Relay and ATM networks. Note that a link may be fully-meshed but not broadcast capable.

Transit and stub shared links Finally, one can distinguish between transit and stub shared links. A *stub* shared link has only one attached router, and a *transit* shared link has two or more attached routers. For example, in the network of Figure 1.1, the wireless link is a stub shared link, and the switched Ethernet connecting all four routers is a transit shared link.

Note that there need not be a direct correspondence between layer-2 technologies and the abstract link types used to represent them. For example, an Ethernet link is a technology that can connect many routers and, therefore, is naturally represented as a shared link. However, it can also be represented as a point-to-point link if it is just connecting two routers. This possibility is illustrated in Section 9.2.1.6.

> *Layer-2 links can be classified as point-to-point or shared links. Point-to-point links connect two, and only two, layer-3 devices. Shared links abstract layer-2 networks and can potentially connect many layer-3 devices. Moreover, shared links can be classified as fully-meshed or partially-meshed, as broadcast or non-broadcast, and as transit or stub links.*

1.7.2   Representing networks through graphs

Graphs provide a simplified way of representing networks, which is useful in many situations. A graph is composed of nodes and arcs. Arcs represent connections between nodes and can be directed or undirected. *Directed* arcs describe a directional relationship between nodes (e.g. the communication in a specific direction). Graphs with directed arcs are called *directed graphs*, and those with undirected arcs are called *undirected graphs*. Both types of graphs are useful when studying routing protocols.

Mapping graphs to networks When studying layer-3 routing protocols, the natural association between network and graph elements is to have nodes representing routers and arcs representing the layer-2 connections between routers. However, in some cases, it may be advantageous to adopt other mappings between network and graph elements. The hosts are not included in the graph if they do not participate in the routing protocol, which is usually the case.

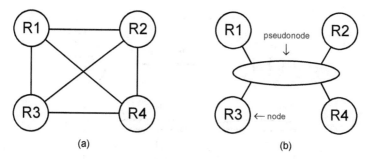

FIGURE 1.17: Connectivity provided by a shared link with four routers represented by (a) arcs connecting each pair of nodes or (b) a pseudonode.

Identifying the network elements Besides the mapping between network and graph elements, the network representation also requires that network elements be identified through unique names. We will assume that the names of routers, point-to-point links, and shared links, are taken from separate name spaces. Router names will be prefixed by R, point-to-point link names by lp and shared link names by ls.

Representing the connections As discussed in Section 1.7.1, one can abstract the layer-2 connections between routers in two types: point-to-point links and shared links. The connectivity provided by a point-to-point link can be represented through an arc. In this case, the arc represents both the communications medium connecting the two routers and the interfaces attaching these routers to the medium.

The pseudonode Similarly, the connectivity provided by a shared link can be represented by a set of arcs, one arc for each pair of routers attached to the link (Figure 1.17.a). However, when the link connects many routers, this type of representation involves many arcs and may become cumbersome. If the link is fully-meshed, a solution to this problem is to represent the connectivity within the link through a single node (Figure 1.17.b). Indeed, this type of shared link can be viewed as a switching element that forwards packets received on one interface to all other interfaces. To distinguish between nodes representing routers and nodes representing shared links, we will refer to the former as nodes and to the latter as *pseudonodes*, following the nomenclature of IS-IS. Moreover, we will draw nodes using circles and pseudonodes using ellipses. The arcs between nodes and pseudonodes have a different interpretation than the ones between nodes: they just represent the interfaces that attach routers to shared links. Note that arcs between pseudonodes are not allowed in this representation, since they have no physical meaning.

An example Figure 1.18 shows the undirected graph that represents the network of Figure 1.1 for studying its IP level routing. Nodes R1 to R4 represent the IP routers. The arcs between nodes R1 and R2, named lp1, and between nodes R2 and R4, named lp2, represent the E1 links. The pseudonode

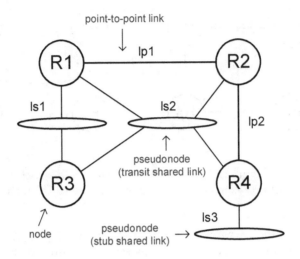

FIGURE 1.18: Graph representing the network of Figure 1.1.

ls1 represents the simple hub-based Ethernet LAN. The pseudonode ls2 represents the switched Ethernet comprising three switches arranged in a triangular topology. These pseudonodes are both transit shared links. The pseudonode ls3 represents the wireless LAN, and is a stub shared link. Of course, there is a routing problem inside the switched Ethernet network represented by shared link ls2. This problem is usually handled through the spanning tree protocol, and could be studied using another graph, now representing the inner details of the switched Ethernet. However, for studying IP routing, the switched Ethernet can be conveniently abstracted through a single entity (in this case, the shared link) since, from this perspective, the switched Ethernet is just a provider of connectivity among the IP routers.

> *Networks can be represented by graphs, and this abstract representation eases the understanding of routing protocols. A graph is composed of nodes and arcs. To study layer-3 routing, a natural mapping between network and graph elements is to have nodes representing routers and arcs the layer-2 connections between routers, which can be point-to-point or shared links. To simplify the representation, the connectivity within a fully-meshed shared link can be summarized through a single node, called a pseudonode.*

1.8 Correctness and performance of routing protocols

The design of a routing protocol must observe several correctness and performance criteria. In particular, a routing protocol must:

- Converge within a reasonable time to a correct state when the following events occur:
  - the attributes of routers or links change;
  - routers are added or removed, possibly due to a failure.

- Deal appropriately with the possibility of corruption of routing messages and of routing information stored at routers, as well as the lack of resources (e.g. memory or processing power) to handle routing information at routers.

- Be efficient, i.e. keep at a minimum the load introduced by control messages, since the resources available to transport control information are subtracted from those of data information.

The correctness of a protocol is usually assessed in terms of *safety* and *liveness* properties, where observation of the first assures that the protocol never produces an incorrect result, and observation of the second assures that the protocol never enters a deadlock state (see Section 2.4.1 of [6]). A distributed routing protocol must ensure that all routers build consistent forwarding tables, each having locally the correct notion of the (global) end-to-end paths.

A particularly challenging scenario for correctness and performance is when all routers are switched on at the same time; this is known as *cold start*.

> The design of a routing protocol must meet several correctness and performance criteria, to assure the protocol converges within a reasonable time to a correct state, and does not impose excessive load on the network.

1.8.1 Reliable transmission of control messages

No communication medium is fully reliable. However, control messages must be transmitted reliably, i.e. they must reach the intended destination without errors and in due time. To achieve reliability, there must be mechanisms (i) to detect if messages contain bit errors and (ii) to ensure that error-free messages are eventually received.

Error detection The usual *error detection technique* is to append to the transmitted message a field computed over its contents, and compare it, at the destination, with the same computation performed over the contents of the received message. There are essentially three types of error detection methods: parity checks, checksums, and cyclic redundancy checks (see, for example, Section 5.2 of [27]).

Periodic repetition of messages To recover from corrupted messages, there are essentially two techniques. One technique frequently used in routing protocols, but not often appreciated as such, is the periodic repetition of messages. In this case, even if some messages get lost or corrupted, others will eventually succeed (if the communication media does not fail completely).

ACK protection Another technique is based on the acknowledgment by the receiver that the message was correctly received. The procedure is the following:

- The sender stores a copy of the message, sets a timer (to wait for the acknowledgment), and transmits the message; the timer, called *ACK timer*, is set to a predefined value called *timeout*.

- The receiver replies with an acknowledgment when it receives the message without errors.

- If the sender does not receive the acknowledgment before the ACK timer expires, it retransmits the message and sets again the timer to the timeout value; this is repeated until an acknowledgment is finally received.

- When the sender receives the acknowledgment, it deletes the copy of the acknowledged message stored initially and clears the ACK timer.

Note that in this interaction both the initial message and the acknowledgment can be corrupted, but the retransmissions only stop when both messages are received correctly.

The acknowledgment can be performed through an explicit control message, usually called ACK, or can be implicit in the response. For example, if the sender requests some information, receiving that information acknowledges that the request was correctly received. We will refer to transmissions protected in this way as being *ACK protected*.

A sender may have to transmit at the same time several messages requiring ACK protection. Moreover, a receiver can receive the same message several times, which may happen when ACKs get lost. Thus, there must be a way to distinguish between a new message and a copy of a previously transmitted one, and there must be a way to correlate each transmitted message with the corresponding acknowledgment. This is usually achieved by numbering the messages with unique sequential numbers, usually just called *Sequence Numbers* (SNs), and numbering the ACKs with the SNs of the messages they acknowledge.

Stop-and-Wait protocol There are several strategies for the retransmission of messages in case of error, usually referred to as Automatic Repeat-reQuest (ARQ) strategies (See Section 2.4 of [6]). The simplest ARQ strategy, often

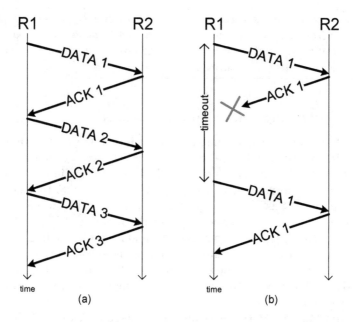

FIGURE 1.19: Stop-and-Wait protocol.

used in LSR, is the *Stop-and-Wait* (SW) protocol. In this case, only one message can be transmitted at a time, and the sender can only move to the transmission of the next message after receiving the acknowledgment that the previous one was correctly received.

Figure 1.19 illustrates the operation of the SW protocol. In Figure 1.19.a, router R1 sends three DATA messages to router R2. DATA messages are numbered sequentially and each ACK repeats the SN of the DATA it acknowledges. There are no transmission errors, but a DATA message can only be transmitted after the acknowledgment of the previous one. In Figure 1.19.b the ACK sent by router R2 in response to the first DATA message is lost. Thus, the timer set by the router (when the message was initially transmitted) expires and, when this happens, router R1 retransmits the DATA message using the same SN. Router R2 discards the second DATA message since it recognizes, based on the SN, that it was already received. However, it still acknowledges its reception. A similar procedure is followed in the case of a DATA message loss.

Reliability on shared links Achieving reliability on a shared link brings an additional challenge: when a router broadcasts a message on a shared link, it must get the confirmation that all other routers attached to the link received it. Thus, the source router must know in advance how many routers are attached to the link and who they are. One way to achieve this is through the Hello protocol, which we describe in Section 2.1.

> *Routing messages need to be transmitted reliably to ensure protocol correctness. Reliability is usually obtained through the periodic repetition of control messages or using ACK protection and a Stop-and-Wait (SW) retransmission strategy.*

1.8.2 Detecting router and link failures

The ability to detect router and link failures is an important feature of routing protocols. In case of a failure, routing protocols need to find alternatives to the damaged routes, as fast as possible. In some cases, failures can be detected through hardware mechanisms, e.g. by electronic circuits that detect loss of signal. However, this is not always possible. To see this, consider the network of Figure 1.20, which reproduces part of the network of Figure 1.1. Suppose that router R3 fails, or that switches Sw1 and Sw3 fail simultaneously (Figure 1.20.a). The first case is a router failure and the second a shared link failure. These failures cannot be sensed by router R2 using hardware mechanisms, since R2 is connected to the shared link through switch Sw2, and this switch remains operational.

Router failures are the most difficult to handle. A router failure can only be detected by its neighbors, since a failed router has no opportunity to inform all others that it is going to fail. This implies the existence of a *keepalive* mechanism, whereby routers periodically announce that they are still operational. This mechanism is included, implicitly or explicitly, in routing protocols.

The failures of shared links may be only partial. Suppose that the links connecting switches Sw3 and Sw1, and switches Sw3 and Sw2, fail simultaneously (Figure 1.20.b). In this case, the shared link can still be used to connect routers R3 and R4 on one side, and routers R1 and R2 on the other. We say that the link has been *partitioned*. A partitioned link is not a completely failed link and, if possible, the network should keep using its surviving parts for routing purposes. As in the case of router failures, a keepalive mechanism can be used to detect the subgroups of routers that can still communicate with each other in case of a shared link partitioning.

> *Routing protocols must be able to handle router and link failures and the partition of shared links. In order to deal with these events, routers must periodically announce to their neighbors that they are still alive; this is called a keepalive mechanism.*

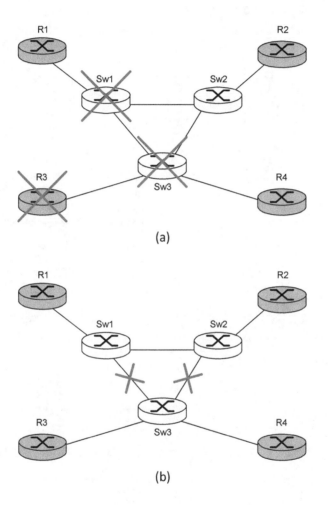

FIGURE 1.20: (a) Failures of routers and links and (b) shared link partition-ing.

Part II

Principles

2

Principles of Single-Area Link State Routing

Link State Routing (LSR) is a routing approach where each router maintains a complete view of the network, i.e. where each router knows the complete network topology and the correspondence between available prefixes and network elements (routers or links). It is based on this view, which we will refer to as the *routing information*, that routers build their forwarding tables. To ensure that routers maintain globally consistent forwarding tables, the routing information must be the same at all routers.

This chapter addresses the principles of LSR, i.e. the fundamental mechanisms and data structures that always must be present in LSR protocols, irrespective of the technology implementing them. The principles of LSR can be organized around three main issues:

- How is the routing information represented?

- What are the mechanisms used to keep this information updated and synchronized at all routers?

- How do the network representation and synchronization mechanisms scale to large networks?

In this chapter, we concentrate on the first two issues; the third issue is related to multi-area networks and will be addressed in chapter 3.

In LSR protocols, each router builds and maintains a *Link State Database* (LSDB) containing the routing information required to construct its forwarding table. There are three types of routing information (see Figure 2.1):

- **Topological information** - Also known as the *network map*, describes the routers, the links between routers, and the costs assigned to router interfaces.

- **Addressing information** - Describes the routable address prefixes (IPv4 or IPv6) and their association with routers and links. One can distinguish between *domain-internal* and *domain-external* addressing information, i.e. information related to prefixes internal and external to the routing domain.

- **Link information** - Describes the addresses used to transport packets in the link between neighboring routers. This information has a local nature and need not be the same across the LSDBs of all routers.

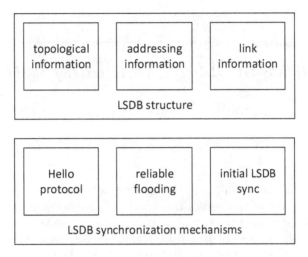

FIGURE 2.1: LSDB structure and synchronization mechanisms.

The LSDB must be kept updated and synchronized at all routers, and for that three mechanisms are required (see Figure 2.1):

- **Hello protocol** - Detects the active neighbors of a router.

- **Reliable flooding procedure** - Disseminates throughout the network the routing information originated by each router.

- **Initial LSDB synchronization process** - Synchronizes the LSDBs of routers that become neighbors.

We will refer to these mechanisms as *synchronization mechanisms*.

In the following sections, we discuss the various aspects concerning the representation of the routing information and the synchronization mechanisms. Section 2.1 introduces the Hello protocol. Section 2.2 discusses the structure of the LSDB and Section 2.3 explains how the forwarding tables are built from the LSDB. Section 2.4 addresses the mechanisms used to disseminate the routing information throughout the network, including the reliable flooding procedure. Section 2.5 describes the initial LSDB synchronization process. Finally, Section 2.6 gives a summary of the control messages that support the various synchronization mechanisms introduced along the chapter.

In this chapter, we adopt our own nomenclature, frequently not adhering to that of OSPF or IS-IS. The goal is to make the nomenclature as expressive as possible and avoid favoring one technology against the other. Later in the book we provide the correspondence between the various terminologies.

> *LSR protocols build and maintain a database, called Link State Database (LSDB), containing topological information (the network map, representing the network topology), addressing information (the address prefixes assigned to routers and links), and link information (the addresses required to transport packets between neighboring routers). The consistency of the LSDB is achieved through three distinct synchronization mechanisms: Hello protocol, reliable flooding procedure, and initial LSDB synchronization process.*

2.1 Hello protocol

The Hello protocol aims at the maintenance of neighborhood relationships among routers. The protocol allows routers to keep track of the neighbors currently active in each of their links. Thus, it plays a central role in keeping the network map updated.

The Hello protocol runs per link and, therefore, a router must run one protocol instance at each of its interfaces. The protocol maintains a list with the identifiers of the neighbors currently active on the link, called *list of active neighbors*.

The protocol must behave correctly even if the network is unreliable, i.e. in case of router and link failures, and loss of control messages. In the next sections, we discuss first the Hello protocol for reliable networks, and then introduce the modifications required to ensure protocol correctness in case of unreliable networks. We also discuss how the protocol is used to verify that the communications between neighbors are bidirectional, what requirements it places in the identification of neighbors, as well as some additional uses of the protocol.

2.1.1 Hello protocol for reliable networks

We start by considering an ideal scenario where both routers and links are reliable, i.e. where (i) routers and links do not fail and (ii) control messages are not lost. This allows a smoother introduction to the issues involved. In this scenario, the protocol requires three control messages that we name HELLO, HELLO REPLY, and BYE, and operates as follows:

- When a router joins the link, it broadcasts a HELLO message containing its identifier;

- Every router receiving a HELLO message replies with a HELLO REPLY message containing its identifier;

- When a router leaves the link, it broadcasts a BYE message containing its identifier;

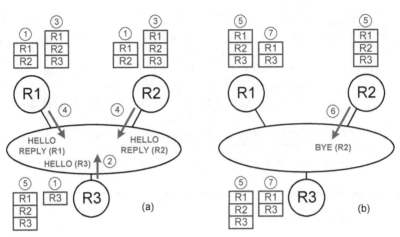

FIGURE 2.2: Hello protocol for reliable networks; (a) router R3 joining the link, (b) router R2 leaving the link.

- When a router receives a HELLO or a HELLO REPLY message, it adds the identifier contained in the message to its list of active neighbors; conversely, when it receives a BYE message, it removes the identifier contained in the message from that list.

We illustrate the operation of this protocol in Figure 2.2. The numbers enclosed in circles indicate the steps of the algorithm, and under each step we show the list of active neighbors of each router. In this example, router R3 joins the link when routers R1 and R2 are already there (Figure 2.2.a), and later router R2 leaves the link (Figure 2.2.b). Initially, (step 1) the lists of active neighbors of routers R1 and R2 contain R1 and R2, and the one of router R3 contains only R3. When router R3 joins the link, it broadcasts a HELLO message with its identifier (step 2). Upon receiving this message, routers R1 and R2 add R3 to their lists of active neighbors (step 3) and each router replies broadcasting a HELLO REPLY message containing its identifier (step 4). When these messages are received, router R3 adds R1 and R2 to its list, and the algorithm terminates (step 5). Later, when router R2 decides to leave the link, it broadcasts a BYE message containing its identifier and deletes its list of active neighbors (step 6). Upon receiving this message, routers R1 and R3 remove R2 from their list of active neighbors and the algorithm terminates again (step 7).

2.1.2 Dealing with unreliable networks

Now suppose that control messages can be lost. This places several problems:

- If a BYE message is lost, the remaining neighbors will keep believing that the departing router is still active;

- If a HELLO message is lost, the neighbors already active on the link will not recognize the presence of the joining router.

The last problem cannot be solved through ACK protection (see Section 1.8.1): when a router joins a link and sends a HELLO message, it doesn't know how many ACKs it should receive since, to begin with, it doesn't know how many routers are active on the link. In fact, the only way to solve this reliability problem is through the periodic repetition of HELLO messages. In this case, even if the first message is lost, one of the subsequent ones will eventually succeed, and the router will become known to all its neighbors. Moreover, if the periodicity of HELLO messages is relatively short, the HELLO REPLY messages can be eliminated from the protocol.

The periodic repetition of HELLO messages can also deal with router and link failures, if combined with an age-lifetime mechanism. Specifically, each entry in the list of active neighbors is assigned an age and a lifetime. When a new entry is installed in the list, meaning that a new neighbor was discovered, it is assigned an age of zero. The age keeps increasing while the entry stays in the list, and is refreshed (reset to zero) whenever a HELLO message sent by the corresponding neighbor is received. When the entry reaches the lifetime, meaning that the communications with the neighbor have failed, it is removed from the list. The lifetime value must be several times the periodicity of HELLO messages. The periodicity of HELLO messages is 10 seconds in both OSPF and IS-IS, and the lifetime is 40 seconds in OSPF and 30 seconds in IS-IS (these are default values).

Note that the Hello protocol does not detect on its own router failures. It just detects loss of connectivity through specific links. The periodic repetition of HELLO messages signals the liveness of the communications with a router through a specific link, and the age-lifetime procedure provides a criterion to decide when communications are no longer possible (due to the lack of received HELLO messages). A router failure is only detected when the failed router becomes isolated in the network map (see Section 2.4.9).

2.1.3   Monitoring bidirectional communications

Receiving HELLO messages from a neighbor means that the neighbor is alive and can communicate with us, but does not guarantee that we can communicate with the neighbor. To see this, consider a point-to-point link consisting of two cables for transmission in either direction, and suppose that one cable fails but the other does not. This is illustrated in Figure 2.3.a. Router R1 receives HELLO messages from router R2, but the communication from router R1 to router R2 is not possible due to the link failure. Thus, the reception of HELLO messages *per se* is not an indication that communication in the outgoing direction is possible. As will be seen later, having this information is required in order to keep the network map updated (see discussion of Section 2.2.1.7).

A router can monitor the outgoing communications with a neighbor, if the

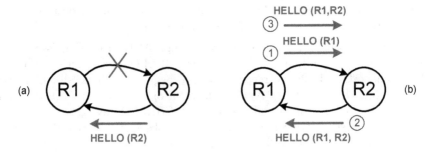

FIGURE 2.3: Monitoring bidirectional communications through the Hello protocol.

neighbor acknowledges the reception of the HELLO messages transmitted by the router. This can be achieved if HELLO messages list not only the identifier of the sending router, but also the identifiers of the active neighbors on the link. The procedure applies to both point-to-point and shared links and is illustrated in Figure 2.3.b. Router R1 sends a HELLO message listing itself (step 1). In the next HELLO message sent by R2, the router lists both itself and R1 (step 2), and when R1 receives this message, it concludes that the communications towards R2 are operational. Likewise, in the next HELLO message sent by R1, the router lists both itself and R2 (step 3), and when R2 receives this message, it concludes that the communications towards R1 are operational. In LSR protocols two routers are only considered neighbors if bidirectional communication between them is verified.

2.1.4   The complete Hello protocol

We can now give a complete description of the Hello protocol:

- Routers maintain a list of the active neighbors on a link, where each entry corresponds to a neighbor and has an age and a lifetime;

- Routers broadcast periodically on a link HELLO messages containing its list of active neighbors;

- When a router receives a HELLO message from a new neighbor, it installs a new entry in the list with the identifier of that neighbor and an age of zero; otherwise, if the neighbor already exists, it simply resets to zero the age of the corresponding entry;

- The age of an entry keeps increasing and, if it reaches the lifetime, it is removed from the list of active neighbors.

The timer that regulates the transmission of HELLO packets is called Hello timer, and the one that is used to detect if a neighbor is still active (i.e. if the

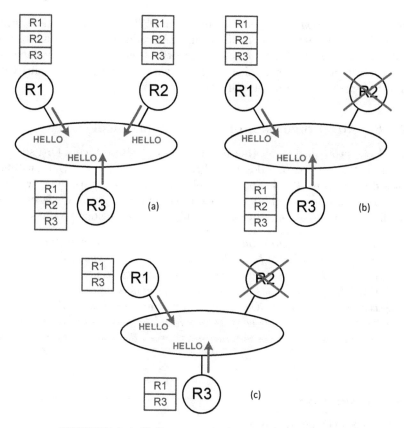

FIGURE 2.4: Hello protocol for general networks.

age of the corresponding entry is still lower than the lifetime) is called Hello liveness timer.

The protocol is illustrated in Figure 2.4. Routers R1, R2, and R3 send periodically on the link HELLO messages and, based on the contents of the received messages, build their lists of active neighbors (Figure 2.4.a). If router R2 fails, it ceases sending HELLO messages on the link, and the entry relative to this router on the lists of R1 and R3 keeps aging (Figure 2.4.b). When the entries reach the lifetime, they are removed from the lists, meaning that R1 and R3 no longer consider R2 as a neighbor (Figure 2.4.c).

2.1.5 Identifying neighbors

The correct operation of the Hello protocol requires that routers have unique identifiers within a link. However, since a router can have more than one interface attached to the same link (in case of shared links), the protocol requires that interfaces have unique identifiers within a link. There are several ways to achieve this, depending on the technology. OSPFv2 identifies each interface using IPv4 addresses, and IS-IS using MAC addresses; these addresses are globally unique. In OSPFv3, routers identify their interfaces by concatenating the router identifier (which is unique within a network domain) with an interface number assigned locally by the router.

2.1.6 Additional uses of the Hello protocol

The Hello protocol assists other mechanisms of LSR protocols. One such mechanism is the election of a designated router at shared links, which will be discussed in Section 2.2.1.4. The protocol is also used to determine if two routers can indeed become neighbors. For example, in OSPF, two interfaces attached to the same link can only become neighbors if they belong to the same area.

> *To keep the network map updated, routers need to maintain a list of the active neighbors on each of its links. This is achieved by the Hello protocol. In this protocol, each router broadcasts periodically on each link HELLO messages containing its list of active neighbors. If a router ceases to send HELLO messages for some time, its neighbors assume that it has been removed from the link. The Hello protocol is also used to determine if the communication between two routers is bidirectional.*

2.2 LSDB structure

The LSDB contains the routing information required to build the forwarding tables. As discussed above, there are three types of routing information: topological, addressing, and link information. Except for the link information, the routing information must be the same at all routers, so that routers can compute consistent forwarding tables. We say that routers must keep the LSDB *synchronized*. A good way to achieve this goal is to use a distributed strategy, where each router *disseminates* to all others information on its *local view* of the network. We call this piece of information the *Network Record* (NR) and the router that creates it the *originating router*. The dissemination of NRs is achieved through a controlled and reliable flooding procedure, which will be explained in Section 2.4. The following sections describe in detail the structure of the LSDB.

2.2.1 The network map

In LSR protocols, each router must build and maintain a *map* of the network, i.e. a representation of the *complete* network topology, containing all its routers

and links. This section discusses the various aspects related to the structure
of the network map.

2.2.1.1 The network representation

The network representation relies on an abstraction of the physical network,
which hides its irrelevant details. To obtain this representation, one needs to
consider three aspects:

- Which types of topological elements are considered in the network representation?

- How are the topological elements identified?

- How is the network topology (i.e. the connection between topological elements) described?

Following Section 1.7.2, we recognize three types of topological elements:
(i) routers, (ii) shared links, and (iii) point-to-point links, and assume that
these elements are identified using separate names spaces. To facilitate the
reading, we will refer to the names of the topological elements as *identifiers*.
Moreover, to distinguish among element types we prefix the router identifiers
by R, the shared link identifiers by ls, and the point-to-point link identifiers
by lp.

To build the network map, each router describes its local view of the
network topology using an NR. The NR includes (i) the router identifier,
(ii) the identifiers of the links the router is attached to, and (iii) the costs
assigned to the corresponding router interfaces. The NR assembled by one
router is disseminated to all others and, to build the complete network map
(to solve the puzzle...), each router joins together the received NRs, as well as
its own.

Figure 2.5 shows, for the network of Figure 1.1, the local view of each
router and the corresponding NRs. To facilitate the analysis, we repeat the
network graph in the upper-part of the figure. For example, the NR of router
R2 tells that the router is attached to links lp1, lp2 and ls2, with costs 10, 20,
and 30, respectively; and the NR of router R3 tells that the router is attached
to links ls1 and ls2, with costs 25 and 15, respectively. In a scenario where
all routers are switched on at the same time (cold start), initially each router
knows only its own NR. At this point in time, the network is still viewed as
four isolated islands!

Figure 2.6 shows the step-by-step construction of the network map at
router R3; all other routers behave similarly. Initially (Figure 2.6.a), router
R3 only knows about itself and the links it is attached to, i.e. ls1 and ls2; it has
only one NR—its own—in its network map. When router R2 disseminates its
NR (Figure 2.6.b), R3 learns about the existence of R2 and its links, lp1, lp2,
and ls2; R3 also learns that it can communicate directly with R2 via link ls2.
However, at this point, R3 still has a partial view of the network: it doesn't

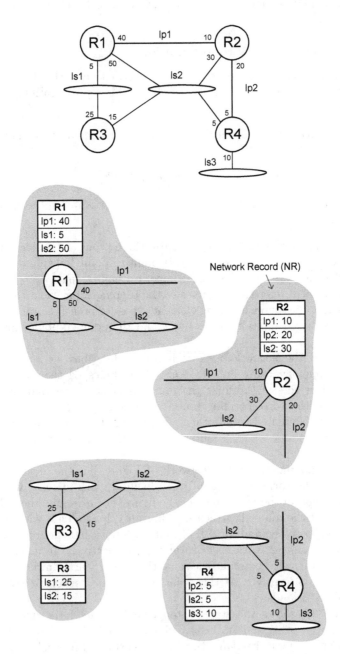

FIGURE 2.5: Local network view of each router and the corresponding Network Records (NRs), in the network of Figure 1.1.

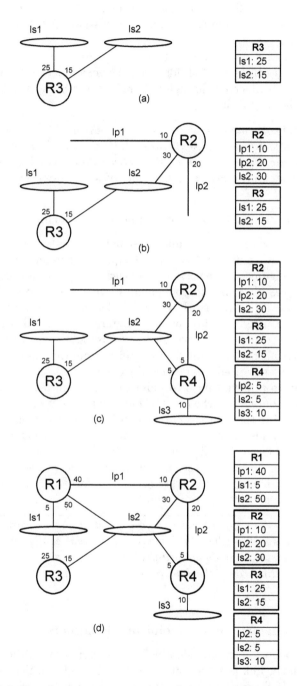

FIGURE 2.6: Step-by-step construction of the network map at router R3.

know about routers R1 and R4 and their links. The complete network map is only obtained when these two routers broadcast their NRs (Figure 2.6.c and Figure 2.6.d).

Note that one considers that two routers are connected through a link if their NRs share the same link identifier. For example, we know that routers R2 and R4 are connected through link lp2 because both their NRs include an entry referring to this link.

> *The network map is built from the individual contributions of all routers: each router disseminates to all others a Network Record (NR) containing its local view of the network, i.e. its identifier, and the identifiers and interface costs of the links it is attached to.*

2.2.1.2   Configuration of router and link identifiers

The identifiers of routers and links must both be configured at routers, since only routers are configurable layer-3 equipment. Router identifiers must be explicitly configured by the network manager; we say they are *static identifiers*. Link identifiers may also be configured statically. However, this solution is repetitive and prone to errors. For example, in the network of Figure 2.5, the network manager would have to configure the same identifier at all four routers attached to link ls2, without any mistakes, to ensure correctness. A better solution is to derive the link identifiers from the router identifiers, such that only the latter need to be statically configured.

One way to achieve this is to build the link identifier based on the identifiers of the routers attached to the link, and to rely on the Hello protocol for disseminating the router identifiers within the link. This solution is more flexible and avoids identifying links through dedicated name spaces. It also brings an additional benefit: due to the involvement of the Hello protocol, the presence or absence of the link identifier reflects directly the link state, as being operational or not operational. The identifiers configured in this way are called *dynamic identifiers*.

> *The router identifiers must be configured by the network manager at the routers. To reduce the amount of manual configuration, the link identifiers can be dynamically obtained from the identifiers of the routers attached to the link, which become known through the Hello protocol.*

2.2.1.3   Dynamic point-to-point link identifiers

A point-to-point link can be identified at a router through the identifier of its neighbor on the link (i.e. the router on the other side), which becomes known through the Hello protocol. However, since there may be several *parallel* point-to-point links between two routers, an additional element is needed to ensure the uniqueness of identifiers. This element can be a tag assigned locally by the

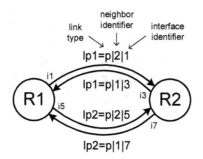

FIGURE 2.7: Point-to-point link identifiers.

router, unique among the point-to-point links leading to the same neighbor; the tag is usually the identifier of the link interface. Thus, a point-to-point link can be uniquely identified at a router using three elements: (i) the link type, (ii) the identifier of the neighboring router, and (iii) the identifier of the interface attaching the router to the link. Note that a point-to-point link is considered a *single* link providing communications in both directions. Thus, according to the above link identification criterion, a point-to-point link is named differently by each of its end routers (we know this seems odd...). Point-to-point link interfaces are also considered bidirectional, i.e. capable of transmitting and receiving traffic. Moreover, we will use the alias lp$i$ to refer to point-of-point link i whenever convenient, bearing in mind that the identifier has a more complex inner structure. This is illustrated in Figure 2.7. Routers R1 and R2 are connected through two parallel point-to-point links. The interface identifiers are i1 and i5 at R1, and i3 and i7 at R2. The upper point-to-point link is identified at router R1 by p|2|1 and at router R2 by p|1|3; the lower point-to-point link is identified at router R1 by p|2|5 and at router R2 by p|1|7.

Note that advertising all parallel links between routers may not be required. For example, if the network map is just being built for computing shortest paths, only the least cost link needs to be advertised.

Point-to-point links are identified using neighbor identifiers in both OSPF and IS-IS. IS-IS has no means to distinguish among parallel links, and only the least cost link is advertised. OSPF distinguishes among these links using the IPv4 address assigned to the interface, in OSPFv2, and the interface identifier, in OSPFv3.

> *A point-to-point link is dynamically identified at a router using three elements: the link type, the identifier of the neighboring router, and a local tag that distinguishes among parallel links.*

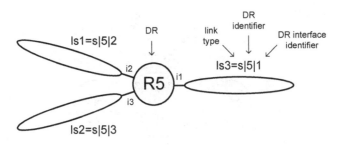

FIGURE 2.8: Shared link identifiers.

2.2.1.4 Dynamic shared link identifiers and designated routers

Shared links can be dynamically identified through the identifier of a router selected among all routers attached to the link. This router is called *Designated Router* (DR). At stub shared links, the DR is the sole router attached to the link. However, at transit shared links, an election must be performed to select the DR.

DR election The DR election can be based on the Hello protocol, since it periodically advertises (through HELLO messages) the identifiers of all routers attached to a shared link. Routers must agree on some criterion to decide which router is the DR. One possible criterion is to select the first router to become active on the link; another criterion is to select the router with the highest (or lowest) identifier. OSPF uses the first criterion and IS-IS the second. For example, in the network of Figure 2.5, the DR would be router R1 in both links ls1 and ls2, if the criterion is lowest identifier first, or router R3 in link ls1 and router R4 in link ls2, if the criterion is highest identifier first.

Uniqueness of link identification The first example above shows that a router can be the DR at more than one shared link. Thus, to ensure uniqueness in the identification of shared links, an additional element is needed. This element can be a tag assigned locally by the DR, unique among the links where the router is the DR; the tag is usually the identifier of the interface attaching the DR to the link. Thus, a shared link is dynamically identified using three elements: (i) the link type, (ii) the DR identifier, and (iii) the identifier of the interface attaching the DR to the link. Similarly to point-to-point links, we will use the alias ls$i$ to refer to shared link i whenever convenient, bearing in mind that the identifier has a more complex inner structure. Figure 2.8 gives an illustration. In the example, router R5 is the DR at three shared links; the identifier of the link attached via interface i1 is s|5|1, the one attached via interface i2 is s|5|2, and the one attached via interface i3 is s|5|3.

Shared links are identified in this way in IS-IS and OSPFv3. In OSPFv2, the DR identifier is not used, and shared links are identified through the IPv4 address assigned to the DR interface attaching to the link.

Advertising the shared link identifier Once the shared link identifier is determined by the DR, it must be advertised to all routers attached to the link. Without this piece of information, the routers cannot finish assembling their NRs. For example, in the network of Figure 2.5, routers R1 to R4 cannot complete building their NRs without knowing that ls2 is the identifier of the shared link they are attached to. Even if the routers can determine on its own which router is the DR (because they all know the election rule), the identifier tag is locally assigned by the DR, and only the DR knows it initially. A simple solution to advertise the shared link identifier is to use the Hello protocol. As before, routers attached to a shared link first become neighbors (i.e. they establish bidirectional communication) and elect the DR. Once the election is complete, the DR composes the shared link identifier and advertises it in the subsequent HELLO messages it transmits on the link. Thus, a field to advertise the shared link identifier must be added to HELLO messages broadcast on shared links.

> *A shared link is dynamically identified through the identifier of a router elected among all routers attached to the link, called Designated Router (DR). The link identifier is composed of three elements: the link type, the DR identifier, and a local tag assigned by the DR to distinguish among the various shared links where it is the DR.*

2.2.1.5 Types of network changes and their consequences

Networks are highly dynamic. When the state of a network changes, the network map must be updated as soon as possible to reflect the new reality. This means that NRs, once created and disseminated, may need to be replaced by *fresher* versions containing newer information. Specifically, if a router senses a change on itself or on the links it is attached to, it must disseminate a new NR to inform all others and update the network map. Thus, NRs must have a *freshness* attribute to express how recent they are, such that fresher NRs can replace older ones in the network map. We will refer to the successive NRs originated by a router as *NR instances*. In Section 2.4.1, we will discuss how freshness attributes can be composed.

Three types of network changes need to be considered:

- Changes in interface costs;
- Additions of routers and links;
- Removals of routers and links (e.g. failures).

Changes in interface costs When the cost of an interface is changed at a router, the router disseminates a new NR instance with the new cost, which will then replace the old one.

Additions of routers and links Similarly, when a new link is attached to a router, the router disseminates a new NR instance with a new entry containing the link identifier and the corresponding interface cost. Moreover, when a router joins the network, it disseminates a new NR identifying itself and describing its links.

Removals of routers and links The removal of routers and links is more difficult to handle. Two types of removals can be identified: *soft* removals, i.e. removals through configuration actions, and *hard* removals, i.e. failures. A link removal (soft or hard) is detected by the routers attached to the link, e.g. through hardware mechanisms or the Hello protocol. When a router detects this event, it disseminates a new NR instance without the entry associated with the removed link. Shared links may fail only partially, keeping two or more separate connectivity islands; we will discuss this problem in Section 2.2.1.6.

The impact of router removals in the network map depends on whether the removal is a soft removal or a failure. When a router is soft removed, it has the opportunity to delete its NR from the LSDBs of the remaining routers prior to being removed (e.g. switched off). This can be achieved by disseminating a delete message throughout the network; we will address this issue in Section 2.4.6. However, when a router fails, it has no opportunity to delete its NR. This shows that the network map may contain outdated NRs and, as will be shown later, these NRs may lead to erroneous conclusions about the network connectivity; this problem will be discussed in Section 2.2.1.7.

The connectivity graph To study the problem of outdated NRs, we resort to a graph expressing the connectivity that derives from the network map; we call it the *connectivity graph*. In this graph, routers are represented as nodes and the connectivity between routers as arcs. There will be an arc between two nodes only if the communication between the corresponding routers is bidirectional, since only in this case can the two routers be considered neighbors (see Section 2.1).

In the next sections, we will analyze in detail how shared link partitioning and router failures impact the network map. The discussion will be carried out using the network of Figure 2.9, which includes one shared link connecting four routers and three point-to-point links. Figure 2.9.a shows the network map and Figure 2.9.b the connectivity graph. In the graph, there is an arc connecting each pair of routers attached to the shared link. The graph also shows that R1 and R2 are directly connected through two different links.

2.2.1.6   Reaction to shared link partitioning

As discussed in Section 1.8.2, shared links may become partitioned in two or more connectivity islands; in this case, the network should keep using the resulting parts for routing purposes, if possible. Fortunately, the mechanism for assigning dynamic shared link identifiers deals correctly with this problem. When a link partitions, the Hello protocol identifies the subgroups of routers that can still communicate among themselves and elects a DR for each subgroup, except for the subgroup of the previous DR. Then, new link

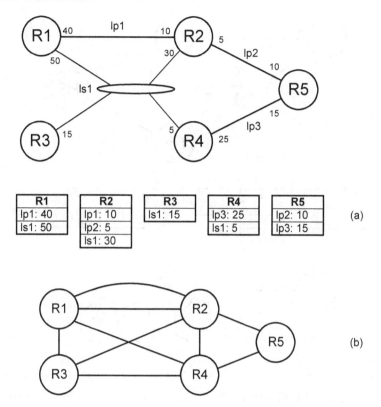

FIGURE 2.9: Example network to study the impact in the network map of link partitioning and router failures; (a) network map and (b) corresponding connectivity graph.

identifiers are obtained (based on the new DRs), and the routers attached to the links where the link identifier changed disseminate new NR instances with the updated information.

This is illustrated in Figure 2.10. Router R2 is initially the DR of the shared link, and the shared link identifier is ls1=s|2; we do not include the interface number in the shared link identifiers since, in this example network, it is not needed to uniquely identify the links. When the link partitions, routers R1 and R3 cease to receive HELLO messages from routers R2 and R4, and vice-versa. Thus, two subgroups get formed, each containing only the routers that recognize themselves as neighbors through the Hello protocol. The subgroup of routers R1 and R3 elects a new DR, which in this example is router R3, and the DR defines the identifier of the new link as ls2=s|3. This identifier is communicated to router R1 through the Hello protocol, and both R1 and R3 disseminate a new NR instance containing the new link identifier. At the subgroup of routers R2 and R4, there is no DR election (router R2 keeps being the DR) and the link identifier stays the same. Figure 2.10.a shows

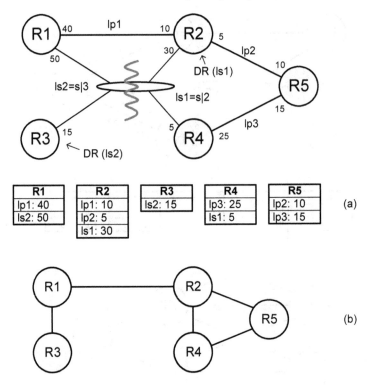

FIGURE 2.10: Shared link partitioning; (a) network map after the partitioning and (b) corresponding connectivity graph.

the network map after the partitioning, where the NRs of routers R1 and R3 already refer to link ls2 (instead of ls1), and Figure 2.10.b shows the corresponding connectivity graph. It can be seen that, despite the partitioning, all routers can still communicate among themselves. For example, routers R3 and R4 cannot communicate directly through the shared link, as before, but can communicate via routers R1 and R2.

2.2.1.7 Reaction to router failures

A router that fails cannot delete its NR from the network map. However, in general, the failure of a router can be inferred from the information disseminated by its neighbors, and this allows updating correctly the forwarding tables. Specifically, when a router fails, its neighbors detect the event through the Hello protocol and disseminate new NR instances where the link with the failed router is no longer listed. Then, in most failure scenarios, the failed router becomes isolated in the network map and, when other routers rebuild their forwarding tables, the new routes will no longer cross the failed router.

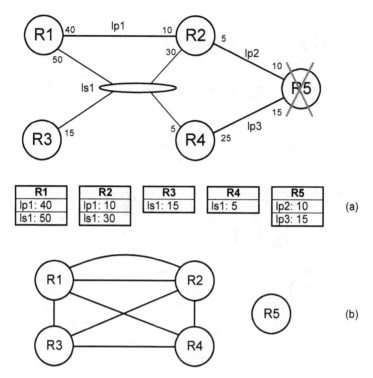

FIGURE 2.11: Failure of router connected with neighbors through point-to-point links (case 1); (a) network map after the failure and (b) corresponding connectivity graph.

Unfortunately, there are some failure scenarios where the failed router does not become isolated in the networks map.

Three cases must be considered regarding the neighbor that detects a failure:

1. The neighbor is connected to the failed router through a point-to-point link;

2. The neighbor is connected to the failed router through a shared link and the failed router was the link DR;

3. The neighbor is connected to the failed router through a shared link and the failed router was not the link DR.

Case 1 The first case is illustrated in Figure 2.11. Router R5 is connected to routers R2 and R4 through point-to-point links. When router R5 fails, routers R2 and R4 detect the failure through the Hello protocol, and both disseminate new NR instances where their link with router R5 is no longer listed. Figure 2.11.a shows that the NR of router R5 remains in the network map after the

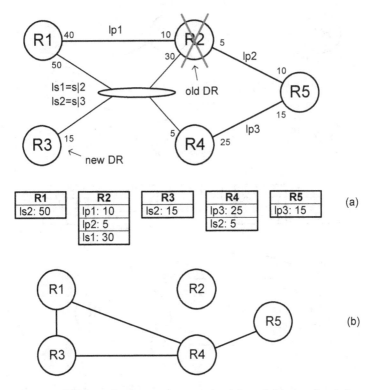

FIGURE 2.12: Failure of router connected with neighbors through a shared link where it is the DR (case 2); (a) network map after the failure and (b) corresponding connectivity graph.

failure. However, as seen in the connectivity graph of Figure 2.11.b, router R5 becomes isolated and, therefore, will no longer be used for routing purposes.

Case 2 The second case is illustrated in Figure 2.12. In this case, router R2 is the DR at the shared link and fails. The failure is detected by its neighbors on the shared link (and on its point-to-point links), through the Hello protocol. Once the failure is detected, the remaining routers attached to the shared link elect a new DR and define a new identifier for the link; this is again performed through the Hello protocol. We assume that the link identifier was initially ls1=s|2 and, after the failure, changed to ls2=s|3. Since the link identifier changed, the routers attached to the shared link disseminate a new NR instance, where the old link identifier is replaced by the new one. In addition, the point-to-point links leading to router R2 are suppressed from the NRs of routers R1 and R5. Figure 2.12.a shows the network map after the failure of router R2, and Figure 2.12.b shows the corresponding connectivity graph. As in the previous example, the NR of the failed router stays in the

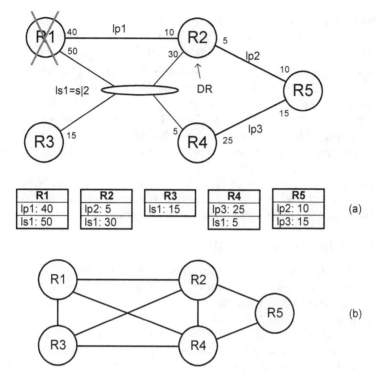

FIGURE 2.13: Failure of router connected with neighbors through a shared link where it is not the DR (case 3); (a) network map after the failure and (b) corresponding connectivity graph.

network map. However, as seen in the connectivity graph, router R2 becomes isolated, and will no longer be used for routing purposes.

Case 3 The third case is illustrated in Figure 2.13. In this case, router R1 fails, but the router is not the link DR. The failure is detected by its neighbors on the shared link and on its point-to-point link. However, since the failed router is not the DR, there is no need to elect a new DR, and the link identifier stays the same. Note that, contrarily to routers R2, R3, and R4, router R5 is not a neighbor of router R1 and, therefore, cannot detect the failure through the Hello protocol: nothing changed, from the perspective of this router! (Indeed, this was the motivation for selecting this network topology to study the problem of outdated NRs). The network map after the failure is shown in Figure 2.13.a, and the corresponding connectivity graph in Figure 2.13.b. Contrarily to previous cases, the failed router (router R1, in this case) remains reachable from all other routers, which is incorrect.

The root problem highlighted by this example is that the shared link identifier abstracts (too much...) the pairwise connectivity between routers attached to a shared link, and this simplification is unable to deal with some failure scenarios. The problem would not exist if a different link identifier was assigned to each pair of routers communicating over a shared link. However, this solution would not scale, as discussed in Section 1.7.1.

Alternative solutions to the problem of case 3 One solution to this problem is to impose a change in the link identifier whenever the failure of a non-DR router is detected. This can be implemented easily: the DR just modifies the tag that, together with the DR identifier, composes the shared link identifier. The drawback of this solution is that, in case of a failure, all routers must disseminate a new NR instance, containing the new link identifier.

Another solution to this problem is to create a new type of NR, describing shared links and their attached routers. This is the solution adopted in OSPF and IS-IS, and we will discuss it in the next section.

2.2.1.8 The shared-link-NR

The failure of a non-DR router attached to a shared link is not reflected in the network map, since the link identifier does not change in reaction to the failure. One way to deal with this problem is to introduce a new type of NR to represent transit shared links and their attached routers. This NR includes (i) the link identifier and (ii) the identifiers of all routers attached to the link. In this way, failures of routers attached to shared links, DR or non-DR, become immediately visible in the network map. This NR will be called *shared-link-NR* (slink-NR for short) and, to avoid confusion, we will call *router-NR* the NR introduced previously to represent a router and its links. Stub shared links need not be represented by slink-NRs, since these links have only one attached router.

One immediate question is what router should be responsible for originating slink-NRs. The natural choice is the DR of the shared link, since this router is already elected among the routers attached to the link for the purpose of defining the link identifier.

Example with router-NRs and slink-NRs Figure 2.14 shows the network map of the network of Figure 1.1, structured with router-NRs and slink-NRs. The router-NRs describe the routers and its links, and are exactly as before. But the network map now includes two slink-NRs, each representing one of the transit shared links, i.e. links ls1 and ls2. For example, the slink-NR of link ls2 lists all four routers attached to the link. Notice that the topological information gathered from the slink-NRs can be completely inferred from the router-NRs. Thus, from this point of view, slink-NRs add no information. However, as discussed above, they provide a way of dealing with router failures.

How slink-NRs solve the problem of non-DR failures at shared links? The use of slink-NRs allows double checking the network connectivity. In particular, two routers are only considered connected to each other through a transit shared link if (i) their router-NRs refer to the link (through the link identifier) and (ii) the slink-NR that describes the link refers to the two routers

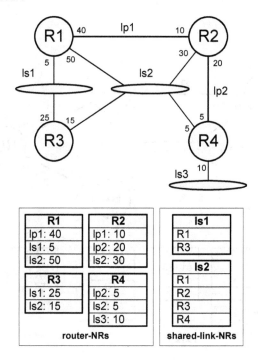

FIGURE 2.14: Network map of the network of Figure 1.1 structured with router-NRs and slink-NRs.

(through their router identifiers). These conditions should be taken into account when building the connectivity graph.

Figure 2.15 illustrates how the use of slink-NRs solves the problem of non-DR failures at shared links, illustrated in Figure 2.13 (case 3). Figure 2.15.a shows the network map before the failure. The network map now includes a slink-NR, listing all routers attached to transit shared link ls1. When router R1 fails, router R2, the DR of link ls1, detects the failure through the Hello protocol and disseminates a new slink-NR without R1 listed, which replaces the previous one. The new network map is shown in Figure 2.15.b. Again, the router-NR of router R1 stays in the network map, indicating that the router might be connected to link ls1. However, the router can no longer be considered connected to this link since it is not listed in the corresponding slink-NR. Thus, as shown in the connectivity graph of Figure 2.15.c, the router becomes isolated in the network map.

Revisiting the case of DR failures at shared links To complete the set of interesting cases, we show in Figure 2.16 how the network map structured with router-NRs and slink-NRs gets formed in the case of a DR failure (case 2). This was already addressed in Figure 2.12, for the case of a network map solely structured with router-NRs. The initial DR is again router R2, and the link identifier is ls1=s|2. When the DR fails, R3 is again elected DR, and

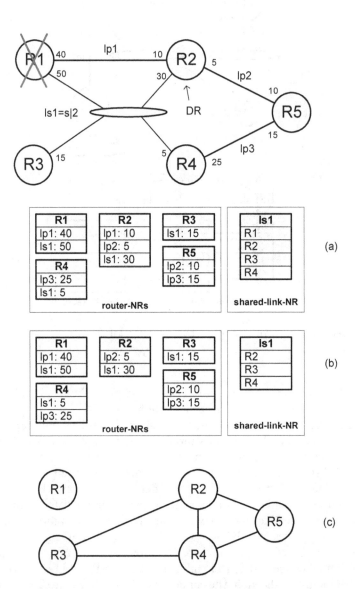

FIGURE 2.15: Failure of router connected with neighbors through a shared link where it is not the DR. Network map with router-NRs and slink-NRs (a) before the failure, (b) after the failure, and (c) corresponding connectivity graph.

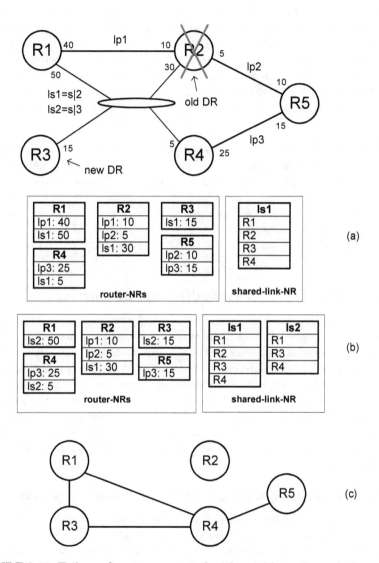

FIGURE 2.16: Failure of router connected with neighbors through shared link where it is the DR. Network map with router-NRs and slink-NRs (a) before the failure, (b) after the failure, and (c) corresponding connectivity graph.

defines ls2=s|3 as the new link identifier. It then disseminates a new slink-NR
that no longer includes router R2. The previous slink-NR, listing routers R1
to R4 as being attached to link ls1, stays at the LSDB since R2 could not
delete it (due to the failure). Thus, as seen in Figure 2.16.b, there are now
two slink-NRs, one new (referring to link ls2) and one outdated (referring to
link ls1). The router-NR of router R2 also remains in the network map. The
two outdated NRs indicate that router R2 might be connected to link ls1.
However, no other router is connected to this link, since the router-NRs of
R1, R3, and R4 have been updated to list ls2 (instead of ls1). Thus, as shown
in the connectivity graph of Figure 2.16.c, router R2 becomes isolated in the
network map.

> *The network map must be kept updated when interface costs change,
> when routers or links are added or removed, and when shared links
> are partitioned. To deal with the problem of router failures, a new
> NR describing shared links and their attached routers is required. It
> is called shared-link-NR (slink-NR) and, to avoid confusion, the NR
> that describes routers and their links is now called router-NR. The
> slink-NR is originated by the DR of the shared link it represents.*

2.2.2   The area-internal addressing information

Up to now we have been only concerned with the representation of the network
topology—the network map. However, in order to build forwarding tables—
the ultimate goal of a routing protocol—routers need to know which *routable*
address prefixes are available at the network and to which topological elements
(routers or links) they have been assigned. There are several possibilities for
combining the topological and addressing information at the LSDB. In the
next sections, we discuss the alternatives for representing the addressing information *internal* to a single-area routing domain. The representation of addressing information *external* to a routing domain will be discussed in Section
2.2.4.

2.2.2.1   Separating topological and addressing information

From the point of view of routing, it is possible and desirable to *separate* the
information related to the network topology, i.e. the *topological information*,
from the one related to routable prefixes, i.e. the *addressing information*.
There are two aspects in this separation: the first is concerned with the name
spaces used by these two types of information, and the second with the way
the information is disseminated. We will discuss these two aspects next.

Name spaces The identifiers of topological elements could, in principle, share
the name space of routable prefixes. However, this is not desirable, since these
two types of identifiers have intrinsically different functions. The identifiers
of routers and links aim at the representation of the intra-domain network:

	Independent name spaces	Independent dissemination
OSPFv2	No	Yes
OSPFv3	Yes	Yes
IS-IS (IPv4 and IPv6)	Yes	No

FIGURE 2.17: Separation of topological and addressing information in OSPFv2, OSPFv3, and IS-IS.

they need only be unique inside a routing domain and need not be present at forwarding tables. Conversely, the routable prefixes represent the ultimate destinations of information: they must be present at forwarding tables and, for Internet-wide communications, they must be prepared to identify unambiguously the whole of Internet interfaces. Thus, the name space of routable prefixes needs to be much larger in size and scope than that of topological identifiers. Significant overhead is incurred if one uses routable addresses to identify routers and links, especially in an era where the number of hosts is increasing very fast, pushing for the rapid deployment of IPv6 addressing. Just as the identifiers of routers, point-to-point links, and shared links are preceded by keywords R, lp, ls, respectively, routable address prefixes will be preceded by keyword ap.

Dissemination The second aspect mentioned above relates to the dissemination of topological and addressing information. These two types of information can be disseminated jointly or separately, but there are good reasons for disseminating them separately. First, prefixes assigned to specific routers or links may change over time, and may change more frequently than the network topology. Second, new addressing schemes may be introduced along time, and this was precisely what happened with the evolution from IPv4 to IPv6.

The importance of separating topological and addressing information has not been recognized, at least initially, by some LSR technologies: OSPFv2 uses the same name space for topological and addressing information (at least in part), but these two types of information can be disseminated independently (although in some cases they are not); IS-IS separates the name spaces but forces topological and addressing information to be disseminated together; finally, OSPFv3 separates the name spaces and allows these types of information to be disseminated independently. Figure 2.17 summarizes this comparison.

> *A good design principle regarding LSR protocols is the separation of topological and addressing information, i.e. to describe these two types of information separately (using independent name spaces) and to disseminate them separately.*

2.2.2.2   The area-internal-prefix-NR

The router-NR and the slink-NR are the elements needed to represent the network topology, and we will refer to them as the *topological NRs*; they form what we have been calling the network map. If topological and addressing information are to be separated according to the principle introduced in the previous section, then a new NR is required to advertise address prefixes and their association with topological elements. It will be called *area-internal-prefix-NR* (aip-NR for short). We use the term *area-internal* since, in this chapter, we are dealing with single-area routing domains; as will be discussed in chapter 3, in multi-area domains there is the need to describe addressing information external to an area (area-external prefixes). Note that only OSPFv3 and IS-IS (IPv4 and IPv6) include the equivalent of an aip-NR; OSPFv2 mixes topological and addressing information, an issue that will be discussed in Section 2.2.2.3.

There are three important issues that need to be addressed regarding aip-NRs: (i) how are prefixes related to the topological elements they were assigned to, (ii) which router originates an aip-NR, and (iii) how many prefixes to advertise in each aip-NR.

How are prefixes related to topological elements? Address prefixes (or even individual addresses) can be assigned to the three types of topological elements—routers, point-to-point links, and shared links—, to provide attachment points for hosts or for management purposes. A simple way to relate prefixes to topological elements is through the topological identifiers, i.e. to establish a correspondence between prefixes and the identifiers of the routers or links they were associated with. Thus, an aip-NR must include (i) the prefix being advertised and (ii) the identifier of the topological element the prefix was associated with (i.e. the *pointer* to the topological element).

Routers advertising prefixes on behalf of links A prefix assigned to a router is naturally associated with the router to which it belongs; thus, an aip-NR advertising a prefix assigned to a router points to the router using the router identifier. However, a prefix assigned to a link may not be directly associated with the link to which it was assigned, but with a router attached to the link. In this case, the aip-NR points to the router, through the router identifier, and not to the link. We may say that the router advertises the prefix *on behalf* of the link.

When prefixes assigned to links are associated with routers, the advertising aip-NR must include, besides the prefix and the router identifier, the cost of the interface attaching the router to the link. This information is required so that the costs of the paths leading to the prefix can be calculated. In fact, this can be seen as a departure from the separation principle, because interface costs are mixed with address prefixes. The path cost calculation performed by a router has two steps: (i) first, calculate the path cost from the router where the calculation is being made to the router associated with the prefix; (ii) second, add to the previous path cost the interface cost declared in the aip-NR that advertises the prefix.

Moreover, in the above solution, all routers attached to the link must advertise the prefix, because each of them provides a path to the link that may have a different path cost. Thus, in this case, a prefix assigned to a point-to-point link is advertised by the two link end routers, a prefix assigned to transit shared links is advertised by all routers attached to the link, and a prefix assigned to a stub shared link is advertised by the (single) link router. In the first and second cases there will be repeated addressing information in the LSDB.

Removing link identifiers from the network map Going one step further, when prefixes assigned to links are associated with routers, the corresponding link identifiers can be removed from the network map. This is the usual practice in the representation of stub shared links. In this case, a prefix assigned to a stub shared link is associated with the (single) router attached to the link and advertised through an aip-NR that includes (i) the prefix, (ii) the router identifier, and (iii) the cost of the interface attaching the router to the link. Moreover, the router-NR advertising the router and its links no longer includes an entry relative to the stub shared link.

Which router originates an aip-NR? Whether advertising prefixes assigned to routers or links, aip-NRs can only be originated by routers, since only routers are layer-3 active equipment. Naturally, an aip-NR advertising a prefix associated to a router is originated by that router. As discussed above, in the case of a prefix assigned to a link but associated with a router, all routers attached to the link must advertise the prefix and, therefore, all of them must originate aip-NRs relative to the prefix.

For prefixes assigned and associated to links there are essentially two alternatives: to have the prefix advertised by (i) all routers attached to the link or (ii) a router selected among all routers attached to the link. In the former solution there will be repeated addressing information in the LSDB, except in the case of stub shared links. The latter solution implies electing a router, which is already done in the case of transit shared links, for the purpose of defining the link identifier and originating slink-NRs (see Sections 2.2.1.4 and 2.2.1.8).

How many prefixes to advertise in each aip-NR? Another issue that needs to be faced is how many address prefixes to advertise in each aip-NR. In fact, a router may be responsible for announcing several prefixes, e.g. assigned to itself, to its point-to-point links, or to the shared links where it is the DR. These prefixes can either be all be advertised in the same aip-NR or in different ones. However, if one of the prefixes originated by a router is changed, it should

Prefix assigned to...	aip-NR originator(s)	Pointer to topological element
router	the router	router id
point-to-point link	the two link end routers	link id
transit shared link	the link DR	link id
stub shared link	the link end router	link end router id[*]

[*]includes cost of router outgoing interface

FIGURE 2.18: The originators and pointers to topological elements of aip-NRs in OSPF and IS-IS.

be possible to update the LSDB without having to disseminate again all the addressing information originated by that router. Thus, it is preferable to advertise only one address prefix in each aip-NR, and we will assume this option hereafter.

How are prefixes advertised in OSPFv3 and IS-IS? In OSPFv3 and IS-IS, prefixes assigned to routers, point-to-point links, and stub shared links are advertised in the same way, but prefixes assigned to transit shared links are advertised differently (see Figure 2.18).

- **Routers -** A prefix assigned to a router is associated with the router (through the router identifier) and is advertised by an aip-NR originated by the router.

- **Point-to-point links -** A prefix assigned to a point-to-point link is associated with the link (through the link identifier) and is advertised by two aip-NRs, each originated by one of the link end routers.

- **Stub shared links -** A prefix assigned to a stub shared link is associated with the (single) link router (through the router identifier) and is advertised by an aip-NR originated by that router. The aip-NR includes the cost of the interface attaching the router to the link, and the link is removed from the network map.

- **Transit shared links -** A prefix assigned to a transit shared link is associated with the link (through the link identifier) in both OSPFv3 and IS-IS. However, in OSPFv3, it is advertised by a single aip-NR originated by the link DR and, in IS-IS, it is advertised by all routers attached to the link, each originating one aip-NR relative to the prefix (and producing repeated addressing information).

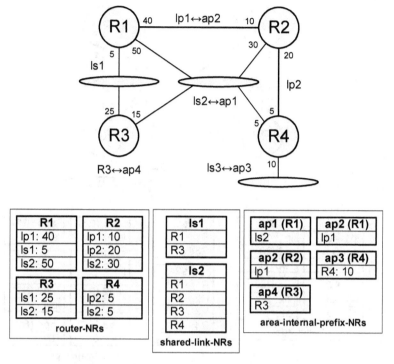

FIGURE 2.19: LSDB of the network of Figure 1.1, with separated topological and addressing information: the OSPFv3 case.

OSPFv3 example In Figure 2.19, we illustrate the LSDB structure of OSPFv3 with topological and area-internal addressing information, using the network of Figure 1.1. We have only assigned four address prefixes: ap1 to transit shared link ls2, ap2 to point-to-point link lp1, ap3 to stub shared link ls3, and ap4 to router R3. The figure shows the various NRs that comprise the LSDB. The topological information is given by the router-NRs and slink-NRs, as before. They coincide with those of Figure 2.14, except for the stub shared link that is no longer represented.

The addressing information is provided by the aip-NRs. In the figure, we have included, in parenthesis, the aip-NR originators. Prefix ap1 is advertised through an aip-NR originated by the DR of link ls2, which we assume being R1. Prefix ap2 is advertised through two aip-NRs, one originated by router R1 and another by router R2. Prefix ap3 is advertised through an aip-NR originated by router R4, and includes the cost of the interface attaching to the link, i.e. a cost of 10. Finally, prefix ap4 is advertised through an aip-NR originated by router R3.

IS-IS example Figure 2.20 repeats the case of Figure 2.19, but for IS-IS. The only difference is that the prefix ap1, assigned to the transit shared link,

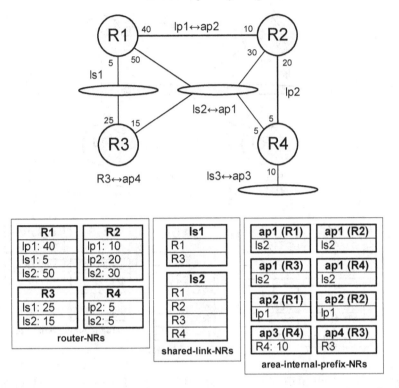

FIGURE 2.20: LSDB of the network of Figure 1.1, with separated topological and addressing information: the IS-IS case.

is advertised by four aip-NRs, each originated by one of the routers attached to the link. There is a lot of repeated information!

We highlight here that not all topological elements need to be assigned routable addresses. This is a common misconception, probably stimulated by the technologies that mix topological and addressing information. Only those topological elements that must be reachable (e.g. for management purposes or because they have hosts attached) need to be assigned addresses. For example, a shared link that is only used as a transit link between routers and has no hosts attached, needs not! We will illustrate this possibility through an experiment described in Section 9.2.1.3.

> When topological and addressing information are separated, the ad-
> dress prefixes internal to an area are advertised through area-internal-
> prefix-NRs (aip-NRs), containing the identifier of the topological ele-
> ment the prefix is associated with. There are several alternatives for
> associating prefixes with network elements and for the origination of
> aip-NRs. The usual practice is to have (i) prefixes assigned to routers
> advertised through an aip-NR originated by the owning router, (ii) pre-
> fixes assigned to point-to-point links advertised through two aip-NRs
> each originated by one link router, (iii) prefixes assigned to transit
> shared links advertised through either one aip-NR originated by the
> link DR, or aip-NRs originated by all routers attached to the link,
> and (iv) prefixes assigned to stub shared links advertised through an
> aip-NR originated by the (single) link router. In the latter case, the
> aip-NR includes the cost of the interface attaching to the link, which
> allows removing stub shared links from the topological representation
> of the network.

2.2.2.3 Mixing topological and addressing information

If the same name space is used for topological and addressing information, then
there is no need to have a separate NR to convey the addressing information:
this information can be included in the router-NRs and slink-NRs. Moreover,
except for prefixes assigned to routers and stub shared links, only the prefix
length needs to be added, since the topological identifier combined with the
prefix length completely defines the prefix. In the context of IP addressing,
mixing topological and addressing information means that the topological
elements (routers and links) are identified through IP addresses.

Figure 2.21 illustrates this solution. In the figure, the keyword *pl* refers
to the prefix length, and pl*i* denotes the length of prefix ap*i*. The prefix assigned to router R3 (ap4) is advertised through a new entry in the router-NR
originated by R3; the prefix definition requires two elements: the prefix base
address and the prefix length. Similarly, the prefix assigned to the stub shared
link (ap3) is advertised through a new entry in the router-NR originated by
R4; this entry needs to include the cost of the interface attaching the router
to the link. The prefix assigned to the point-to-point link (ap2) is advertised
by adding the prefix length pl2 to the entries of the router-NRs of R1 and
R2 that announce the link. Finally, the prefix assigned to the transit shared
link (ap1) is advertised by associating the prefix length pl1 with the shared
link identifier of the slink-NR that announces the link. Instead, the prefix
length could have been added to the router-NRs of the routers attached to
the link (as in the point-to-point links), but this would unnecessarily repeat
this information and waste memory resources.

The solution just described for advertising the topological and addressing
information is very close to the one adopted by OSPFv2.

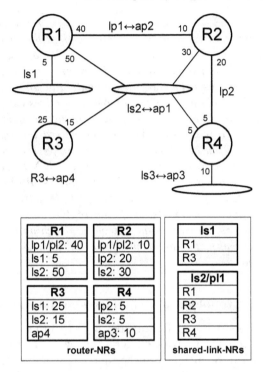

FIGURE 2.21: LSDB of network of Figure 1.1 with mixed topological and addressing information: the OSPFv2 case.

> *When the topological and addressing information are mixed, the router-NRs and slink-NRs describe both types of information.*

2.2.3 The link information

We have already introduced three NR types: the router-NR, the slink-NR, and the aip-NR. The first two describe the network topology, and the third advertises the address prefixes available at the network and their association with topological elements. From this information, a router can compute the shortest paths from itself to every prefix in the network. The path determines, locally at a router, the outgoing interface and, except for the last-hop router, the next-hop router leading to the prefix. However, when a router needs to forward a packet through a layer-2 network to a next-hop router, it still needs to find the layer-2 address of the next-hop router interface that should receive the packet. For example, if a router needs to send a packet over an Ethernet LAN to another router, the packet must be encapsulated into an Ethernet frame containing the MAC address of the destination router interface. Layer-

2 addresses are not needed in the case of point-to-point links, but are required for shared links, since more than one next-hop router may be available.

We will call *link address* a layer-2 address, or an address that might be resolvable into a layer-2 address at which the next-hop router can be reached within the link. Link addresses need not be layer-2 addresses, if there is some address resolution mechanism to resolve the link address into the corresponding layer-2 address. For example, the next-hop router may provide an IPv4 address, to be resolved into a MAC address using ARP.

Forwarding tables must include, in each entry, the link address of the next-hop interface that leads to the corresponding destination. In the forwarding tables of commercial routers, the next-hop link addresses are layer-3 addresses (IPv4 or IPv6) and are included in all entries relative to non-directly attached links, even in entries relative to point-to-point links, where they are not needed; we will also adopt this approach. To distinguish link addresses from other types of identifiers, we will prefix them by *la*.

To provide information on link addresses, we introduce the *link-address-NR* (la-NR for short). An la-NR includes (i) the link address, (ii) the topological identifier of the originating router, and (iii) the topological identifier of the link where the address is to be used. A router originates one la-NR per attached link, and each la-NR advertises the link address at which the router can be reached on a link. The la-NRs need only be sent to neighboring routers in a given link, i.e. la-NRs need only have link flooding scope.

We note that the link information is of local nature, since it needs only to be known within a link: a router only needs to know the layer-2 addresses of its neighbors on each of its links. Thus, unlike the topological and the addressing information, the link information is not disseminated throughout the whole network and, therefore, is not the same at the LSDBs of all routers.

OSPFv3 is the only technology that provides link addressing information through la-NRs. In OSPFv2, this information is provided through the equivalent of router-NRs, and in IS-IS it is provided through the Hello protocol.

Figure 2.22 shows the network of Figure 1.1, including the link addresses of all interfaces and the la-NRs of links ls1 and ls2. Thus, this is the complete LSDB of router R3. For example, router R3 learns from the la-NRs present in its LSDB that it can use link address la6 to communicate with router R1 via link ls2, and link address la8 to communicate with the same router via link ls1. The figure also illustrates that link addresses can be the same at different links. For example, router R3 provides link address la2 to its neighbors on links ls1 and ls2.

> *When a router wants to transmit a packet on a layer-2 link to a next-hop router, it needs to determine first the layer-2 address of the interface that should receive the packet, or an address that might be resolvable into it. This address is called a link address. Link addresses are provided through link-address-NRs (la-NRs), which are disseminated only within the link.*

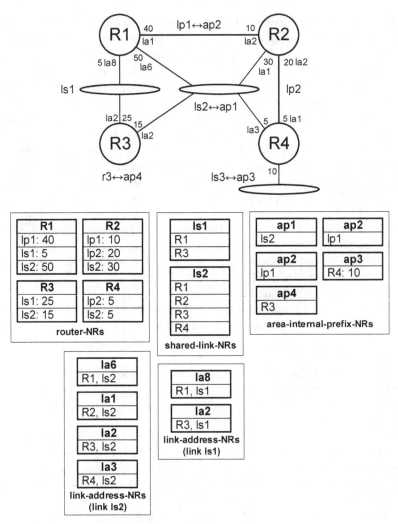

FIGURE 2.22: LSDB of router R3 in the network of Figure 1.1, with separate topological, addressing, and link information.

2.2.4   Interconnection with other routing domains

LSR protocols provide routing inside routing domains. However, users inside one routing domain must be able to communicate with users located in other routing domains. The problem of communication among different routing domains can be solved through an inter-domain routing protocol that exchanges the address prefixes owned by each domain and determines paths between them. The inter-domain routing protocol runs on special routers located in the border between routing domains, called Domain Border Routers (DBRs).

We may say that DBRs and their interconnections form a routing overlay for the exchange of inter-domain routing information. This issue was discussed in Section 1.4.3.

Besides the structure in routing domains, the Internet is organized in administrative domains called Autonomous Systems (ASes). Each AS can contain one or more routing domains, but routing domains are strictly contained within ASes. Thus, at the worldwide Internet level, ASes can be seen as single routing domains. In this case, the DBRs are called Autonomous Systems Borders Routers (ASBRs) and there must be one ASBR on either side of the border between ASes. Moreover, the inter-domain routing protocol that runs on the ASBRs and is used for inter-AS communications is currently the BGP protocol.

From the perspective of a router internal to a routing domain that needs to contact an external destination, it is irrelevant whether the domain is connected to another domain of the same AS or to another domain of a different AS. Thus, we will discuss this problem in the generic context of routing domains and DBRs, bearing in mind that routing domains may coincide with ASes. Figure 2.23 shows a network with several routing domains.

DBRs run both the inter-domain and intra-domain routing protocols. They are responsible for *exporting* to other domains the addressing information collected internally through the intra-domain routing protocol, and for *importing* the addressing information of other domains collected through the inter-domain routing protocol. This is a two-way process. We assume that the topological information is not exchanged across routing domains.

From the point of view of an internal router, there are now two types of address prefixes: *domain-internal* prefixes and *domain-external* prefixes. Moreover, the path from an internal router to an external destination, located in another routing domain, now has two parts: the internal sub-path, from itself to the outgoing DBR, and the external sub-path, from the DBR to the external destination. The end-to-end path need not be the same in both directions, i.e. the outgoing and ingoing DBRs of a domain need not be the same for a given origin-destination pair.

Routing across different domains brings new challenges. A routing decision involves determining both the internal and external sub-paths. However, different domains may use different types of path metrics, which can be difficult to combine into an end-to-end metric. For example, one domain may use minimum-hop routing, another shortest path routing based on arbitrary link costs, and a third one widest-path routing. Moreover, when routing between ASes, the performance is no longer the single path selection criterion, and is often not the most important one: political, economic, and security factors must also be accounted for in routing decisions. For example, in Figure 2.23, if domain d is not trusted or charges too much, it should not be used to route information, e.g. between domains a and c. To address these issues, inter-domain routing protocols associate each external prefix with a set of attributes that characterize the external sub-path. We will call them *external*

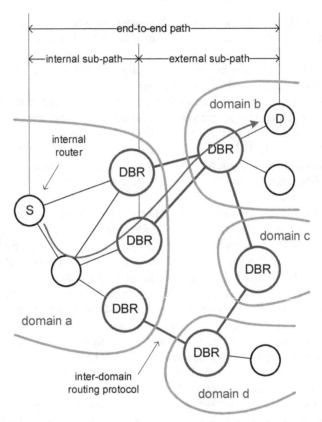

FIGURE 2.23: Routing across routing domains.

attributes. For example, in BGP [46], one of the most important external attributes is the AS PATH, which describes the sequence of ASes in the path towards the associated external prefix. Thus, a BGP router can perform routing decisions while knowing the identity of all ASes in the candidate paths to a destination and, in this way, it can avoid unwanted ASes. The information that gathers the external prefixes with their attributes is sometimes called *external reachability information.*

2.2.4.1 Exporting and importing addressing information

Exporting addressing information is a simple task. DBRs are part of the routing domain and, therefore, know all addressing information of their domain, which they obtained through the intra-domain routing protocol. Thus, they can easily inject this information into the inter-domain routing protocol and deliver it to their external neighboring DBRs.

The import issue can be more complex and depends on whether the routing domain has a single or multiple DBRs (see Figure 2.24). If the domain has

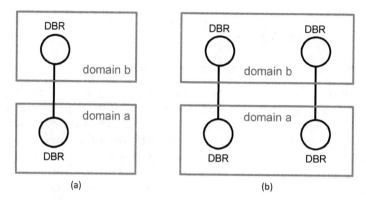

FIGURE 2.24: Routing domains (a) with one DBR or (b) multiple DBRs.

a single DBR, then there is only one entry/exit point for external domains. In this case, there is no need to inject external addressing information into the domain. One can simply install a special entry in the forwarding tables of internal routers pointing to the (single) DBR, to be used whenever the intended destination does not match any other entry. This entry is called *default route* or *gateway of last resort*. To support this routing solution one additional feature is required: internal routers must be able to identify the DBR in the network map. This can easily be achieved by including a flag in router-NRs, to signal whether the originating router is a DBR. The issue of default routes will be further discussed in Section 2.3.4.

If the domain has more than one DBR, then it becomes relevant to determine the best DBR for each external destination. Different DBRs may receive information on the same external address prefix with different external attributes, which can then be used for a decision. There are several alternatives to deal with this issue:

- Let each DBR inject the external reachability information in the domain and leave the decision up to internal routers.

- Allow the various DBRs of a domain to exchange external reachability information among themselves to determine which DBR is the best for each external destination, and let only the best DBR inject in the domain the corresponding external reachability information. This approach is allowed by the BGP protocol; it can be implemented through internal BGP sessions established among the BGP routers of the same AS.

- Let internal routers decide which DBR to use without accounting for external reachability information. In this case, one selection criteria could be to use the closest DBR.

In the first and second alternatives, there must be a way to advertise external reachability information inside a routing domain. This issue will be addressed in the next section.

2.2.4.2 The domain-external-prefix-NR

External reachability information can be advertised inside a routing domain through a new NR, called *domain-external-prefix-NR* (dep-NR for short). The dep-NR is originated by a DBR and includes (i) the advertised external prefix and (ii) the identifier of the originating DBR. We will assume it also includes a single cost attribute, called *external cost*, that summarizes the external attributes associated with the advertised prefix. This attribute is only relevant in the first alternative described above, where the routing decision is left to internal routers. In solutions that mix topological and addressing information, the domain-external prefixes could also be advertised through the router-NRs originated by the DBRs. However, we will not consider this possibility since it is not used in practice.

Consider the example of Figure 2.25, where we omitted the topological NRs; the example relates to the second alternative described above, where the routing domain is an AS connected to the Internet through two ASBRs (routers R1 and R2), and the ASBRs communicate with each other and with external ASBRs using BGP. Suppose that both ASBRs receive reachability information regarding external prefixes ap1 and ap2 (step 1), and that each prefix comes with an AS PATH attribute indicating the number of hops (ASes) in the path that leads to it. Suppose that routers R1 and R2 are allowed to exchange this attribute (step 2) and are programmed to select the ASBR that provides the lowest number of hops. In this case, router R1 verifies it is the best ASBR for reaching ap2 and router R2 verifies is the best for reaching ap1. Thus, router R1 floods internally the dep-NR advertising ap2, and router R2 floods the one advertising ap1 (step 3).

> *A routing domain communicates with other routing domains through its DBRs, and learns the address prefixes of other domains through the inter-domain routing protocol. If the domain has a single DBR, there is no need to inject these prefixes into the domain; default routes pointing to the DBR should be used instead. If the domain has more than one DBR, the DBRs may inject information on the prefixes into the domain, to allow internal routers deciding on the best outgoing DBR to reach each prefix. This information is disseminated through domain-external-prefix-NRs (dep-NRs), originated by the DBRs.*

2.2.5 NR identifiers

The NRs must be uniquely identified within the LSDB to avoid confusion in the construction of the network map. One could consider the possibility of identifying NRs through the identifiers of the elements they represent, e.g.

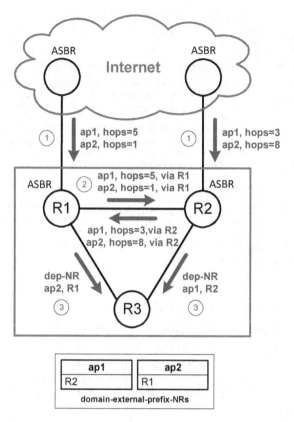

FIGURE 2.25: Injecting external reachability information in a domain with multiple ASBRs.

using the shared link identifier, in the case of slink-NRs, or the address prefix, in case of aip-NRs. However, this would result in lengthy and non-uniform identifiers. Moreover, in the case of prefixes assigned to point-to-point links, the same prefix is advertised by two different NRs.

Thus, we will adopt a generic way of identifying NRs, also used in OSPF and IS-IS, where NRs are identified using the following three elements: (i) the identifier of the originating router, (ii) the NR type indicator, and (iii) a locally assigned tag that distinguishes among the NRs of the same type originated by the same router. We will refer to this set of three elements as the *NR identifier*.

> *NRs are uniquely identified using three elements: the identifier of the originating router, the NR type indicator, and a locally assigned tag to distinguish among NRs of the same type originated by the same router.*

NR	role	originator
router-NR	describes router and attached links	own router
slink-NR	describes transit shared link and attached routers	link DR
aip-NR	describes domain-internal prefix and its association with network element	varies with network element
dep-NR	describes domain-external prefix and its association with injecting DBR	DBR
la-NR	describes link address to transport packets between neighbors	own router

FIGURE 2.26: NRs that form the LSDB of a single-area LSR network, their role and originating router.

2.2.6   Summary of LSDB structure

We summarize now the NR types introduced in previous sections. The actual LSDB structure, namely which NRs are used and the way they are used, depends on whether the topological, addressing, and link information are described separately, and on the way the addressing information is related to the topological elements. The role and the originator of each NR is summarized in Figure 2.26.

- **router-NRs and slink-NRs** - The router-NR and the slink-NR provide the topological information, i.e. the network map. The router-NR describes a router and its attached links, and is originated by the router being described. The slink-NR describes a transit shared link and its attached routers, and is originated by the link DR. Stub shared links are not represented in the network map. In solutions mixing topological and addressing information, these NRs also provide addressing information.

- **aip-NR** - The aip-NR provides addressing information internal to the routing domain. The aip-NR originator depends on the network element that the prefix is associated with (see discussion in Section 2.2.2.2). This NR is not used in solutions mixing topological and addressing information.

- **dep-NR** - The dep-NR provides addressing information external to the routing domain, and is originated by the DBR that injects the prefix into the domain.

- **la-NR** - The la-NR provides link information. It is originated by a router on each of its links to advertise the link address at which the router can be reached on the link.

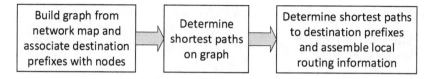

FIGURE 2.27: Steps in the construction of a forwarding table.

2.3 From the LSDB to the forwarding table

As discussed in Section 1.4.2, the forwarding table of a router has one entry per destination prefix, and each entry contains information on the path towards the prefix, which includes the outgoing interface, the link address of the next-hop interface, and the path cost.

The forwarding table can be obtained at a router from the topological, addressing, and link information contained in its LSDB. This involves three steps (see Figure 2.27):

1. Represent the network topology through a graph and associate the destination prefixes with nodes of the graph.

2. Run a centralized shortest path algorithm over the graph, to determine the shortest paths between the node representing the router where the forwarding table is being built (the source node) and the other nodes of the graph. Several shortest path algorithms are available, but Dijkstra's algorithm is the one used in LSR protocols.

3. Determine the shortest path to the destination prefixes and assemble the local routing information, i.e. the outgoing interface and the link address of next-hop interface that leads to each prefix.

We detail each of these steps in the next sections.

2.3.1 Building the network graph and associating destination prefixes with nodes

How to set up the network graph The graph on which Dijkstra's algorithm will be used to find the end-to-end shortest paths is set up based on the network map (router-NRs and slink-NRs). However, a decision must be made on which network elements are represented by nodes of the graph. Naturally, routers are represented by nodes. Regarding links, the decision depends on the association between prefixes and links, as defined in the aip-NRs. Recall that a prefix assigned to a link may be associated with a router attached to the link (through the router identifier), and not directly with the link. Some general guidelines are:

- Routers are represented by nodes.

- Stub shared links are not represented by nodes since they are not part of the network map.

- In general, transit shared links are represented by nodes, since this simplifies the graph structure and eases the computation of shortest paths. A possible exception is when the link is only connecting a few routers (e.g. two routers).

- Point-to-point links may or may not be represented by nodes. It is not worth doing it if, in the LSDB structure, a prefix assigned to a point-to-point link is not directly associated with the link.

In this graph, the outgoing arcs of nodes representing routers correspond to their outgoing interfaces and are labeled with the corresponding interface costs. The outgoing arcs of nodes representing links are labeled with cost zero, since links are not layer-3 switching equipment.

Association between prefixes and nodes Once the graph is set up, the prefixes can be associated with the nodes of the graph, using the information contained in the aip-NRs and dep-NRs. In case of a prefix assigned to a link that is not represented by a node of the graph, the prefix must be associated with each node that represents a router attached to the link, and must include the cost from the router to the prefix.

Example Figure 2.28 shows the graph of the network of Figure 2.22 where, besides the routers, only the transit shared links are represented by nodes of the graph. Nodes n1 to n4 represent the routers R1 to R4, and nodes n5 and n6 represent the transit shared links ls1 and ls2. To simplify the graphical presentation, we use undirected arcs and place the labels near the nodes of the corresponding outgoing arcs. However, the computation of shortest paths is performed over the directed graph. For example, the label 40 placed near node n1 is the label of the outgoing arc from n1 to n5, and the label 5 placed near the same node is the label of the outgoing arc from n1 to n7. Note that the labels of the outgoing arcs of nodes representing links are all zero.

Regarding the association between prefixes and nodes, prefix ap1 is assigned to the transit shared link ls2 represented by node n6. Prefix ap2 is assigned to the point-to-point link between R1 and R2; since this link is not represented by a node of the graph, it must be associated with the nodes representing the link end routers, i.e. it is associated to node n1 with a cost of 40 and to node n2 with a cost of 10. Prefix ap3 is assigned to the stub shared link; it is associated to node n4, representing the link router, with a cost of 10. Finally, prefix ap4 is assigned to router R3 represented by node n3.

Note that in the LSDB structure of Figure 2.22, (i) the point-to-point links are represented in the network map and (ii) the aip-NRs advertising prefixes assigned to these links associate the prefixes directly with the links, through the link identifiers. In this case, the graph could include nodes representing point-to-point links and, as will be seen in Section 2.3.3, these would facilitate

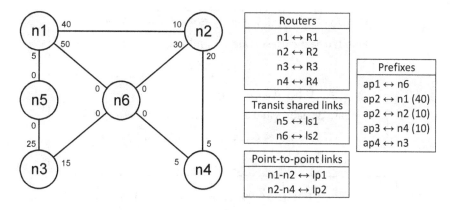

FIGURE 2.28: Graph of the network of Figure 2.22, and correspondence between nodes, network elements, and prefixes.

the determination of the shortest paths to the prefixes. This solution would not be possible in the LSDB structure of Figure 2.20, since the aip-NRs associate the prefixes assigned to point-to-point links with the link end routers.

2.3.2 Determining shortest paths on a graph - Dijkstra's algorithm

LSR protocols use Dijkstra's algorithm (Figure 2.29) to determine the shortest paths between the nodes of the network graph. This algorithm can determine in a single run the shortest paths from the source node to every other node, i.e. the *shortest path tree* rooted at the source node. Dijkstra's iterates in the order of increasing path cost. Each node is assigned a *label*, updated in each iteration, corresponding to the cost of the path from the source node to the node itself. The labels become *permanent* when the true value of the shortest path cost has been found; otherwise they are said to be *temporary*. The algorithm ends when all labels become permanent. The algorithm also records the parent (predecessor) of each node in the shortest path tree, to enable the determination of the actual paths towards destination nodes.

Let s denote the source node, S the set of permanently labeled nodes, c_{ij} the cost of the arc from node i to node j, p_i the parent of node i, and C_j the path cost estimate from the source node to node j (i.e. the label of node j). If there is no arc between i and j, then $c_{ij} = \infty$; nodes i and j for which $c_{ij} \neq \infty$ are said to be *neighbors*. The label of the source node is always $C_s = 0$ since it corresponds to the cost from a node to itself. Initially (step 0 of Figure 2.29), only the source node is placed in S and, for every other node j, the labels are set to the costs of the arcs leading to j, and the node parents are all set to the source node s. Then, the algorithm starts iterating. The first step in each iteration (step 1 of Figure 2.29) determines the node not yet in S that is

---

0. Initially, set $S = \{s\}$, $C_s = 0$, $C_j = c_{sj}$, and $p_j = s$ for all $j \notin S$.

1. Find the next closest node, i.e. find $k \notin S$ such that

$$C_k = \min_{m \notin S} C_m.$$

   Set $S := S \cup \{k\}$. If S contains all nodes, stop.

2. Update the labels and the parent nodes, i.e. for all $j \notin S$,

   If $C_k + w_{kj} < C_j$, set $C_j = C_k + c_{kj}$ and $p_j = k$.

   Go to step 1.

FIGURE 2.29: Dijkstra's algorithm.

closer to the source, i.e. the one with lowest label. This node is now considered permanently labeled and placed in S; we will refer to it as *node k*. The second step (step 2 of Figure 2.29) updates the labels of all nodes still not in S. At this point, the algorithm only needs to examine the paths where node k is the penultimate node. Whenever a node label is updated, i.e. the cost estimate decreases, its parent node is set to node k. Then, the algorithm returns to the first step and proceeds from there, until all nodes have been included in S. When the algorithm ends, all nodes are labeled with the true shortest path cost from the source node to themselves.

Figure 2.30 illustrates the transition from step 1 to step 2. Node k has just been given a permanent label and included in S. At this point, the algorithm updates the labels of the neighbors of node k that are not in S, i.e. nodes a and b. Nodes not in S that are not neighbors of k, i.e. nodes c and d, need not be considered at this stage. After this relabeling operation, the algorithm selects among all nodes not in S, i.e. nodes a to d, the one with the lowest label.

The parent node information allows determining the actual path from the source node to all other nodes. A path is usually defined at a node through its next-hop node. The next-hop node can be obtained by backtracking on the graph from the destination node to the source node, using the parent node information.

We illustrate the step-by-step operation of Dijkstra's algorithm in Figure 2.31, using a network graph simpler than the one of Figure 2.28. Actually, the graph of this example is a representation of the network of Figure 1.1 where only routers and shared links with more than two connections are represented as nodes. We show how the shortest path tree rooted at node n2 is determined. In each graph of Figure 2.31, nodes represented with a dashed line are nodes not yet permanently labeled. The figure also shows, at each iteration, the shortest path table of node n2, which includes an entry for each destination (dest), indicating the next-hop node (nh) and shortest path cost towards the destination (cost). The algorithm iterates as follows:

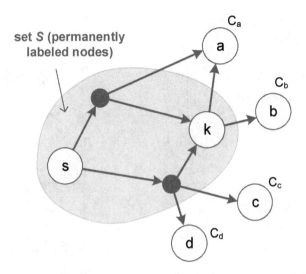

FIGURE 2.30: Transition from step 1 to step 2 of Dijkstra's algorithm.

- **Initialization** - Initially (Figure 2.31.a), only node n2 is in S, i.e. $S = \{2\}$. Each node is labeled with the cost of the directed arc from the source node to the node itself; note that node n3 is assigned a label of ∞, since there is no arc between nodes n2 and n3. Moreover, the parent nodes are all set to node n2.

- **1st step** - Following the initialization, node n1 becomes the node with the lowest label, $C_1 = 10$. Thus, it is given a permanent label and placed in S, leading to $S = \{1, 2\}$ (Figure 2.31.b). This means that the shortest path from n2 to n1 (and its cost) has been found. The shortest path cost equals the node label, i.e., it is 10. The next-hop node is determined from the parent node information: since the parent of n1 is n2, the next-hop node is set to n1. Now, the labels of nodes not in S that are neighbors of node n1, the node k at this step, must be updated. Specifically, C_3 decreases to 15, and the labels of other nodes remain unmodified. Moreover, since the label of n3 decreased, its parent must be changed to node n1, because n1 is the current node k, i.e. $p_3 = 1$.

- **2nd step** - Then, the algorithm moves to a new iteration where node n3 is given a permanent label, leading to $S = \{1, 2, 3\}$ (Figure 2.31.c). Thus, the shortest path from n2 to n3 (and its cost) has been found. The next-hop node is determined by backtracking from n3 to n2. The parent of n3 is n1, and the parent of n1 is n2; thus, the next-hop node is n1. At this point, no relabeling occurs and the algorithm proceeds to the next iteration.

- **3rd step** - At this iteration, node n4 is included in S, leading to $S = \{1, 2, 3, 4\}$ (Figure 2.31.d), and the next-hop node is set to n4. The label of

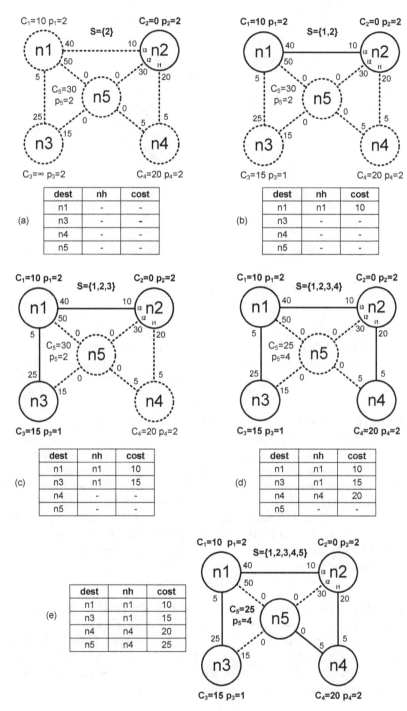

FIGURE 2.31: Dijkstra's algorithm example.

node n5 decreases to $C_5 = 25$, since it can now be reached via node n4 and, consequently, its parent changes to node n4, i.e. $p_5 = 4$.

- **4th step** - Finally, in the last iteration (Figure 2.31.e), node n5 is included in S and the algorithm stops, since S now spans all nodes. The shortest path cost from n2 to n5 is set to 25 and the next-hop node is set to n4. The next-hop node is again determined by backtracking from the destination node: the parent of n5 is n4, and the parent of n4 is n2, leading to a next-hop node of n4.

2.3.3 Determining shortest paths to destination prefixes and assembling local routing information

Once the shortest paths between the source node and all other nodes of the network graph have been computed, one needs to determine the actual shortest paths towards each destination prefix and assemble the forwarding tables. Figure 2.32 shows the forwarding tables of the network of Figure 2.22. The tables include, for each destination prefix (dest), the link address of the next-hop interface (nh), the outgoing interface identifier (int), and the path cost (cost).

Two issues need to be considered, the first related to the location of the prefix relative to the source router, and the second related to the association between nodes of the graph and prefixes.

Location of destination prefix relative to source router There are two cases regarding the location of the destination prefix relative to the source router:

1. The prefix is assigned to a link directly connected to the router;

2. The prefix is assigned to some link or router beyond a directly connected link and is reachable through a next-hop router.

Prefix assigned to directly connect link In the first case, and as discussed in Section 1.5, the directly connected route takes preference over any other path (e.g. shortest path). Thus, the outgoing interface installed at the forwarding table must be the one attaching to the link. In these types of entries, forwarding tables include information stating that the destination prefix is at a directly connected link, e.g. the sentence "is at a directly connected link" or the keyword "dc".

For example, in the network of Figure 2.32, the path from R2 to ap1 is through interface i2 (the directly connected route), even if the path R2 → R4 → ap2 has lower cost.

Prefix assigned to link or router beyond directly connect link In the second case, one needs to determine the shortest path from the source router to the prefix, using the results of running Dijkstra's algorithm over the

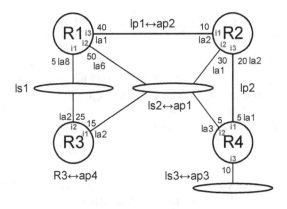

dest	nh	int	cost
ap1	dc	i2	-
ap2	dc	i3	-
ap3	la2	i1	30
ap4	la2	i1	5

Router R1

dest	nh	int	cost
ap1	dc	i2	-
ap2	dc	i1	-
ap3	la1	i3	30
ap4	la1	i1	15

Router R2

dest	nh	int	cost
ap1	dc	i1	-
ap2	la1	i1	25
ap3	la3	i1	25
ap4	local	-	-

Router R3

dest	nh	int	cost
ap1	dc	i2	-
ap2	la2	i1	15
ap2	la1	i2	15
ap3	dc	i3	-
ap4	la2	i2	5

Router R4

FIGURE 2.32: Forwarding tables of the network of Figure 2.22.

network graph. The shortest path is defined locally at a router through the outgoing interface and the link address of the next-hop interface.

Given a path on the network graph, the first node of the path (the source node) represents a router, but the second node can either represent a link or a router. The next-hop router corresponds to the node that represents the next router on the path and it can be either the second or the third node. The outgoing interface is the interface with the second node of the path. The link address (i.e. the link information) is provided by the la-NR originated by the next-hop router on the corresponding link, or some equivalent means (see discussion in Section 2.2.3).

Computing shortest paths towards prefixes assigned to network elements represented by nodes of the graph The computation of the (final) shortest path towards a prefix depends on the association between prefixes and nodes. Dijkstra's algorithm determines the shortest paths between the source node and all other nodes of the network graph. If the destination prefix is assigned to a network element represented by a node of the graph, then the shortest path is directly determined by Dijkstra's algorithm.

For example, in the network of Figure 2.32, consider the path from router R2 to prefix ap4, assigned to router R3. In the corresponding network graph (see Figure 2.28), router R2 is represented by node n2 and router R3 by node n3. When Dijkstra's algorithm is run over the graph having n2 as the source node, it indicates that the shortest path is n2 → n1 → n3, with a cost of 15. Since the next-hop node in the path is node n1, the link address of the next-hop interface is the one provided by router R1 on link lp1, i.e. la1.

Computing shortest paths towards prefixes assigned to network elements not represented by nodes of the graph There can be prefixes assigned to links not represented by nodes of the network graph. This is precisely the case of the point-to-point link and the stub shared link in the graph of Figure 2.28. Specifically, in the case of prefixes assigned to links, (i) a prefix may be associated with a node that does not represent the link to which the prefix was assigned, and (ii) a prefix may be associated with more than one node. In these cases, the computation of the final shortest path requires additional steps.

A node representing a router that is associated with a prefix assigned to a link will be called *proxy node* (relative to the prefix). Proxy nodes must include the cost from the router represented by the node to the prefix. For example, in the graph of Figure 2.28, n4 is a proxy node for ap3, with cost 10, and n1 and n2 are proxy nodes for ap2, with costs 40 and 10, respectively.

Computing the shortest path to a prefix associated with one proxy node The cost of the shortest path from a router to a prefix associated to a proxy node must be determined in two steps. First, using Dijkstra's algorithm, determine the shortest path, and the corresponding path cost, from the source node to the proxy node of the prefix. Then, add this path cost to the prefix cost.

For example, in the network of Figure 2.32, consider the path from router R1 to prefix ap3, assigned to the stub shared link. In the network graph of Figure 2.28, router R1 is associated with node n1 and prefix ap3 with proxy node n4. When Dijkstra's algorithm is run over the graph having n1 as the source node, it indicates that the shortest path from n1 to n4, is n1 → n7 → n3 → n8 → n4, with cost 20. From this information, one determines that the outgoing interface of router R1 is the one attaching link ls1 (represented by node n5), and the next-hop router is router R3 (represented by node n3). The link address of the next-hop interface is the one provided by router R3 on link ls1, i.e. la2. Finally, since the proxy node has a cost of 10 relative to ap3, it follows that the path cost from R1 to ap3 is 20+10=30.

Computing the shortest path to a prefix associated with more than one proxy node When a prefix is associated with more than one proxy node, then one needs to determine the shortest paths to the prefix via each proxy node, and the actual shortest path is the least cost one.

Again, in the network of Figure 2.32, consider the path from router R4 to prefix ap2, assigned to the point-to-point link. In the network graph of Figure 2.28, R4 is associated with n4 and ap2 is associated with two proxy nodes: n1 with cost 40 and n2 with cost 10. One needs to determine first the shortest path costs from n4 to ap2 via n1 and n2. The shortest path from n4 to n1 is n4 $\rightarrow$ n1 with cost 5 and, therefore, the shortest path cost to ap2 via n1 is 5+40=45. Moreover, there are two equal cost shortest paths from n4 to n2, n4 $\rightarrow$ n2 and n4 $\rightarrow$ n6 $\rightarrow$ n2 both with cost 5; therefore, the shortest path cost to ap2 via n2 is 5+10=15. Finally, when comparing the path costs via n1 and n2, we conclude that the shortest path is via n2 with a cost of 15.

This means that there are two equal cost shortest paths from R4 to ap2. In such a case, the usual practice is to include both in the forwarding table. The analysis of the network graph shows that one path is via the transit shared link ls2 and the other is via the point-to-point link lp2. Thus, the link addresses to be used in each path are the ones provided by R2 in these links, i.e. la1 for the path via the transit shared link and la2 for the path via the point-to-point link.

2.3.4   Default routes

Forwarding tables may contain an entry to be used when the destination prefix of an incoming packet does not match any other entry. As discussed in Section 2.2.4.1, this entry, called default route or gateway of last resort, is useful in routing domains with only one DBR, i.e. where there is a single exit point to other routing domains. In this case, there is no need to disseminate external addressing information inside the domain, and all traffic destined to external prefixes can internally follow the default route.

We give an example in Figure 2.33. A domain is connected to the Internet through router R1. In this example, there is only one internal address prefix: ap1 assigned to link ls1. We assume that the interface costs are all 10. When analyzing the LSDB, router R2 learns, through the router-NR of R1, that R1 is a DBR (the router-NR includes a flag to indicate if the originating router is a DBR). It then installs a default route in its forwarding table corresponding to the shortest path from itself to the DBR, i.e. having i1 as the outgoing interface, la1 as the link address of the next-hop interface in link lp2, and a path cost of 10. The default route entry is the last one to be checked for any incoming packet. For example, if a packet arrives at router R2 and is destined to ap1, it will be forwarded to R3 via interface i2; however, if it is destined to any other address prefix (e.g. a domain-external prefix), it will not match the first entry of the forwarding table and will be sent towards the DBR via interface i1.

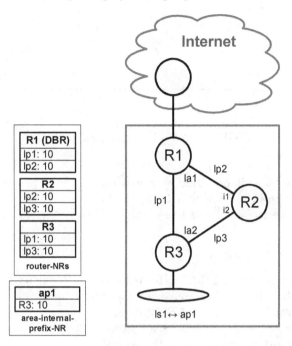

Forwarding table of router R2

destination	next hop	interface	cost
ap1	la2	i2	20
default	la1	i1	10

FIGURE 2.33: Default routes.

Forwarding tables are built from the routing information contained in the LSDB in three steps: (i) obtain a graph of the network from the network map and associate destination prefixes with nodes of the graph, (ii) run Dijkstra's algorithm over the network graph to determine the shortest paths between the nodes of the graph, and (iii) determine the shortest paths to destination prefixes and assemble the local routing information, i.e. the outgoing interface and the link address of the next-hop interface that leads to each prefix.

2.4   Dissemination and updating of routing information

LSR protocols require a mechanism to disseminate throughout the network the routing information originated by one router and allow the LSDB update

FIGURE 2.34: Linear sequence numbers.

by replacing older information with new one. Three types of operations are required: (i) to install a new NR, (ii) to update an existing NR, and (iii) to delete an existing NR. These issues will be discussed in the next sections.

2.4.1   NR freshness

As discussed in Section 2.2.1.5, NR instances must have a freshness attribute expressing how recent they are, such that fresher instances can replace older ones. This attribute is called *NR freshness*.

There are several possibilities for expressing freshness (see Section 2 of [40]). The most common way is to use a linear space of sequence numbers (SNs). When the NR is first created, it is assigned an initial SN value, and this value is incremented by one every time a new instance of the NR is created (Figure 2.34). In this case, the rule for replacing an NR instance stored at a router is to allow substitution only when the incoming NR instance has a higher SN, i.e. is fresher than the stored one. One advantage of using SNs is that they can also be used to control the flooding process, as will be discussed in the next section.

Using linear sequence numbers poses the problem of what to do when the final SN is reached. This issue will be addressed in Section 2.4.8.

> *NRs have a freshness attribute to reflect how recent they are, such that fresher NR instances can replace older ones. Linear sequence numbers (SNs) are usually used for this purpose.*

2.4.2   Identification of NR instances

An NR can have several instances, each with a different freshness value. Thus, NR instances must be uniquely identified by two elements: the NR identifier and the NR freshness. As discussed in Section 2.2.5, the NR identifier includes (i) the identifier of the router that originates the NR, (ii) the NR type indicator, and (iii) a locally assigned tag to distinguish among NRs of the same type originated by the same router. As will be seen later, there are circumstances where only the NR identifier is needed, but others where the more complete NR instance identifier is required.

> *NR instances are uniquely identified by two elements: the NR identifier*
> *and the NR freshness.*

2.4.3   Controlled flooding

We start by discussing the dissemination process in a generic context. Consider a network where each router originates successive messages that must reach all other routers. The dissemination of these messages can be achieved through the following procedure, named *flooding*:

- The originating router transmits the message to be flooded through all its interfaces;

- Any router receiving the message on one interface retransmits it on all other interfaces (i.e. the message gets retransmitted on all interfaces except the one where it was received).

Using this simple procedure, the message initially transmitted by the originating router is surely delivered to all other network routers. However, if the network contains cycles, the message will keep circulating indefinitely, eventually causing network congestion and breakdown. We say that this flooding procedure is *uncontrolled*.

The solution to this lack of control is to make routers remember if they have already transmitted a message that is being flooded. To accomplish this goal, messages need to be labeled with unique identifiers and these identifiers must be stored at routers when the messages are received. Unique message identifiers can be obtained using two elements: the identifier of the originating router (ID) and a unique number generated locally by the originating router. An example of unique numbers is the sequence numbers (SNs) introduced in Section 2.4.1. With these ingredients, the *controlled* flooding procedure works as follows:

- The originating router assigns each message to be flooded a unique identifier composed of two elements, i.e. the (ID, SN) pair, and transmits the message through all its interfaces.

- When a message is received at a router, the router verifies if the received (ID, SN) pair is already stored in memory. If yes, the message is discarded; if not, the pair is stored and the message is transmitted through all interfaces, except the one where the message was received.

Note that all (ID, SN) pairs must stay stored at routers unless, in addition to being unique numbers, SNs also express the message freshness, such that older messages become outdated and no longer need to be flooded. This is precisely the case of LSR protocols, which will be detailed in Section 2.4.5.

2.4.4   Reliable flooding

The controlled flooding procedure described in the previous section provides
a mechanism to disseminate messages throughout the network. However, to
ensure that messages are delivered to all routers, without exception, an additional mechanism is needed. One way to achieve *reliable flooding* is to ACK
protect the transmissions between neighboring routers (see Section 1.8.1). In
this case, whenever a router sends a message being flooded to a neighbor, the
neighbor replies with an ACK message. The reception of a message must be
acknowledged even if the message is discarded due to the controlled flooding
procedure.

Figure 2.35 shows an example of the reliable flooding mechanism. Router
R5 wants to disseminate a message to all other routers. It assigns SN=1 to
the message and transmits it on the shared link using a broadcast address
(Figure 2.35.a). The message sent by router R5 is received by routers R3 and
R4, which store the corresponding (ID, SN) pair, and confirm the correct
reception with an ACK. Then, router R3 sends the message to router R1, and
router R4 to routers R1 and R2. Again, the receiving routers store the (ID,
SN) pair and reply to the transmitting routers with ACK messages (Figure
2.35.b). The figure assumes that the two messages received by router R1 arrive
exactly at the same time; thus, router R1 will only send the received message
to R2. Finally, routers R1 and R2 send the message to each other, but both
messages are discarded since the (ID, SN) pair they carry is already stored at
the receiving routers (Figure 2.35.c). Despite discarding the messages, routers
R1 and R2 confirm their correct reception with an ACK. From then on, no
further messages circulate in the network (Figure 2.35.d).

2.4.5   The dissemination of NRs

The dissemination of NRs can be based on the controlled and reliable flooding
process described in the two previous sections. As explained in Section 2.4.3,
the messages to be disseminated must be labeled with unique identifiers. In
LSR protocols, (i) the messages are the NR instances, (ii) the NR instances
are uniquely identified using the NR identifier and the NR freshness, and (iii)
the NR freshness is expressed using SNs. Moreover, when a new NR instance
is created, all other instances of the same NR (which must have a lower SN)
become outdated and irrelevant to the routing process. Thus, an NR instance
arriving at a router with an SN lower than an existing instance is outdated
and needs not be retransmitted, even if it is arriving at the router for the first
time.

The rules for the controlled and reliable flooding of an NR instance arriving
at a router are then the following:

- If no instance of the incoming NR is stored at the router, store the incoming
  instance and retransmit it on all interfaces except the one where it was
  received;

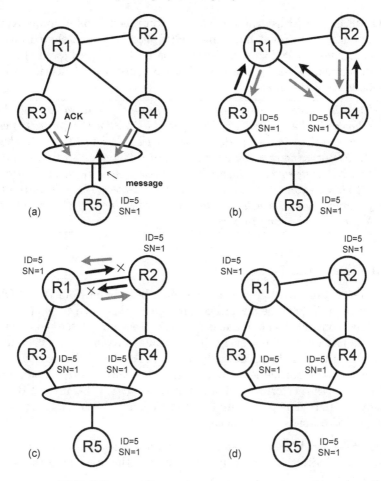

FIGURE 2.35: Controlled and reliable flooding.

- If there is a stored instance with a lower SN, replace it by the incoming instance and retransmit the incoming instance on all interfaces except the one where it was received;

- Otherwise, discard the incoming instance;

- Always acknowledge the reception of an incoming NR to the neighboring router that transmitted it.

The acknowledgment of an NR reception needs not carry the full NR contents; only the NR instance identifier is needed to fully identify the NR being acknowledged.

Depending on the NR type and network structure, NRs may need to be flooded only in some zones of the network; the network zone where an NR is to be flooded is called the *NR flooding scope*.

OSPF and IS-IS use the reliable and controlled flooding process just described on point-to-point links, and OSPF also uses it on shared links, with minor changes. In IS-IS, reliability on shared links is achieved through the periodic transmission, by the DR, of the NR instance identifiers of the last NRs flooded on the link.

> *The dissemination of NRs can be performed through a controlled and reliable flooding procedure. The control of the flooding process is based on the NR instance identifier, which uses SNs to express freshness. An NR arriving at a router is discarded if there is a stored instance of that NR with equal or higher SN; otherwise it is retransmitted on all interfaces except the one where it was received. The reliability of the flooding process is achieved by ACK protecting the NRs transmitted between neighboring routers.*

2.4.6   Deleting NRs

Several situations exist in which NRs must be deleted from the LSDB. The deletion of an NR can be performed through a *delete indication*, disseminated using the controlled and reliable flooding procedure described above. The delete indications need not carry the full contents, and not even the freshness attribute, of the NR being deleted: only the NR identifier is needed.

When disseminating delete indications, the control of the flooding process is done through the verification of the intended action (i.e. if the NR has been already deleted). The procedure followed when an indication to delete an NR arrives at a router is the following:

- If the referenced NR is still stored in the LSDB, delete this NR and transmit the delete indication on all interfaces except the one where the indication was received;

- Otherwise, discard the indication;

- Always acknowledge the reception of a delete indication to the neighboring router that transmitted it.

The need for a guard time A router can delete an NR and create it again shortly after. Thus, an indication to delete an NR can be followed by an indication to create a new NR with the same identifier, which will then have an SN equal to the initial value. In this case, there is the danger that the indication to create arrives earlier at a router than the indication to delete. If this happens, the new NR will not be installed since it has an SN lower than the stored one. The solution to this problem is to impose a guard-time at the originating router between the transmissions of these two indications. The guard-time must be enough to allow the propagation of the delete indication to all network routers; it is called RESTART-TIME in [40], where this problem is also discussed.

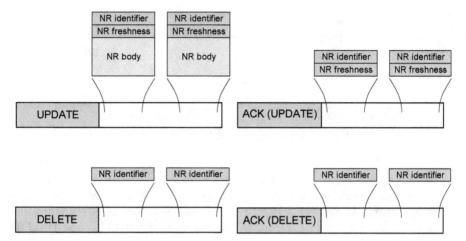

FIGURE 2.36: UPDATE, ACK (UPDATE), DELETE, and ACK (DELETE) messages, and their contents.

> *NRs can be deleted by disseminating a delete indication using a controlled and reliable flooding procedure. The indication only needs to carry the NR identifier of the NR being deleted. There must be a guard-time between flooding an indication to delete an NR and flooding a new NR having the same identifier as the previously deleted one.*

2.4.7 Control messages

The NRs, the acknowledgment NRs, and the indication to delete NRs must be transmitted between neighboring routers encapsulated in control messages. The type of control message provides the semantics of the intended action. We will name UPDATE the messages that transmit full NR instances, DELETE the messages that carry delete indications, ACK (UPDATE) the messages that acknowledge the reception of full NRs, and ACK (DELETE) the messages that acknowledge the reception of delete indications.

As highlighted previously, a DELETE message needs only to include the identifier of the NR being deleted, and the same is true for ACK (DELETE) messages. Likewise, an ACK (UPDATE) message needs only to include the instance identifier of the NR instance being acknowledged. This is illustrated in Figure 2.36.

To add flexibility, we consider that these messages can transport information relative to more than one NR. For example, an UPDATE message can transport more than one NR instance, and an ACK (UPDATE) message can transport more than one acknowledgment. Thus, the control messages are just containers that transport information relative to NRs between neighbors.

On shared links, there is no need to transmit one UPDATE message (or one

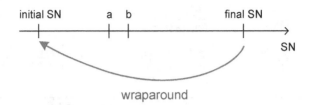

FIGURE 2.37: Wraparound linear sequence numbers.

DELETE) message to each individual neighbor on the link: a single message transmission suffices if sent to a broadcast or a multicast address known by all neighbors. All routers receiving an NR instance (or NR delete indication) must acknowledge its reception. Unlike the case of UPDATE and DELETE messages, there is no benefit in transmitting ACK messages using broadcast or multicast addresses.

> *NR instances are transported between neighboring routers using UP-DATE messages, NR delete indications using DELETE messages, the acknowledgments of NR instances using ACK (UPDATE), and the acknowledgments of NR delete indications using ACK (DELETE) messages. These messages can transport information relative to more than one NR.*

2.4.8 Wraparound problem

The use of a linear space of SNs to express freshness, growing from an initial to a final value, poses the problem of deciding what to do when the final value is reached. Usually, the space of SNs is made sufficiently large to make this event rare; anyway, LSR protocols must consider this possibility. One solution to this problem is to *wrap around* the SN space and restart from the initial value, as illustrated in Figure 2.37. In this case, the NR instance disseminated after the one with the final SN value is transmitted with the initial SN value. This introduces another problem: after the wrap around, the fresher NR instance will have an SN lower than the older one (the fresher instance has SN=1 and the older one has the final SN value), and the flooding rules prevent its dissemination. One simple solution is to delete first the old NR instance, using the deletion mechanism introduced in Section 2.4.6, and only afterwards flood the new NR instance with the initial SN. As pointed out in Section 2.4.6, there must be a guard-time between the transmission of the delete indication and the transmission of the subsequent NR, to ensure that the delete indication arrives first at all routers.

> *When the SN reaches its final value, the SN space wraps around, i.e.*
> *the next NR instance will have an SN equal to the initial value. How-*
> *ever, to enable the dissemination of the instance with the initial SN*
> *value, the instance with the final SN value must be deleted first.*

2.4.9 Removal of outdated NRs

The outdated NRs originated by a failed router can stay indefinitely in the
LSDB. This does not harm the routing process since, as discussed in Section
2.2.1.7, the failed router becomes isolated in the network map. However, it
unnecessarily wastes memory resources. One solution to save memory is to let
routers delete NRs based just on topological considerations. For example, if
a router analyses its network map (e.g. when building the forwarding table)
and observes that a router has become isolated, the NRs originated by that
router can be safely removed from the LSDB. In the example of Figure 2.16,
this process would delete from the LSDB the router-NR and the slink-NR
originated by router R2.

An alternative to the process described above is to let outdated NRs be
removed using an age-lifetime mechanism. In this case, each NR is assigned
an age that keeps increasing while the NR is stored in memory and, when
the age reaches a predefined lifetime, the NR is deleted from the LSDB. In
this case, some measure must be taken to avoid deleting legitimate NRs under
stable operation. In particular, the age of a legitimate NR must be refreshed
(reset to zero) before its lifetime is reached. Refreshing can be implemented by
periodically generating, with a period lower than the lifetime, a new NR instance with an age of zero. This NR instance is then disseminated throughout
the network and replaces the older instance (with higher age). Thus, in this
alternative, new NR instances must be created and disseminated periodically
even if the network remains unchanged. This is precisely the solution adopted
by both OSPF and IS-IS to delete outdated NRs. The lifetime is 1 hour in
OSPF and 20 minutes in IS-IS, with a refresh period of 30 minutes in OSPF
and 15 minutes in IS-IS.

> *When a router fails, the NRs it originated must be deleted from the*
> *LSDB, to avoid wasting memory resources. One alternative is to delete*
> *the NRs originated by routers that have become isolated in the network*
> *map. Another alternative is to use an age-lifetime mechanism, where*
> *NRs are assigned an age and are deleted from the LSDB when their age*
> *reaches a predefined lifetime. In this case, to avoid deleting legitimate*
> *NRs under stable operation, new NR instances with zero age must be*
> *periodically disseminated.*

2.5 Initial LSDB synchronization

When two routers become neighbors (through the Hello protocol), they may need to synchronize their LSDBs. The two routers, even if they have been attached to the network for a long time, may not anticipate whether the contents of their LSDBs is the same. Moreover, when a router joins a network, it may do so by connecting several neighbors. If the network is in a stable condition, all these neighbors should have the same LSDB. The question is then if the joining router needs to synchronize its LSDB with all the contacted neighbors, or if it suffices synchronizing with a single one.

In any case, a protocol must be devised for the LSDB synchronization between two neighbors. Since the size of the LSDB may be large, the protocol should try to minimize the amount of exchanged information.

The LSDB synchronization process must be coupled with the flooding procedure. A router that synchronizes its LSDB with a neighbor may obtain new NRs or fresher instances of existing NRs from the neighbor. In this case, the router must disseminate the new information throughout the network, using the procedure of Section 2.4.5.

We discuss these aspects in the next sections.

2.5.1 Which neighbors to synchronize with?

To discuss which neighbors a router must synchronize with when joining a network, we consider several scenarios with increasing complexity. In these scenarios, the joining router attaches to the network through:

- a single neighbor;

- multiple neighbors located in the same network;

- multiple neighbors located in different networks parts (i.e. with different LSDBs), via point-to-point links;

- multiple neighbors located in different network parts, via a shared link.

These scenarios are illustrated in Figure 2.38.

Attaching the network through a single neighbor In the first (and simpler) scenario, the joining router attaches to the network through a single neighbor (Figure 2.38.a). In this case, the two routers must exchange their LSDBs. Note that even a router that has just been switched on will already have an LSDB, consisting at least of its self-originated router-NR. In Figure 2.38.a, router R1 sends LSDB1 to router R2, and router R2 sends LSDB2 to router R1. Then, each router merges the LSDB received from the neighbor with its own LSDB, and both arrive at the same LSDB, i.e. they become synchronized. This example shows that the initial LSDB synchronization between

two routers must always be bidirectional, i.e. either router must send to the
other a copy of its LSDB, the simpler it might be.

**Attaching the network through multiple neighbors located in the
same network** In the second scenario, the joining router attaches to the network through multiple neighbors located in the same network (Figure 2.38.b).
In this case, one may ask if it suffices synchronizing the LSDB with just one
neighbor. The contacted neighbors may have exactly the same LSDB if the
network is in a stable state, but may also have different LSDBs, none of them
fully updated, if the network is in a changing state. The LSDB of a router
is not fully updated in the period between the origination of routing information that will change the LSDB (e.g. insertion of a new NR or deletion of
an existing one) and the arrival of this information at the router. However,
if the LSDB of a contacted neighbor is not fully updated, that neighbor will
certainly receive later the missing information, and will then have the opportunity to deliver it to the joining router through the flooding procedure. Thus,
according to this scenario, a joining router only needs to contact one neighbor.

**Attaching the network through multiple neighbors located in dif-
ferent network parts, via point-to-point links** In the third scenario, the
joining router attaches to the network through neighbors located in different
network parts, each with its own LSDB, via point-to-point links. We say that
the joining router is solving a *network partition problem*. In this scenario, if
the router synchronizes only with one neighbor, it will get only the LSDB of
one network part, and will not be able to communicate with the other parts, or
provide connectivity among the various disconnected parts. In Figure 2.38.c,
router R1 joins the network, becomes neighbor of R2 and R3, but synchronizes only with R2. In this case, the router will not be able to communicate
with the network part of router R3, and the network remains partitioned.
Note that the flooding procedure does not help in this case. Since router R1
obtains LSDB2 from router R2, which is new information, it must disseminate
it to router R3. Thus, R3 obtains LSDB2; however, neither R1 nor R2 obtain
LSDB3. As shown in Figure 2.38.d, the solution to this problem is to have
R1 synchronizing with both R2 and R3. In this case, R1 can inject on each
side the LSDB received from the other, so that the LSDBs become merged.
Since a joining router has initially no information on whether two neighbors
belong to the same network part, this scenario shows that a joining router
must synchronize the LSDB with all its neighbors.

**Attaching the network through multiple neighbors located in dif-
ferent network parts, via shared links** The procedure defined above can
be relaxed in the case of neighbors attached to the same shared link. These
neighbors are directly connected to each other and, under stable conditions,
belong necessarily to the same network part and have the same LSDB. Therefore, it suffices synchronizing with a single neighbor at a shared link. This
neighbor is usually the link DR. When the routers attached to a shared link
are all switched on at the same time (cold start), achieving full LSDB syn-

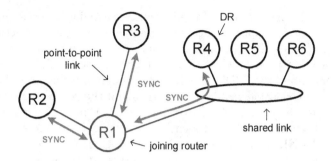

FIGURE 2.39: Neighbors that must synchronize with a joining router during the initial LSDB synchronization process.

chronization involves the pairwise LSDB synchronization between neighbors and the flooding procedure. Consider the example of Figure 2.38.e where four routers, belonging to four disjoint network parts and having four distinct LS-DBs, attach to a shared link at the same time. Suppose that the DR has the role of disseminating new routing information on the link. First, the routers establish neighborhood relationships among themselves (on a pairwise basis) and elect R2 as the link DR. Then, they start synchronizing their LSDBs with the DR. Suppose that R2 synchronizes first with R1. In this case, R1 and R2 exchange LSDB1 and LSDB2 and, since LSDB1 is new routing information for R2, R2 disseminates it on the link. This process is repeated for R3 and R4 and when it ends, all routers share a common LSDB, merging the four previous ones. This is precisely the solution adopted by OSPF and IS-IS. In IS-IS, the DR broadcasts periodically the LSDB at shared links, to make sure it is kept fully synchronized.

We conclude from the discussion above that a joining router must synchronize its LSDB with all its neighbors on point-to-point links and with the DR of each shared link it attaches to. This is illustrated in Figure 2.39. This means that, on shared links, the LSDB synchronization can only start after the DR is elected.

> *Two routers that become neighbors may need to synchronize their LS-DBs. LSDB synchronization is required when the routers become neighbors over a point-to-point link. On shared links, a router needs only to synchronize the LSDB with the link DR.*

2.5.2   Protocol for exchanging LSDBs

The protocol for exchanging LSDBs between two neighbors can, in principle, be a simple request-response interaction, where each router requests the LSDB to its neighbor and the neighbor replies by sending it. However, the LSDBs of the two neighbors may not be completely different, i.e. they may contain several equal NR instances. In this case, exchanging the complete LSDBs becomes inefficient, since duplicate information is retrieved unnecessarily. One way to deal with this problem is to introduce a preliminary phase in the synchronization process where routers retrieve, not the full LSDB contents, but a *summary* of the LSDB. The LSDB summary must contain the minimum amount of information required to determine if the NR instances stored at a neighbor are fresher than those stored locally. This information is just the NR instance identifier, i.e. the NR identifier and the NR freshness attributes; for example, the NR instance identifier of a router-NR does not include the list of links the router is attached to and the corresponding interface costs. Thus, the introduction of the preliminary phase can lead to significant savings in the amount of information exchanged.

Based on the above observations, a possible protocol for initial LSDB synchronization is the following (see Figure 2.40):

- When two routers establish a neighborhood relationship, each of them sends to the other a message, called LSDB SUMMARY REQUEST, requesting the LSDB summary.

- A router that receives an LSDB SUMMARY REQUEST replies with an LSDB SUMMARY message, including the NR instance identifiers of all NRs contained in its LSDB.

- Upon receiving an LSDB SUMMARY, a router compares the contents of its own LSDB with the NR instance identifiers received from the neighbor. It may then send to the neighbor a PARTIAL LSDB REQUEST message asking for the full contents of all NRs that are missing or for which the neighbor has a fresher instance. This request is made using the NR instance identifiers.

- Finally, a router that receives a PARTIAL LSDB REQUEST replies with the full contents of the requested NRs, transported in an UPDATE message.

The protocol described above includes two request-response interactions: the first interaction is the exchange of LSDB SUMMARY REQUEST and LSDB SUMMARY messages, and the second the exchange of PARTIAL LSDB REQUEST and UPDATE messages. Each of these interactions can be protected using ACK protection controlled by the requester, where the response implicitly acknowledges the reception of the request. Note that Figure 2.40 includes only the request-response interactions initiated by router R1.

2.5.3   Dealing with outdated self-originated NRs

An important scenario related to the initial LSDB synchronization process is when a router is switched off and later switched on, before the deletion of its self-originated NRs at the LSDB of its neighbor. In this case, the neighbor could have at its LSDB older instances of the self-originated NRs with equal

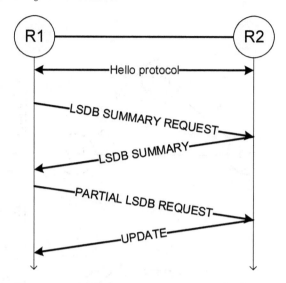

FIGURE 2.40: Control messages exchanged during the initial LSDB synchronization process (only the requests initiated by router R1 are shown).

or higher SNs, since the NRs created by the newly reborn router would have SN=1. If no measure is taken, flooding would be prevented until the SNs of the new NR instances exceed the ones being used immediately before the router was switched off.

Figure 2.41 illustrates this problem, using a network with three routers. In this example, we assume that (i) all interface costs are initially 10, (ii) the router-NR of R1 has initially a large SN, say 50, and (iii) link lp3 is much slower than the other two (Figure 2.41.a). When router R1 is switched off, routers R2 and R3 detect the failure through the Hello protocol and disseminate new router-NRs where links lp1 and lp2 are no longer listed (Figure 2.41.b). At this point, and before each router receives the router-NR sent by the other, R2 and R3 still believe that router R1 is alive and reachable. For example, the connectivity graph of router R2 (Figure 2.41.c) shows that router R1 can be reached through R3. The arrival of the two router-NRs would provide information to both R2 and R3 that router R1 became isolated in the network, which could trigger the deletion of its router-NR from the LSDB.

However, suppose that before this happens, (i) router R1 is switched on again, (ii) disseminates a new router-NR (which must have SN=1) with different interface costs and, since link lp3 is much slower than the other two links, (iii) this router-NR arrives at routers R2 and R3 before the ones sent previously by these routers to each other (Figure 2.41.d). In this case, the old router-NR of R1 (with SN=50) is still at the LSDB of routers R2 and R3 when the new router-NR (with SN=1) arrives, and the new router-NR will be discarded, since it has a smaller SN. Thus, outdated information regarding

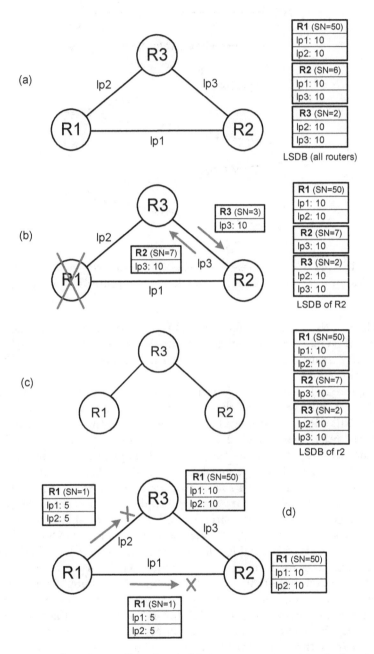

FIGURE 2.41: Switching off and switching on a router again, before its self-originated-NRs are deleted from the LSDB of neighbors.

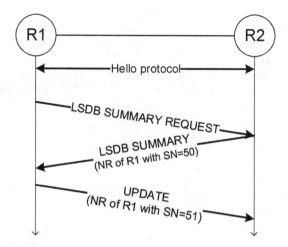

FIGURE 2.42: Dealing with self-originated NRs during the initial LSDB synchronization process.

router R1 remains at the LSDB of routers R2 and R3, and this situation will be kept until the SN of the router-NR sent by router R1 becomes larger than 50.

The problem described above is easy to solve. During the initial LSDB synchronization process, if a router obtains a self-originated NR instance from a neighbor, it floods a new instance of that NR with the SN incremented by one (in relation to the SN received from the neighbor). This is illustrated in Figure 2.42. When router R1 is switched on and synchronizes with router R2, it receives an LSDB summary from router R2, indicating that R2 still has an NR instance of R1 with SN=50. Then, router R1 floods a new instance of that NR with SN=51, which replaces the previous one at the LSDB of all other routers.

> *To avoid retrieving the same LSDB from several neighbors, the initial LSDB synchronization process between two routers can be divided into two phases: in the first phase, both routers advertise a summary of their LSDBs; in the second one, each router requests to the other the NRs it is missing or for which the neighbor has fresher instances, but only those. When a router finds that the neighbor has NRs originated by itself, it disseminates new instances of those NRs with the SN incremented by one.*

message	role	body of message
HELLO	maintains neighborhood relationships and elects DR	list of active neighbors and of DR
UPDATE	transports NR instances between neighbors	full NR instances
ACK (UPDATE)	transports acknowledgments of the reception of NR instances	NR instance identifiers
DELETE	transports NR delete indications between neighbors	NR identifiers
ACK (DELETE)	transports acknowledgments of the reception of NR delete indications	NR identifiers
LSDB SUMMARY REQUEST	requests LSDB summary from neighbor	none
LSDB SUMMARY	sends LSDB summary to neighbor in response to request	NR instance identifiers
PARTIAL LSDB REQUEST	requests specific NR instances from neighbor	NR instance identifiers

FIGURE 2.43: Control messages of LSR protocols, their role and contents.

2.6   Summary of control messages

In this section, we summarize the various control messages required by LSR protocols that have been introduced in this chapter. The list of messages, together with their role and contents, is shown in Figure 2.43.

HELLO messages are part of the Hello protocol. Each router transmits these messages periodically, in all its links, to determine which neighbors are active in each link. Thus, it plays a central role in detecting router and link failures. The message is also used to elect the DR and to determine if two routers can indeed become neighbors.

UPDATE messages carry the NR instances that are being disseminated throughout the network to update the LSDB. These messages are transmitted between neighboring routers, and one message can transport more than one NR instance. The acknowledgments that the NR instances were correctly received are carried in ACK (UPDATE) messages and, as in the case of UPDATE messages, one message can carry more than one acknowledgment. The acknowledgments only need to include the instance identifier of the NR instances being acknowledged (i.e. the NR identifier and NR freshness attributes).

DELETE messages carry the NR delete indications that are being disseminated throughout the network to update the LSDB. Like the UPDATE messages, DELETE messages are transmitted between neighboring routers, and one message can transport more than one delete indication. The delete indications need only to include the NR identifier of the NR being deleted (the NR freshness attribute is not needed). The acknowledgments of the correct receptions of NR delete indications are carried in ACK (DELETE) messages, which can again carry more than one acknowledgment. The acknowledgments only need to include the NR identifier.

The LSDB SUMMARY REQUEST, LSDB SUMMARY, PARTIAL LSDB REQUEST messages are used in the initial LSDB synchronization process, which occurs between pairs of neighbors (see Section 2.5). The LSDB SUM-MARY REQUEST asks the neighbor to send the summary of its LSDB, and the LSDB SUMMARY carries this summary. The LSDB summary contains the NR instance identifiers of all NRs present in the LSDB. The LSDB SUM-MARY message implicitly acknowledges the reception of the LSDB SUM-MARY REQUEST. Finally, the PARTIAL LSDB REQUEST asks the neighbor to send the complete contents of selected NRs, specifically the NRs that the router is missing or for which the neighbor has fresher instances. These messages carry only the NR instance identifiers of the full NRs being requested. The full NR instances are sent in UPDATE messages. Thus, receiving a full NR (encapsulated in an UPDATE message) implicitly acknowledges the reception of the request (encapsulated in a PARTIAL LSDB REQUEST message); a router will keep requesting an NR instance until it receives it from the neighbor, under the control of an ACK timer.

OSPF and IS-IS give different names to their control messages (with the exception of the HELLO), and the names are also different from the ones introduced in this chapter. However, the roles played by the messages are the same in all three cases. In Figure 6.1, we provide a detailed correspondence between the control messages of OSPF, IS-IS, and of our own protocol (called *generic* in the figure).

3

Principles of Multi-Area Link State Routing

LSR protocols require the storage of a full instance of the LSDB at each router. However, when the network is large, containing many routers and links, the LSDB also becomes large, and some routers may lack memory resources to store it completely. One way to overcome this problem is to structure the network in smaller areas, such that routers only need to keep the network map of the area they belong to. Significant memory savings can be achieved in this way.

Areas are delimited by special routers called Area Border Routers (ABRs). Each area builds and maintains its own LSDB, called *area LSDB*. The LSDB contains the network map of the area, which needs only to be known inside the area, and the prefixes internal to the area. However, in order to communicate with other areas, an area still needs to know the prefixes available outside its boundaries and the paths that should be used to reach them. This information is exchanged among areas through an *inter-area routing protocol* running among the ABRs. There are several possible approaches available to inter-area routing.

As in the case of Chapter 2, we want to keep the discussion in this chapter more centered on concepts rather than on technologies. As will be seen in Chapter 7, OSPF and IS-IS constrain the inter-area network topology to a two-level hierarchical structure and use a distance vector approach in their inter-area routing protocol. In this section, we will consider arbitrary inter-area topologies and both the distance vector and the link state approaches to inter-area routing.

Chapter structure Section 3.1 introduces the main aspects related to the multi-area network structure and the inter-area routing protocol. Section 3.2 presents the distance vector and link state approaches to inter-area routing. Section 3.3 discusses the consequences of restricting the routing information available at areas. Finally, Section 3.4 presents alternatives for advertising domain-external addressing information across areas.

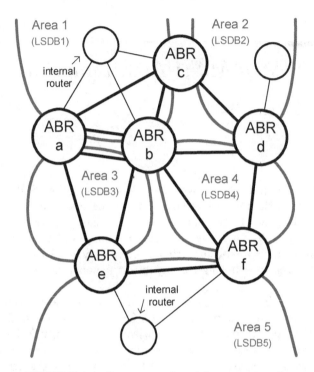

FIGURE 3.1: Structure of multi-area network.

3.1 Multi-area network structure and inter-area routing

In a multi-area network, the network is partitioned in several areas that communicate through ABRs. Figure 3.1 depicts a network structured in five areas (1 to 5) connected by six ABRs (a to f). It also includes some routers internal to areas.

The routing information internal to an area, i.e. the network map of the area and the internal prefixes, is obtained through the routing protocol that runs inside each area, i.e. the *intra-area routing protocol*. This is an LSR protocol and, as such, the information is obtained using the procedures of single-area networks discussed in Chapter 2.

The ABR overlay The set of ABRs and their interconnections form a routing overlay, i.e. a logical network over the physical network utilized for the exchange of inter-area routing information. The inter-area routing protocol runs on this overlay, and it is through this protocol that ABRs learn about the external destinations and compute the shortest path costs towards them. In Figure 3.1, the ABR overlay is represented by thicker lines.

It is assumed that (i) each ABR interface is associated with an area and

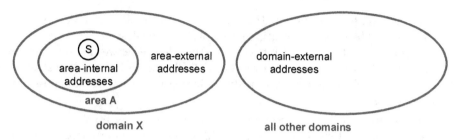

FIGURE 3.2: Address types from the perspective an area-internal router (router *S*).

that (ii) (direct) connections between ABRs can only exist between two interfaces of the same area. Each connection then becomes associated with an area. The connections between ABRs are performed through the flooding mechanism provided by the intra-area LSR protocol. ABRs attached to the same area are called neighbor ABRs. The connections between neighbor ABRs form a full mesh within each area. For example, in the network of Figure 3.1, ABRs *a*, *b*, and *e* are neighbors on area 3, and are interconnected on this area through a full mesh; the same is true for ABRs *b*, *d*, and *f* on area 4. Note that two neighbor ABRs may not be directly connected through a layer-2 link, since there may be several internal routers interposed in the path between them.

The area LSDB In a multi-area network, each area builds and maintains its own LSDB. Routers internal to an area maintain only one LSDB, whereas ABRs maintain as many LSDBs as areas they directly attach to. For example, the network of Figure 3.1 has five different LSDBs, and ABR *b* maintains the LSDBs of areas 1, 3, and 4.

A router is unaware of the network topology beyond its area. Thus, the topological information contained in the area LSDB is restricted to the area. Regarding the addressing information, within an area one can now distinguish three types of prefixes: *area-internal* (prefixes internal to the area), *area-external* (prefixes external to the area but internal to the routing domain), and *domain-external* (prefixes external to the domain). We will refer to the information that is either area-external or domain-external simply as *external* addressing information. Figure 3.2 shows the various types of addresses from the point of view of router *S*.

The area-external destinations that need to be known inside an area are usually address prefixes but, in some special circumstances, can also be routers. This is not required, but some technologies do it. For example, in OSPF, the identifiers of the DBRs are advertised across areas through the inter-area routing protocol. We will discuss this feature in Section 3.4.

The inter-area routing protocol The external address prefixes and the paths to reach them are learned through the inter-area routing protocol running on the ABR overlay. The role of the inter-area routing protocol is then to determine paths that support the communications with destinations external to an area. The inter-area routing protocol should be designed to achieve globally optimal routing, i.e. such that the paths selected between two routers are always the shortest ones. However, both OSPF and IS-IS include restrictions in the way routing information is disseminated that prevents this possibility in some cases.

Inter-area routing approaches The main inter-area routing approaches are link state and distance vector; we will discuss these alternatives in Section 3.2. We could also have considered the *path vector* approach (see Chapter 3 of [33]). However, this approach seems more tailored to inter-domain routing and shares many common aspects with the distance vector approach. Note that, in our case, the intra-area routing protocol is considered to be of LSR type, irrespective of the inter-area routing approach.

Comparison with inter-domain routing Inter-area routing is, in many aspects, similar to inter-domain routing (which we discussed in Section 2.2.4). From the perspective of an area or a domain, the problem in both cases is to learn external prefixes and determine which ABR or DBR should be used to reach those prefixes. In both cases, the routing protocol that obtains this information runs on some routing overlay. In inter-area routing it is the overlay of ABRs and in inter-domain routing it is the overlay of DBRs. The main difference between these two routing types is the path metrics. In inter-area routing, the path metrics of different areas are *comparable*, which allows computing end-to-end shortest paths between routers located in different areas. In inter-domain routing, the path metrics of different domains are usually not comparable, and routing decisions usually involve routing attributes of various types (e.g. the preference of one domain over the others and the hop count).

> *A routing domain can be structured in multiple areas. Each area builds and maintains an LSDB containing the network map of the area, and three types of addressing information: area-internal, area-external (external to the area but internal to the routing domain), and domain-external (external to the domain). Routers have no knowledge of the network map of other areas. The exchange of routing information across areas is performed through an inter-area routing protocol running between the routers located in the frontier between areas, called Area Border Routers (ABRs).*

3.1.1 Steps in communicating external information across areas

Consider the communication between router S, located in a given area (the source area), and prefix ap1, located in a different area (the destination area). Instead of a prefix, the destination can, in special cases, be a router, but the discussion is similar. Before being able to send data messages to ap1, router S

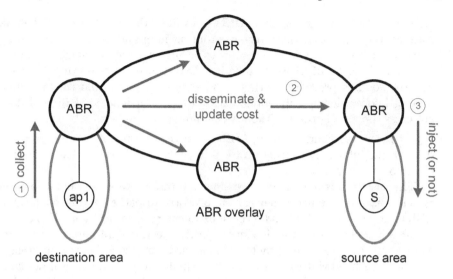

FIGURE 3.3: Steps in communicating external information across areas.

needs to learn that the prefix exists and needs to determine a path (hopefully a shortest path) to it. As highlighted in Section 1.4.3, the routing information starts being advertised by the destination and flows from destination to source. There are three steps in this process, which we represent in Figure 3.3.

Collect First, the ABR of the destination area collects information relative to prefix ap1 and injects this information into the inter-area routing protocol. This information has two elements: the prefix and the shortest path cost from the ABR to the prefix; we will refer to it as the (prefix, cost) pair. ABRs obtain this information from the LSDB of the destination area.

Disseminate and update cost Second, the inter-area routing protocol disseminates the routing information on the ABR overlay and updates the shortest path costs to the advertised prefix.

As mentioned above, the inter-area routing protocol can take either a link state or a distance vector approach. These approaches differ on the type of routing information that is exchanged among ABRs and on how it is disseminated. In the first approach, the routing information originated by one ABR is delivered unmodified to all other ABRs, i.e. it is transmitted directly from each originating ABR to all other ABRs. In the second approach, the routing information is transmitted only between neighbor ABRs, and the estimated costs towards the advertised prefixes are updated by each intermediate ABR as the information travels away from the originating ABR. These two inter-area routing approaches will be studied in detail in Section 3.2.

For the exchange of routing information within the ABR overlay, some communication mechanism between ABRs is necessary. The flooding mechanism intrinsic to LSR protocols provides a natural way of disseminating routing information among neighbor ABRs. As will be discussed later, in the link state routing approach the inter-area routing information uses the flooding procedure with domain flooding scope, to ensure that the information is delivered unmodified to all ABRs. Contrarily, the distance vector routing approach uses the flooding procedure with area flooding scope, such that the inter-area routing information is only communicated among neighbor ABRs and does not cross the area borders without being updated.

Inject In the third and last step, ABRs inject the routing information received through the inter-area routing protocol in the source area, so that it reaches router S.

This last step is not mandatory, since internal routers need not know the available external prefixes: they simply forward packets to one of the area ABRs and let the ABRs handle the subsequent routing decisions. If the area is a *stub area*, i.e. an area with a single ABR, injecting external routing information into the area is even not recommended, since it unnecessarily wastes memory and bandwidth resources; this is similar to the injection of domain-external addressing information in stub domains, discussed in Section 2.2.4. Otherwise, if the area has more than one ABR, injecting routing information is required to achieve globally optimal routing. We will further discuss this issue in Section 3.3.

3.1.2   Determining end-to-end shortest paths

When the source and destination are in different areas, the determination of the end-to-end shortest path is done with the intermediation of the area ABRs. From the point of view of the source router, this computation is equivalent to selecting the *outgoing* ABR that leads to the destination, since routers are unaware of the network topology beyond their area.

Figure 3.4 illustrates the point of view of the source router. An end-to-end path has two sub-paths, one internal and another external to the area. The *internal sub-path* is from the source router (router S) to one ABR, and the *external sub-path* is from that ABR to the destination prefix (prefix ap1). The source router knows its area in detail: based on the topological information present in its LSDB, it knows what routers are ABRs and how to compute the path cost from itself to each of the ABRs. Let c_{Si} denote the cost from router S to ABR i, i.e. the cost an internal sub-path. The costs of the external sub-paths are provided by the ABRs. ABRs obtain this information through the inter-area routing protocol and disseminate it inside the areas they are directly attached to. Let C_{iD} denote the cost from ABR i to destination D, i.e. the cost of an external sub-path.

Thus, the process of determining, at a router, the end-to-end shortest path to a destination comprises three steps:

1. The router obtains the costs of the external sub-paths, i.e. it obtains the costs from each ABR to the destination (the C_{iD} cost

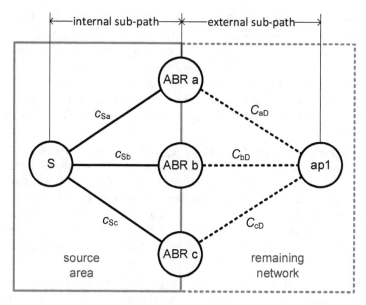

FIGURE 3.4: View of source router when determining the shortest path to an area-external destination.

component); these costs are injected inside the source area by the area ABRs, and are disseminated within the area using the flooding procedure.

2. Using the network map of the area, the router determines the costs of the internal sub-paths, i.e. the costs of the shortest paths from itself to each ABR (the c_{Si} cost component).

3. The router determines the end-to-end path cost through each ABR, i.e. it computes $c_{Si} + C_{iD}$ for each ABR and selects the least cost path and the corresponding outgoing ABR.

> *To determine the shortest path to a destination located in another area, a router adds the cost of the shortest paths from itself to each of its area ABRs, obtained through the area network map, to the cost of the shortest paths from those ABRs to the destination, obtained through the inter-area routing protocol and injected into the area by the ABRs; it then selects the best among these paths, i.e. the least cost one.*

3.1.3 Area LSDB

The area LSDB includes the network map of the area and the addressing information internal and external to the area. As in the case of single-area networks, the network map is formed by router-NRs and slink-NRs. These

NRs are now flooded only inside areas, since the topological information is of no interest outside areas; we say that these NRs are flooded with *area flooding scope*. Moreover, a router-NR of an area only describes the interfaces of the represented router that belong to that area. Thus, an ABR will have its interfaces described in different router-NRs, belonging to different LSDBs.

The addressing information internal to an area, i.e. the prefixes assigned to the area network elements, is provided by the aip-NRs, as in the case of single-area networks.

The area-external-prefix-NR To represent area-external addressing information within the area LSDB, we introduce the *area-external-prefix-NR* (aep-NR for short). These NRs are originated by the ABRs and disseminated inside the areas they are directly attached to, to accomplish the third step of Figure 3.3. Thus, the aep-NR includes (i) the identifier of the originating ABR and (ii) the (prefix, cost) pairs obtained through the inter-area routing protocol. Like router-NRs and slink-NRs, aep-NRs are only disseminated within an area, i.e. they are flooded with area flooding scope.

Additions to the domain-external-prefix-NR The domain-external addressing information also needs to be disseminated inside areas. In Section 2.2.4.2, we introduced the dep-NR for this purpose. In single-area networks, the dep-NR is originated by a DBR and includes (i) the advertised domain-external prefix, (ii) the identifier of the originating DBR, and (iii) a domain-external cost attribute (relative to the sub-path external to the routing domain). In multi-area networks, a router trying to communicate with a domain-external prefix may be located outside the area of the DBR that injected it in the routing domain. Thus, domain-external addressing information must be advertised across areas. To this end, we give dep-NRs a role similar to aep-NRs: the dep-NRs are injected inside areas by the ABRs to advertise domain-external prefixes and the cost of shortest path from the ABR to the DBR that injected the prefix in the domain (this cost is obtained through the inter-area routing protocol). Thus, dep-NRs must be added with a (domain-internal) cost attribute that expresses this path cost value. However, we want to keep the initial role of dep-NRs. Thus, as in the case of single-area networks, the dep-NRs can be injected in the routing domain by the DBRs to advertise the domain-external prefixes (which they obtain through the inter-domain routing protocol); in this case, the domain-internal cost is set to zero. To sum up, dep-NRs can be originated either by ABRs or DBRs and include (i) the advertised domain-external prefix, (ii) the identifier of the originating router, (iii) a domain-external cost attribute, and (iv) a domain-internal cost attribute corresponding to the cost of the shortest path from the originating router to the DBR that inject the prefix in the domain.

Summary of additional NRs required in multi-area networks Figure 3.5 lists the NRs that need to be introduced or modified for the support of multi-area networks. The dbr-NR will be introduced in Section 3.4.

Example Figure 3.6 gives an example of a multi-area network and shows

NR	role	originator
aep-NR	describes area-external prefix and its association with ABR	ABR
dep-NR	describes domain-external prefix and its association with ABR or DBR	ABR or DBR
dbr-NR	describes ASBR and its association with ABR	ABR

FIGURE 3.5: NRs introduced or modified for the support of multi-area networks, their role and originating router.

the LSDB of two of its four areas. Routers R2 to R7 are ABRs, routers R1, R8, and R9 are area-internal routers, and router R8 is also a DBR. Area 1 includes a shared link (ls1). All other links are point-to-point. To facilitate the example, the costs of point-to-point link interfaces are the same in both directions. In this example, only four address prefixes are to be advertised: (i) ap1, assigned to router R1, (ii) ap2, assigned to link ls1, (iii) ap3, assigned to router R9, and (iv) ap4, injected into the domain through router R8. We assume that the domain-external cost associated with ap4 is zero. ABRs know the LSDBs of all areas they directly attach to. For example, router R4 knows the LSDBs of areas 1 and 3. Figure 3.6 shows the LSDBs of areas 1 and 4.

The LSDB of area 1 has four router-NRs and one slink-NR relative to routers R1, R2, R3, and R4, and to link ls1; these are the topological NRs that describe area 1. The remaining NRs convey addressing information. There are two aip-NRs describing the address prefixes internal to the area, ap1 and ap2, and their association with router R1 and link ls1, respectively. The area-external address prefix ap3 is described by three aep-NRs, each injected by one of the area ABRs, indicating the cost of the shortest path from the originating ABR to ap3. For example, based on this information, router R1 determines that its shortest path to ap3 is via router R2 with a cost of 4. Note that the cost to ap3, advertised inside area 1 by router R2, is not the lowest one among the area ABRs: router R4 advertises a cost of 2 and router R2 a cost of 3. However, the internal cost from R1 to R4 is much higher than the one from R1 to R2, and thus router R2 gets selected as the outgoing ABR to ap3. The LSDB of area 1 also includes three dep-NRs describing ap4, each injected in the area by one of its ABRs.

The LSDB of area 4 has the same structure as that of area 1, although with different details. It contains (i) four router-NRs relative to routers R6, R7, R8, and R9, (ii) one aip-NR describing ap3 and its association with router R9, (iii) four aep-NRs describing the external address prefixes ap1 and ap2, as injected by the area ABRs (routers R6 and R7), and (iv) one dep-NR describing ap4 and its association with router R8. For example, based on this information, router R8 selects router R6 as the outgoing ABR to reach both ap1 and ap2.

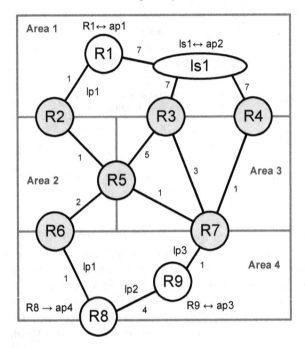

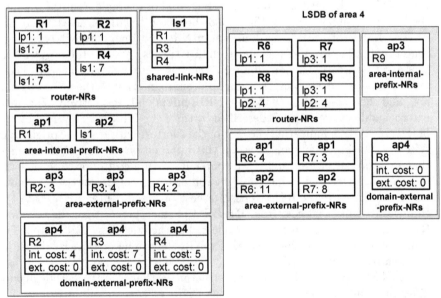

FIGURE 3.6: Multi-area network with four areas and the area LSDBs of areas 1 and 4.

Note that the router-NRs of an area only include the interfaces that belong to that area. For example, in area 1 the router-NR of R3 has only one entry, relative to the interface with link ls1. The other two interfaces, which connect to routers R5 and R7, respectively, are included in the router-NR of area 3.

> *The area LSDB includes the topological information of the area, and the addressing information. The topological information is provided by router-NRs and slink-NRs, as in single-area networks, the area-internal addressing information by area-internal-prefix-NRs (aip-NRs), the area-external addressing information by area-external-prefix-NRs (aep-NRs), and the domain-external addressing information by dep-NRs.*

3.1.4 Graph of the ABR overlay

A convenient abstraction to study inter-area routing is the graph of the ABR overlay. Figure 3.7 shows the overlay graph of the network of Figure 3.6. In this type of graph, nodes correspond to ABRs, arcs to intra-area shortest paths between neighbor ABRs, and arc weights correspond to their costs. The internal routers are not part of the graph. Moreover, each ABR interface is labeled with information on the prefixes available in its attached area. The prefixes internal to the domain are characterized by (prefix, cost) pairs, where the first element identifies the prefix and the second is the cost of the intra-area shortest path from the ABR to the prefix. The prefixes external to the domain are characterized by (prefix, internal cost, external cost) triplets, where the first element identifies the prefix, the second is the cost of the intra-area shortest path from the ABR to the DBR that injected the prefix in the domain, and the third element is the external cost associated with the prefix.

In the graph of Figure 3.7, routers R2 to R4 are labeled with the (prefix, cost) pairs relative to prefixes ap1 and ap2, assigned to area 1 network elements, and routers R6 and R7 with the pairs relative to prefixes ap3 and ap4, assigned to area 4 network elements. For example, router R6 has a label (ap3, 5), since the intra-area shortest path cost from itself to ap3 is 5, through path R6 → R8 → R9, and router R2 is labeled (ap2, 8), since the intra-area shortest path cost from itself to ap2 is 8, through path R2 → R1 → ls1. The figure also shows the costs assigned to each arc. For example, the arc between router R3 and router R5 represents the shortest path between these two routers on area 3 and is assigned a cost of 4, corresponding to path R3 → R7 → R5.

Note that the graph is fully meshed within each area, since there is always some path within an area between each pair of neighbor ABRs. Note also that, in general, the graph is directed, since the shortest path costs may differ in each direction between two neighbor ABRs.

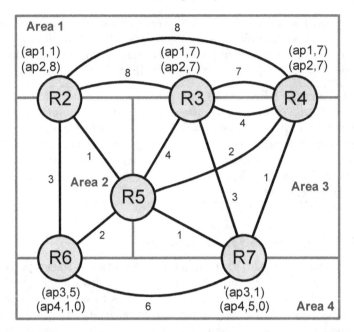

FIGURE 3.7: ABR overlay graph of the network of Figure 3.6.

> *The graph of the ABR overlay is a useful abstraction to study inter-area routing. In this graph, nodes represent ABRs, arcs the intra-area shortest paths between neighbor ABRs, and arc weights their costs. The area-internal routers are not represented in the graph.*

3.1.5   Synchronization mechanisms

The synchronization mechanisms of multi-area networks are those of single area networks with a few modifications. Recall that these mechanisms are the Hello protocol, the initial LSBD synchronization process, and the flooding procedure. One addition to the Hello protocol is the inclusion of the area identifier in Hello messages. This ascertains that neighborhood relationships can indeed be established based on area information. Moreover, the flooding procedure needs to be scoped, since some types of routing information may need to be flooded only inside areas, while other types may need to be flooded across the entire routing domain.

3.2 Inter-area routing approaches

The inter-area routing protocol can take either a link state or a distance vector approach. We describe these alternatives in the next sections.

3.2.1 Link state routing approach

The graph representation introduced in the previous section immediately suggests an LSR approach to inter-area routing. In this approach, each ABR builds and maintains the graph of the ABR overlay. To this end, each ABR floods to all other ABRs its local view of the overlay topology, i.e. who its neighbor ABRs are and what are the intra-area shortest path costs towards them; it also floods information on the prefixes available at the areas it directly attaches to (i.e. it floods the ABR labels). Recall that ABRs maintain the LSDBs of all areas they are directly attached to, and that each LSDB contains the network map of the corresponding area. One requirement of this approach is that LSDBs include information on whether a router is an ABR or just an internal router. This can easily be achieved by introducing a flag in router-NRs.

The overlay LSDB The representation of the ABR overlay topology and of the routable prefixes is made by *the overlay LSDB*, containing three new types of NRs: (i) the *overlay-router-NRs* (orouter-NRs for short) for the topological information, (ii) the *overlay-domain-internal-prefix-NRs* (odip-NRs for short) for the information regarding prefixes internal to the routing domain, and (iii) the *overlay-domain-external-prefix-NRs* (odep-NRs for short) for the information regarding prefixes external to the domain. The orouter-NR includes the identifier of the originating ABR and of its neighbor ABRs, and the costs of the intra-area shortest paths towards them. The odip-NR includes (i) the domain-internal prefix being advertised, (ii) the identifier of the originating ABR and (iii) the cost of the intra-area shortest path from the ABR to the prefix. The odep-NR includes (i) the domain-external prefix being advertised, (ii) the identifier of the originating ABR, (iii) the cost of the intra-area shortest path from the ABR to the DBR that injected the prefix in the domain, and (iv) the domain-external cost associated with the prefix.

The overlay LSDB needs only to be stored at the ABRs; it can be ignored by the area-internal routers. The overlay NRs can be exchanged between ABRs using the flooding procedure (see Section 2.4), but with domain scope, since the NRs must be communicated as originated, i.e. unmodified, to all ABRs. The overlay LSDB corresponding to the network of Figure 3.6 is shown in Figure 3.8.

How do ABRs compute shortest paths and communicate this information to area-internal routers? Through the overlay LSDB, each ABR can build the graph of the ABR overlay and compute the shortest paths from

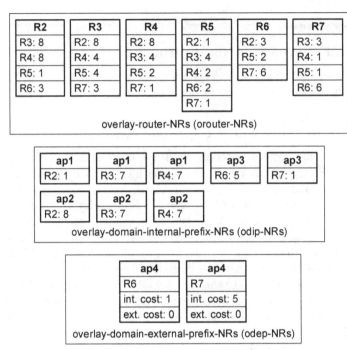

FIGURE 3.8: Overlay LSDB of the network of Figure 3.6.

itself to each domain-internal prefix or DBR, using Dijkstra's algorithm. ABRs then have to inject this information into the areas they directly attach to (the third step of Figure 3.3). They do this using the aep-NRs and the dep-NRs introduced in Section 3.1.3, both flooded with area flooding scope. Recall that both aep-NRs and dep-NRs are originated by ABRs to advertise area-external and domain-external prefixes and the (intra-domain) shortest path costs to reach them.

To illustrate how shortest paths are computed through the overlay graph, consider the determination of the shortest path from router R2 to ap3 using the overlay graph of Figure 3.7. There are several candidate paths, for example:

- R2 → R6 → ap3 with cost 8;

- R2 → R5 → R7→ ap3 with cost 3;

- R2 → R3 → R7→ ap3 with cost 12.

Each (end-to-end) path cost is computed by adding two cost components: (i) the cost of the shortest path from router R2 to an ABR that is labeled with ap3, and (ii) the cost from that ABR to ap3 (present in the ABR label). The path with lowest cost is the second one above. Router R2 can now inject this information into its directly attached areas (area 1 and area 2) using aep-NRs.

Specifically, the aep-NRs advertise that R2 can provide a path to ap3 with a cost of 3.

Note that, in this kind of approach, two types of flooding scope are used: the overlay NRs, i.e. the orouter-NRs, the odip-NRs, and the odep-NRs, are flooded with domain scope, whereas all other NRs are flooded with area scope.

To the best of our knowledge, there is still no implementation of this link state multi-area routing approach.

> *In the LSR approach to inter-area routing, ABRs build and maintain the graph of the ABR overlay. Each ABR floods to all other ABRs its local view of the overlay topology, using overlay-router-NRs (orouter-NRs), and information on the prefixes available inside the areas it directly attaches to, using overlay-domain-internal-prefix-NRs (odip-NRs) for prefixes internal to the domain, and overlay-domain-external-prefix-NRs (odep-NRs) for prefixes external to the domain. The routing information computed by the ABRs using the overlay NRs, i.e. the shortest path cost to each prefix, is injected in the areas using aep-NRs and dep-NRs.*

3.2.2 Distance vector routing refresher

Before describing in detail the Distance Vector Routing (DVR) approach to inter-area routing, we review here its basic principles (see also Section 5.2 of [6]). DVR is based on the distributed and asynchronous version of the Bellman-Ford algorithm. In this algorithm, each router sends to all its neighbors its estimate of the shortest path cost to every network destination. A router updates its estimate of the shortest path cost to a destination via a specific neighbor by adding two cost components: (i) the cost estimate received from that neighbor and (ii) the cost of the link from itself to the neighbor. It then selects the neighbor that provides the shortest path cost (i.e. the next-hop router). When a neighbor is removed (e.g. because of a failure), the cost through that neighbor must be set to ∞. The messages broadcast to neighbors are called distance vectors, and each vector element is a (destination, cost) pair. Distance vectors need to be sent at startup and when the cost estimate towards a destination changes.

Which information to store at routers For each destination, routers can either store the best information received from each neighbor or simply the information received from the best neighbor (the next-hop router); the stored information is the neighbor identifier and the shortest path cost it provides. The second approach is used by the RIP protocol [30], and is also the one used in OSPF and IS-IS. In this case, upon receiving a distance vector from one of its neighbors, a router updates the stored shortest path cost and the next-hop estimate according to the following rules:

- If the next-hop router increases its cost estimate towards a destination, the stored cost is increased accordingly, but the stored next-hop remains unchanged;

- If any neighbor provides a cost to a destination lower than the one currently stored, the stored cost is updated to the lower one, and the next-hop router is set to the neighbor that provides that cost.

Reliable transmission of distance vectors Correct operation of DVR protocols requires (i) that distance vectors are transmitted reliably and (ii) that routers know who their neighbors are. The last requirement ensures that routers can detect new neighbors and neighbor failures. The two requirements can be met by the periodic transmission of distance vectors among neighboring routers, which is precisely the solution adopted in RIP (distance vectors are sent every 30 seconds). In the EIGRP protocol [4, 13], the neighborhood relationships are established through a Hello protocol, as in LSR, and the reliability of message transmissions is ensured through ACK protection. Note that in DVR protocols, routers only need to know their neighbors and not the complete network map, as in LSR.

Count-to-infinity problem DVR protocols suffer from the well-known problem of *count-to-infinity*, which we illustrate with the help of Figure 3.9. Let D_{i1} denote the estimate of the distance from router i to router R1; in the figure, we also indicate the next-hop (nh) router leading to router R1. Assume that, at $t = 0$, the protocol is in a stable state, where each router knows the correct distance and correct next-hop to R1. Suppose that the routers exchange distance vectors synchronously (i.e. at the same time) and that, shortly after $t = 0$, the connection between routers R1 and R2 fails. Following the failure, at $t = 1$, router R2 sends to R3 a distance of ∞, and router R3 sends to both its neighbors, R2 and R3, a distance of 2, as usual. Based on this information, router R2 is led to believe that it can now reach R1 through R3 with a distance of 3 and router R2 believes that it can reach R1 through R4 with a distance of 4. After the next exchange of distance vectors, at $t = 2$, routers R2 and R3 become engaged in a *routing cycle*, from which they will never get out. The distance from R2 to R1 keeps increasing by one every two iterations, slowly counting to infinity, but R2, and indeed all other routers, will never completely understand that router R1 is no longer reachable.

One way to overcome this problem is to "redefine infinity"; for example, in RIP, infinity is set to 16. However, this solution imposes a topological restriction on the network. In fact, networks running RIP cannot have paths longer than 15 hops. Besides this solution, other techniques can be used to reduce the impact of count-to-infinity, such as *triggered updates* and *split horizon*. The DUAL technique [13], used in the EIGRP protocol, completely eliminates this problem.

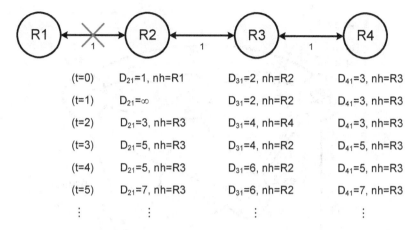

FIGURE 3.9: Count-to-infinity problem.

3.2.3 Distance vector routing approach

In the distance vector approach to inter-area routing, ABRs run a DVR protocol among themselves. The area-internal routers do not participate in this process, except as intermediaries in the communication between ABRs. Note that, in this approach, ABRs have no knowledge of the complete ABR overlay; they only know their neighbor ABRs. Each ABR broadcasts distance vectors to all its neighbor ABRs, containing (prefix, cost) pairs. When an ABR receives a distance vector from a neighbor ABR regarding some prefix, it updates its estimate of the shortest path cost and next-hop ABR towards this prefix using the usual DVR rules, which we reviewed in the previous section. For that, it needs to compute the shortest path from itself to the neighbor ABR, which it can do through the analysis of the LSDB it has in common with the neighbor. As in the inter-area LSR approach, each ABR is responsible for originating information relative to the prefixes available inside its directly attached areas, which it learns from the corresponding LSDBs; this information corresponds to the ABR labels of the overlay graph.

Example In Figure 3.10 we give an example of the operation of the DVR approach related to the ABR overlay of Figure 3.7. The figure shows one of the many possible sequences of (asynchronous) steps whereby ABRs get to know the shortest path cost to ap3 and the next-hop ABR that provides this cost. The originators of this information are the ABRs of area 4, i.e. routers R6 and R7. Note that the shortest path costs and next-hop ABRs from any ABR to ap3 can be anticipated by inspecting the overlay graph. For example, the shortest path cost from router R2 to ap3 is certainly 3 via router R5 (the next-hop). However, in the DVR approach, routers are not aware of the overlay graph; each ABR only knows in detail the areas it is directly attached

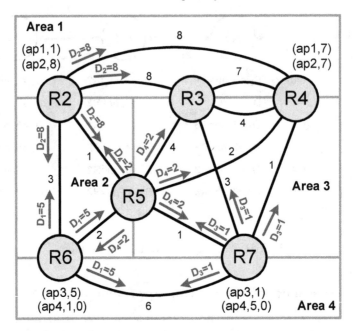

FIGURE 3.10: Example of DVR in the ABR overlay of Figure 3.7.

to. Let, for each router, D_i denote its estimate at step i of the distance (cost) to reach prefix ap3.

- **Step 1** - Initially, router R6 advertises to all neighbors a distance of 5 to ap3; this cost was obtained through the analysis of the LSDB of area 4.

- **Step 2** - Next, router R2 advertises a distance of 8, which corresponds to the sum of the distance received from R6 with the cost of the shortest path between R2 and R6; the shortest path cost is obtained through the analysis of the LSDB of area 2, which is kept by R2. Note that, at this point of the (distributed) computation, router R2 still has an incorrect estimate of the shortest path cost to ap3.

- **Step 3** - At this step, router R7 advertises a distance of 1 to ap3 and, based on this information, routers R3, R4, and R5 learn their correct shortest path costs to ap3, which are 4, 2, and 2, respectively; they also learn the next-hop router to ap3, which is R7 for all of them. Again, the shortest path costs are obtained through the analysis of an LSDB, the LSDB of area 3 in this case.

- **Step 4** - Finally, router R5 advertises a distance of 2 to ap3 and, based on this information, router R2 updates its distance to 3 and the next-hop to R5. After this step, all ABRs have learned the correct shortest path cost and next-hop ABR to ap3. Any other sequence of steps would lead to the same result.

How are distance vectors communicated between ABRs and how is the routing information computed by ABRs communicated to area-internal routers? The distance vectors are exchanged between neighbor ABRs. Two questions remain to be answered:

- How are the distance vectors communicated between ABRs?

- How is the routing information computed by ABRs communicated to area-internal routers?

It turns out that both the communication process and the control messages that communicate the routing information can be the same.

Communication process The communication (i) between neighbor ABRs and (ii) between ABRs and area-internal routers uses the flooding procedure with area flooding scope (see Section 2.4). Recall that the intra-area routing protocol is of the LSR type, even in the inter-area DVR approach.

The communication between ABRs contrasts with the inter-area LSR approach, where the overlay NRs (orouter-NR, odip-NR, and odep-NR) are communicated between ABRs with domain flooding scope (see Section 3.2.1).

Moreover, since the flooding mechanism is reliable, the control messages are transmitted without loss or corruption. Thus, unlike RIP where there is no explicit protection mechanism for the transmission of distance vectors, in this approach distance vectors need not be repeated periodically.

Note also that the underlying intra-area LSR protocol is what implicitly establishes neighborhood relationships between ABRs attached to the same area. It is through the area LSDB maintained by the intra-area LSR protocol that an ABR knows who its active neighbor ABRs are on that area, and it is through this LSDB that it can detect eventual ABR failures. We may say that the intra-area LSR protocol plays the role of the Hello protocol in the ABR overlay.

Control messages As in the case of the inter-area LSR approach, the aep-NRs and the dep-NRs introduced in Section 3.1.3 are used to communicate the routing information computed by the ABRs to the area-internal routers. Moreover, since these NRs have the (prefix, cost) semantics required by distance vectors, they can also be used to communicate routing information between neighbor ABRs. Recall from Section 3.1.3 that aep-NRs and dep-NRs are originated by ABRs to advertise (i) a prefix (an area-external prefix, in the case of aep-NRs, and a domain-external prefix, in the case of dep-NRs), and (ii) the path cost from the originating ABR to the prefix (in the case of aep-NRs) or to the DBR that injected the prefix in the routing domain (in the case of dep-NRs).

Note that two different types of control messages could have been used to accomplish the two functions described above, but this is not needed in this case. In fact, flooding a distance vector within an area serves two purposes simultaneously, and this is an important property. First, it communicates routing information between ABRs (the second step of Figure 3.3). Second, it injects into the whole area the information computed by ABRs, since the distance vectors are disseminated to all internal routers (the third step of Figure 3.3).

> *In the DVR approach to inter-area routing, neighbor ABRs exchange distance vectors among themselves containing (prefix, cost) pairs and, based on this information, estimate their path costs to the advertised prefixes using the usual DVR rules. The aep-NRs and the dep-NRs are used both to communicate the routing information between neighbor ABRs (i.e. as distance vectors) and to communicate the routing information computed by the ABRs to the area-internal routers.*

3.3   Is it still possible to have globally optimal routing?

An important question is whether globally optimal routing is achievable in a multi-area network, i.e. if the paths selected between any two routers are still shortest paths. Problems may arise when routers are restricted in the amount of routing information available to determine shortest paths, and in fact they are in both OSPF and IS-IS. We will give two examples.

Preference to intra-area paths First, routers may give preference to intra-area paths. An *intra-area* path is a path completely contained within one area; an *inter-area* path is a path that crosses more than one area. If preference is given to intra-area paths, the routing between a source and a destination belonging to the same area will not be optimal whenever there is an inter-area path with smaller cost than the least cost intra-area path. This case is illustrated in Figure 3.11. The cost of the intra-area path S → D is 4 and the cost of the inter-area path S → ABR 2 → ABR 1 → D is 3. Note that ABRs may be even forbidden to inject into an area inter-area paths relative to a source and a destination located on that area; this restriction is made by both OSPF and IS-IS.

Short-sighted ABRs The second case is more intricate. It happens when ABRs have no sense of the ABR overlay, and only summarize "what they see" in one area to the other areas they are attached to. We call them *short-sighted* ABRs. Consider the example of Figure 3.12.a. The source and destination are in different areas, each with its own LSDB. It is clear from the figure that the shortest path from S to D is S → ABR 3 → ABR 2 → ABR 1 → D with a cost of 4. Suppose that the ABRs inject information in area 2 based solely on the information contained in the LSDB of area 1; note that the link between ABR 1 and ABR 2 is not included in this LSDB, since it is not part of area 1. In this case, ABR 1 advertises a cost to D of 1, ABR 2 advertises a cost of 4 and ABR 3 a cost 5. Based on this information, router S decides that the best path is S → ABR 1 → D with a cost of 5, which is not optimal.

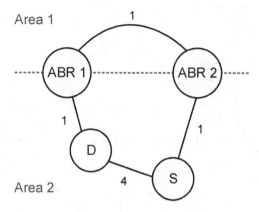

FIGURE 3.11: Giving preference to intra-area paths.

This problem happened because ABRs deal in isolation with each area. To see how the problem could be solved, Figure 3.12.b shows the graph of the ABR overlay. The graph indicates the shortest paths between all pairs of ABRs in each area, and the ABR labels related to the destination, determined through the LSDB of area 1. A label corresponds to the cost of the shortest intra-area path from the ABR to the destination; the labels are 1 for ABR 1, 4 for ABR 2, and 5 for ABR 3. From this information, ABR 3 can determine that the shortest path from itself to D follows the path ABR 3 $\rightarrow$ ABR 2 $\rightarrow$ ABR 1 with a cost of 3; ABR 3 can then inject this information into area 2, allowing S to determine that the shortest path to D is S $\rightarrow$ ABR 3 $\rightarrow$ ABR 2 $\rightarrow$ ABR 1 $\rightarrow$ D with a cost of 4.

> *Routers may be restricted in the amount of information available to compute shortest paths, in which case globally optimal routing might not be achieved.*

3.4 Advertising domain-external addressing information

In previous sections, we already discussed the dissemination of domain-external addressing information across areas. Domain-external prefixes are injected into a routing domain by DBRs using dep-NRs with an internal cost of zero. This information is captured by the ABRs of the areas where the DBRs are located, which disseminate it throughout the ABR overlay. In the LSR approach to inter-area routing, the information is disseminated using odep-NRs, as discussed in Section 3.2.1. In the DVR approach, it is disseminated using dep-NRs distributed as distance vectors, as discussed in Section 3.2.3. This is the solution adopted in IS-IS.

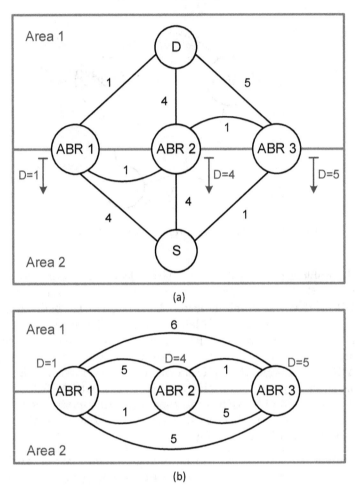

FIGURE 3.12: Short-sighted ABRs; (a) example network, (b) graph of the ABR overlay.

Figure 3.13.a illustrates the DVR approach described above. The zone "other areas" represents the various areas that may exist between the source and destination areas; the arcs represent the shortest paths between routers and the arc labels represent the corresponding shortest path costs. The domain-external prefix ap1 is injected in the domain by DBR 1 through a dep-NR carrying an internal cost of zero. When ABR 2 receives this dep-NR, it injects it into the inter-area routing protocol with an internal cost of 5, which corresponds to the cost of the shortest path from itself to DBR 1. Similarly, when the dep-NR reaches ABR 1, ABR 1 computes the shortest path to ap1, which is 8, and injects this information into the source area. Router S, upon receiving the dep-NR, determines that the cost to ap1 is 10 via outgoing

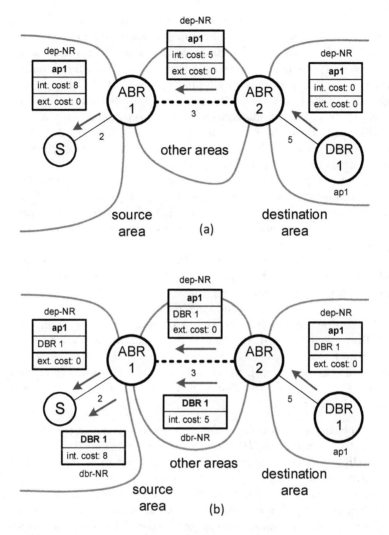

FIGURE 3.13: Alternatives for advertising domain external-information across areas; (a) with dep-NRs only, (b) with dep-NRs and dbr-NRs.

ABR 1. The domain-external cost travels unmodified from the DBR to router S.

There is an alternative to disseminate domain-external addressing information across areas, which involves two NR types. In this solution, a domain-external prefix is injected into a routing domain by a DBR using a dep-NR, as in the previous solution, but the dep-NR is disseminated throughout the complete routing domain, i.e. with domain flooding scope. Thus, the dep-NR by-passes the inter-area routing protocol, reaches all domain routers unmodified and, in this way, it carries no information on how to route to the originating DBR. This information must be provided by a new type of NR, disseminated through the inter-area routing protocol. Different NR types must be used in the DVR and LSR inter-area routing approaches. We will concentrate on the former, since it is the solution adopted in OSPF. The NR type of the DVR approach is called *domain-border-router-NR* (dbr-NR for short), is originated by the ABRs and disseminated as a distance vector, and contains (i) the DBR identifier and (ii) the shortest path cost from the originating ABR to the DBR. In this solution, the dep-NR must include the identifier of its originating DBR, such that the two NR types can be related to each other.

One advantage of this solution is that dbr-NRs act as gathering elements of all domain-external prefixes originated by the same DBR. Thus, in case a DBR is removed, deleting all its originated domain-external prefixes can be done through a single delete indication, i.e. the indication to delete the corresponding dbr-NR.

We give an example in Figure 3.13.b. In this case, the domain-external prefix ap1 is advertised through a dep-NR that is disseminated using domain flooding scope and includes the identifier of DBR 1; the dep-NR includes no internal cost information. At the same time, ABR 2 injects into the inter-area routing protocol a dbr-NR describing DBR 1 and the cost of the shortest path from itself to DBR 1, which is 5. When the dbr-NR reaches ABR 1, ABR 1 updates the cost to 8 and injects it into the source area. Combining the information received in the dep-NR and the dbr-NR, router S concludes that it can reach the DBR that injected ap1 using ABR 2 as the outgoing ABR, with a cost of 10.

> *In multi-area networks, a domain-external prefix is injected into the routing domain by a DBR and is advertised within the area where the DBR is located using a dep-NR; this information is then disseminated across areas using the inter-area routing protocol. An alternative solution involves two NR types: the domain-external prefix is advertised through a dep-NR disseminated with domain flooding scope, i.e. bypassing the inter-area routing protocol, and the DBR that originated the dep-NR is advertised using a domain-border-router-NR (dbr-NR), which is disseminated across areas through the inter-area routing protocol.*

Part III

Technologies

4

Overview of OSPF and IS-IS

4.1 A bit of history

The first LSR protocol was developed for the ARPANET in the late 1970's [32]. It was by then a very simple protocol, since the ARPANET itself was a very homogeneous network, consisting of single-vendor routers connected through synchronous serial lines. Several improvements to this protocol were later proposed in [40], and efforts to standardize LSR technologies started in the 1980's.

There have been many proposals for the use of LSR protocols in both fixed and wireless networks (see, for example, Chapter 7 of [38], for wireless networks proposals). However, their main success is undoubtedly in fixed networks. LSR protocols are currently the preferred ones for intra-domain routing in the Internet. Over the years there were several implementations of LSR technologies, e.g. LSP (for IPX), PNNI (for ATM), IS-IS (for CLNP, IPv4, and IPv6), and OSPF (for IPv4 and IPv6). An introductory description of these technologies can be found in Chapter 14 of [41].

4.2 Today's Link State Routing

OSPF and IS-IS are the main LSR players in today's Internet. They are implemented by virtually all vendors of routing equipment and have been widely deployed in most large IP networks worldwide. Moreover, they keep being adapted to modern routing technologies, such as in the recent adoption of IS-IS as the routing protocol of IEEE 802.1aq shortest path bridge networks [5].

OSPF and IS-IS standards The current versions of OSPF are OSPFv2 (for IPv4), defined in RFC 2328 [36], and OSPFv3 (for IPv6), defined in RFC 5340 [8]. Unlike OSPF, IS-IS was not born as a routing protocol for IP networks: it was initially defined for ISO networks in the ISO/IEC 10589 specification [1] and only later extended to IPv4, in RFC 1195 [7], and IPv6, in RFC 5308 [20]. Besides these specifications, which serve as its basis, there are many others providing various types of enhancements and extensions. For example, OSPF and IS-IS were extended to support MPLS traffic engineering [19, 21, 23, 29], multi-topology routing [44, 45], and multicast routing [34, 44].

Main aspects of LSR protocols As discussed in the previous part of the book dedicated to the principles of LSR, LSR protocols involve three main aspects: (i) the structure of the LSDB, (ii) the synchronization mechanisms required to keep the LSDB updated and synchronized, and (iii) the extensions of the previous two aspects in order to deal with large networks. The first two issues were covered in Chapter 2 and the third one in Chapter 3. In this part of the book, we will devote one chapter to each of these aspects, in the context of OSPF and IS-IS technologies.

Recall that the LSDB includes three types of routing information: (i) topological information, (ii) addressing information, and (iii) link information. The addressing information can be internal or external to the routing domain. The synchronization mechanisms are (i) the Hello protocol, (ii) the flooding procedure, and (iii) the initial LSDB synchronization process. To deal with large networks, the network is usually structured in multiple areas.

Differences between OSPF and IS-IS OSPF and IS-IS differ in several fundamental aspects but also share many common features. The common features are sometimes hidden behind different nomenclature, an issue to which we give particular attention in this book.

When comparing the IPv4 and IPv6 versions of each technology (OSPF and IS-IS), the main differences are in the LSDB structure. The synchronization mechanisms are essentially the same within a given technology. OSPFv3 uses the same Hello protocol, the same dissemination and update procedures, and the same initial LSDB synchronization process as OSPFv2; and this is also true for the IPv4 and IPv6 versions of IS-IS.

The LSDB structure adopted initially by IS-IS facilitated the evolution from IPv4 to IPv6. In OSPF, the changes were more profound. OSPFv2 mixes topological and addressing information by using IPv4 addresses as topological identifiers. OSPFv3, certainly impelled by the enormous growth of the address space (from 32 to 128 bits), redesigned substantially the LSDB structure and clearly separated topological and addressing information.

4.3   The structure of OSPF and IS-IS routing domains

The structure of OSPF and IS-IS routing domains is illustrated in Figure 4.1. Both OSPF and IS-IS routing domains can be structured in multiple areas, to keep the size of the LSDB manageable when networks become large. However, neither OSPF nor IS-IS support arbitrary inter-area topologies (which we discussed in Chapter 3). The inter-area topology is restricted to a two-level hierarchy, where the upper level can only have one area, and all traffic between

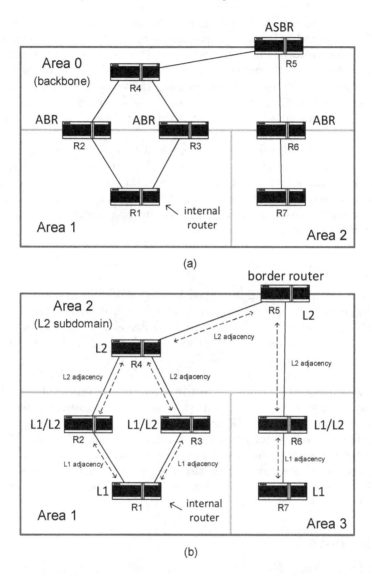

FIGURE 4.1: Structure of (a) OSPF and (b) IS-IS routing domains.

lower level areas is forced to go through the upper one. The upper level area is called *backbone* in OSPF and *L2 subdomain* in IS-IS; we will mainly use the OSPF designation. Because of these topological restrictions, OSPF and IS-IS networks structured in multiple areas are usually called *hierarchical networks*, and not multi-area networks. We will adopt the latter designation hereafter. A proposal for an OSPF extension supporting arbitrary multi-area topologies can be found in [51].

In both hierarchical OSPF and IS-IS, each area maintains a separate LSDB. Areas exchange routing information among themselves using an inter-area routing protocol which, in both cases, is a distance vector protocol. Distance vector protocols suffer from well-known convergence problems, which were probably the root cause for limiting the inter-area topology to a (less flexible) two-level hierarchical structure. However, as discussed in Chapter 3, there exist technical solutions for structuring multi-area networks with arbitrary inter-area topologies.

The domain-external addressing information is injected into a routing domain through its domain border routers. This issue was discussed in Section 2.2.4. A domain border router can be connected to (and inject external prefixes from) a different AS or another routing domain of the same AS. The process of transferring addressing information across different routing domains is usually called *redistribution.*

4.3.1   OSPF

In OSPF, each router within a routing domain belongs to one of three types (see Figure 4.1.a): (i) Autonomous System Border Router (ASBR), which establishes the connection with other routing domains, (ii) Area Border Router (ABR), which interconnects the backbone with one or more lower level areas, and (iii) routers internal to areas. The ASBR can be located at any area, and not just at the backbone. Unfortunately, the designation "ASBR" is misleading since the ASBR can interconnect two routing domains located in the same AS (e.g. it can interconnect a routing domain running another intra-domain routing protocol, such as RIP or EIGRP).

In OSPF, each router interface is assigned to one area. Thus, ABRs have one interface assigned to the backbone area, and one or more interfaces assigned to lower level areas. This also means that ABRs do not belong to a single area (we may say they belong to the backbone and to all lower level areas they directly attach to).

4.3.2   IS-IS

In IS-IS, the hierarchical structure is based on two types of LSDB, called L1 LSDB and L2 LSDB. These LSDBs are independent, and each LSDB is built and maintained using specific control packets (L1 and L2 control packets). For that purpose, the router interfaces can establish two types of relationships, called L1 and L2 adjacencies. Moreover, interfaces can be configured in three types: (i) L1-only, (ii) L2-only, and (iii) L1/L2. The interface type determines the type of routing information and control packets that the interface can exchange: L1-only interfaces can only exchange information relative to the L1 LSDB (i.e. they can only establish L1 adjacencies), L2-only interfaces can only exchange information relative to the L2 LSDB (i.e. they can only establish L2 adjacencies), and L1/L2 interfaces can exchange information relative to both

LSDBs (i.e. they can establish both types of adjacencies, using different control packet types).

IS-IS routers are then classified according to the interface types they support: (i) L1 routers have only L1-only interfaces, (ii) L2 routers have only L2-only interfaces, and (iii) L1/L2 routers can have the three types of interfaces. In IS-IS, areas are not assigned to router interfaces as in OSPF, but are assigned to routers.

As in OSPF, an IS-IS routing domain structured in areas has a distinct LSDB for each area. The LSDBs of lower level areas must be of the L1 type and the LSDB of the backbone area (L2 subdomain) must be of the L2 type. The exchange of routing information between the backbone and one lower level area or, saying another way, between the L2 LSDB and one L1 LSDB, is done at an L1/L2 router. Thus, L1/L2 routers are the ABRs of IS-IS. Because a router is assigned to a specific area, L1/L2 routers must be assigned to lower level areas. Thus, lower level areas are formed by L1 routers and the L1/L2 routers that interconnect with the backbone (which establish L1 adjacencies among themselves for the construction of the L1 LSDB), and the backbone is formed by L2 routers (which establish L2 adjacencies among themselves for the construction of the L2 LSDB). On the surface, this seems very different from the OSPF routing domain structure but, as will be discussed in Chapter 7, it is not!

The routers that interconnect to other routing domains (i.e. the equivalent of the ASBR of OSPF) are simply called *border routers*.

4.3.3   OSPFv3 instances

OSPFv3 includes an interesting feature, which is the possibility of overlaying multiple logical networks (called instances) on top of the same physical infrastructure. Each instance can have its own LSDB and, for that purpose, OSPFv3 routers establish a distinct adjacency per instance. This is similar to the L1 and L2 adjacencies of IS-IS (and the corresponding L1 and L2 LSDBs), although the motivation was not the support of multiple areas. One application scenario referred to in the OSPFv3 specification [8], is when several providers connect their networks to a NAP (Network Access Point) for the exchange routing information and want to run separate OSPF routing domains over some set of common equipment (routers or links). However, this feature is yet to be fully exploited in real networks!

4.4   Architectural constants and configurable parameters

Both OSPF and IS-IS make a clear distinction between architectural constants and configurable parameters. Architectural constants are parameters that were fixed by the specification and cannot be changed by the network manager. They are listed in Appendix B of RFC 2328 [36] and in Section 7.5 of ISO/IEC 10589 [1]. The configurable parameters of OSPFv2 are listed in Appendix C of RFC 2328, and those of OSPFv3 in Appendix C of RFC 5340 [8]. For example, in both OSPF and IS-IS, the lifetime of the LSDB elements is an architectural constant, and the periodicity of HELLO transmissions is a configurable parameter. In this book, we will refer to the architectural constants by appending "*" to their names.

4.5   Goals and structure of this part of the book

In the principles chapters (Chapters 2 and 3), we sketched the fundamental aspects of an LSR protocol, discussing several design alternatives but leaving aside the technological details. Our objective was to build a framework from where the essential features of OSPF and IS-IS—their commonalities and dissimilarities—could be understood in a more solid way. From now on, we will refer to this protocol as the *generic* protocol.

In this part of the book, we dive into the technological details of OSPF and IS-IS. Bearing in mind that OSPF and IS-IS are still evolving technologies, with constant additions and enhancements, our aim is to give the reader enough information so that he/she can fully operate real life OSPF and IS-IS networks (and easily learn any specific feature that we might not have covered).

Part structure In Chapter 5, we explain the LSDB structure of OSPF and IS-IS for the case of single-area networks and, in Chapter 6, we introduce the synchronization mechanisms required to keep the LSDB updated and synchronized. Finally, in Chapter 7, we discuss the extensions for the support of hierarchical routing. In part IV of the book, we illustrate through many examples obtained with real equipment the various features of OSPF and IS-IS introduced in the present part of the book.

5

LSDB Structure of OSPF and IS-IS Single-Area Networks

The LSDB contains the routing information required to build the forwarding tables at routers. Routing information falls in three types: (i) topological information, describing the routers and the way they are connected through links, (ii) addressing information, describing the routable layer-3 addresses of the actual network destinations (where hosts reside), (iii) link information, describing the layer-2 addresses required to transport packets between routers. The addressing information can be internal or external to the routing domain (domain-internal or domain-external). The description of the network topology, which we also call *network map*, involves abstracting the physical network in a suitable way. In particular, the links between routers range from simple point-to-point links to relatively complex layer-2 networks. The abstraction needs to define how routers and links are identified and how links are typified.

OSPF and IS-IS use different philosophies and nomenclature to structure their LSDBs. In Chapter 2, we introduced a generic LSDB structure that will now be used as reference. The basic building blocks of this LSDB are the *Network Records* (NRs), which have two important properties: (i) they provide elementary routing information and (ii) they can be flooded independently. Moreover, NRs can convey *separately* topological and addressing information. We will devote particular attention to the ability of current technologies to achieve this separation.

Except for the link information, which has a local nature, the routing information must be the same in the LSDBs of all routers located in the same area. The mechanisms required to ensure this synchronization will be discussed in Chapter 6.

Chapter structure In this chapter, we concentrate on the LSDB structure of routing domains containing a single area, but possibly connected to other routing domains. We start by introducing, in Section 5.1, the basic elements of the OSPF and IS-IS LSDBs. Then, in Section 5.2 we explain how routers and links are identified, and in Section 5.3 how links are typified. The general structure of the LSDB elements is presented in Section 5.4. We next discuss how these elements are used to describe the topological and domain-internal addressing information in OSPFv2 (Section 5.5), in IS-IS (Section 5.6), and in OSPFv3 (Section 5.7). The representation of the domain-external addressing information is introduced in Section 5.8. Finally, in Section 5.9, we discuss the representation of the link information.

5.1   LSDB elements

OSPF In OSPF, the LSDB elements are called *Link State Advertisements* (LSAs). There are different types of LSAs to convey different types of routing information, and LSAs can be flooded independently. However, in OSPFv2, some LSAs mix the topological and addressing information.

IS-IS In IS-IS, the LSDB is first divided into *Link State PDUs* (LSPs). There are only two types of LSPs: (i) *Nonpseudonode-LSPs*, to convey routing information associated with routers, and (ii) *Pseudonode-LSPs*, to convey routing information associated with transit shared links. The different types of routing information contained in each LSP are provided through a variable number of Type-Length-Value (TLV) records. Thus, from a routing information perspective, the TLVs are the equivalent of the LSAs. However, unlike LSAs, TLVs cannot be flooded independently: they are forced to be flooded together with all other TLVs belonging to the same LSP. In IS-IS, the LSPs are indivisible from a flooding perspective. For example, the TLVs describing the topological and addressing information associated with one specific router cannot be flooded separately, since they are part of the same Nonpseudonode-LSP.

Originating routers The routers that create LSAs and LSPs have an important role in maintaining the LSDB properly updated. We will refer to them as *originating routers* or as *Advertising Routers* (ARs), following the OSPF nomenclature.

Comparing the generic protocol, OSPF, and IS-IS In Figure 5.1, we summarize the correspondences between the LSDB elements of the generic protocol, OSPF, and IS-IS from the routing information and flooding perspectives. From the routing information perspective, there is an equivalence between NRs, LSAs, and TLVs; from the flooding perspective, the equivalence is between NRs, LSAs, and LSPs.

Revisiting the separation principle IS-IS and OSPFv2 somewhat compromise the separation principle discussed in Section 2.2.2. In IS-IS, the routing information elements (the TLVs) cannot be flooded independently. In OSPFv2, some routing information elements (the LSAs) mix topological and addressing information. OSPFv3 is the only technology that separates topological and addressing information both from the routing information and the flooding perspectives.

In Figure 5.2, we show the correspondence, from the routing information perspective, between the NRs of the generic protocol introduced in Chapter 2

routing information perspective

flooding perspective

FIGURE 5.1: Correspondence between the LSDB elements of the generic protocol, OSPF, and IS-IS.

and the LSDB elements of OSPF and IS-IS. These elements will be discussed in the next sections.

> *In OSPF, the LSDB elements are the Link State Advertisements (LSAs). Different LSA types convey different types of routing information, and can be flooded independently. In IS-IS, the LSDB is first divided into Link State PDUs (LSPs), and each LSP includes a variable number of Type-Length-Value (TLV) records to convey different types of routing information. The TLVs contained in one LSP cannot be flooded independently. There are two types of LSPs: Nonpseudonode-LSPs describe routing information associated with routers, and Pseudonode-LSPs describe routing information associated with transit shared links.*

5.2 Router and link identifiers

The network representation requires suitable router and link identifiers. In the next sections, we discuss the structure of these identifiers in OSPFv2, OSPFv3, and IS-IS. In Figure 5.3 we summarize their main characteristics.

5.2.1 Router identifiers

OSPF router identifiers In OSPFv2 and OSPFv3, the router identifier is a 32-bit number unique within the routing domain, called *Router ID* (RID), and expressed in dotted-decimal notation as an IPv4 address. In OSPFv2, it is not mandatory to explicitly configure a RID at a router. If not configured,

generic	OSPFv2	OSPFv3	IPv4 IS-IS	IPv6 IS-IS
router-NR	Router-LSA	Router-LSA	IS Neighbors TLV or Extended IS Reachability TLV of Nonpseudonode-LSP	IS Neighbors TLV or Extended IS Reachability TLV of Nonpseudonode-LSP
shared-link-NR (slink-NR)	Network-LSA	Network-LSA	IS Neighbors TLV or Extended IS Reachability TLV of Pseudonode-LSP	IS Neighbors TLV or Extended IS Reachability TLV of Nonpseudonode-LSP
area-internal-prefix-NR (aip-NR)	Router-LSA and Network-LSA	Intra-Area-Prefix-LSA	IP Internal Reachability Information TLV or Extended IP Reachability TLV	IPv6 Reachability TLV
link-address-NR (la-NR)		Link-LSA		
domain-external-prefix-NR (dep-NR)	AS-External-LSA	AS-External-LSA	IP External Reachability Information TLV or Extended IP Reachability TLV	IPv6 Reachability TLV

FIGURE 5.2: Types of routing information elements of the generic protocol, OSPF, and IS-IS.

Identifier	OSPFv2	OSPFv3	IS-IS
router	RID	RID	SID
point-to-point link	RID of neighbor | IPv4 address of router interface	RID of neighbor | Interface ID of router interface	SID of neighbor
transit shared link	IPv4 address of DR interface	RID of DR | Interface ID of DR interface	SID of DIS | non-zero Pseudonode ID

FIGURE 5.3: Router and link identifiers in OSPF and IS-IS.

the RID is set as one of the IPv4 addresses assigned to the router interfaces, and actual implementations use the highest one. Thus, in OSPFv2, the RID can be a routable IPv4 address.

IS-IS router identifiers In IS-IS, routers are uniquely identified by the *Network Entity Title* (NET), shown in Figure 5.4. The NET uses ISO Network Service Address Point (NSAP) addresses and is represented in hexadecimal notation. It includes both the area identifier, called Area ID (AID), and the identifier of the router within its area, called System ID (SID). The AID can have a length between 1 and 13 octets. However, a common approach in IP networks is to use 3 octets. The first octet is called AFI (Authority and Format Identifier) and tells how to interpret the remaining part of the AID. IP networks usually use the AFI of private addressing, i.e. 0×49. The SEL field (NSAP Selector) is never examined in IP networks; it can take any value and is usually set to 0×00.

5.2.2 Link identifiers

Links are dynamically identified through the Hello protocol. The link identifier is based on the identifier of one of the routers attached to the link: the router on the other side of the link, in case of point-to-point links, or the link DR, in case of transit shared links.

Point-to-point link identifiers Point-to-point links are represented at a router through the identifier of its neighbor on the link, in both OSPF and

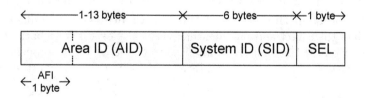

FIGURE 5.4: Structure of the Network Entity Title (NET).

IS-IS (see Section 2.2.1.3). The identifier is the RID in case of OSPF, and the SID in case of IS-IS.

The DR, the BDR, and the DIS A shared link is identified and represented in the LSDB through its DR (see Section 2.2.1.4). In IS-IS, the DR is called *Designated Intermediate System* (DIS). In OSPF, a second router is elected at transit shared links, called the *Backup Designated Router* (BDR), with the role of replacing the DR in case of failure. However, the BDR plays no role in the identification of shared links.

Shared link identifiers OSPFv2 uses a different method than OSPFv3 and IS-IS to identify transit shared links. In OSPFv2, a shared link is identified through the IPv4 address of the DR interface attaching to the link; note that this identifier is an interface address and may not coincide with the RID of the DR.

OSPFv3 and IS-IS use a concatenation of the DR identifier with a local tag generated by this router, which must be unique for each shared link it represents. The tag is called `Interface ID` in OSPFv3 and `Pseudonode ID` in IS-IS. The `Interface ID` has four octets and the `Pseudonode ID` has one octet. Moreover, the `Pseudonode ID` must have a non-zero value in the case of transit shared links.

Figure 5.5 gives an example of shared link identifiers. Router R1 is the DR at both the left and right shared links. The figure shows how the shared link identifiers of IS-IS, OSPFv2, and OSPFv3 are obtained from the `Pseudonode IDs` (IS-IS), IPv4 interface addresses (OSPFv2), and `Interface IDs` (OSPFv3).

Distinction among parallel links Moreover, to distinguish among parallel links at a router, OSPFv2 concatenates the neighbor's RID with the IPv4 address of the interface attaching the router to the link, and OSPFv3 concatenates it with the interface identifier. IS-IS has no provision to distinguish among parallel links.

Use of the Pseudonode ID The `Pseudonode ID`, besides being a component of the transit shared link identifier, is also used to discriminate between the identifiers of routers and transit shared links. The `Pseudonode ID` must be zero in the first case and non-zero in the second one. In OSPF, the distinction between the element an identifier applies to (router, point-to-point link, or transit shared link) is made through the context where the identifier is used, e.g. the type of LSA and the so-called link description.

A comment on the length of the link identifiers We highlight that the length of the router and link identifiers is clearly excessive in both OSPF and IS-IS, given the fact that they only need to be unique inside a routing domain: OSPF uses 4 octets (an IPv4 address) and IS-IS uses 6 octets (the SID).

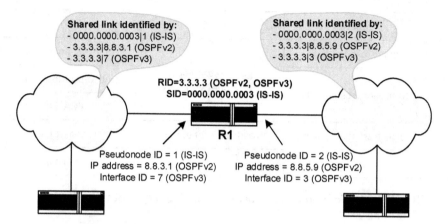

FIGURE 5.5: Identification of shared links in IS-IS, OSPFv2, and OSPFv3.

> *Routers are identified by 32-bit numbers expressed as IPv4 addresses,
> in both OSPFv2 and OSPFv3, and by NSAP addresses in IS-IS.
> Links are dynamically identified using the Hello protocol. Point-to-
> point links are identified through the RID of the neighboring router on
> the link which, in case of OSPF, is concatenated with an IPv4 address
> (OSPFv2) or a locally generated tag (OSPFv3), to distinguish among
> parallel links. Shared links are identified through the IPv4 address of
> the DR interface attached to the link (OSPFv2) or through the RID of
> the DR concatenated with a locally generated tag (OSPFv3 and IS-IS)
> to distinguish among routers designated in more than one shared link.*

5.3 Link types

The link types are abstractions of the actual link technologies for the purpose
of network representation. In Section 1.7, we classified layer-2 links in two
types: point-to-point links and shared links. Furthermore, shared links were
classified according to their degree of connectivity and broadcast capability.
The degree of connectivity impacts the way shared links are represented in the
network map, and the broadcast capability impacts the way control packets are
transmitted on shared links. According to the degree of connectivity, links can
be classified as *fully-meshed* or *partially-meshed*. In particular, fully-meshed
shared links can be represented by the equivalent of a slink-NR. According to
the broadcast capability, links can be classified as *broadcast* or *non-broadcast*.
The link types considered by OSPF and IS-IS do not follow exactly this classi-

generic	IS-IS	OSPF
point-to-point link	general topology subnetwork	point-to-point link
non-broadcast shared link		point-to-multipoint link
		NBMA link
broadcast shared link	broadcast subnetwork	broadcast link
		virtual link

FIGURE 5.6: Correspondence between the link types of the generic protocol, OSPF, and IS-IS.

fication and differ slightly between them. Figure 5.6 shows the correspondence between the link types of the generic protocol, OSPF, and IS-IS.

IS-IS link types IS-IS considers only two link types, called *general topology subnetwork* and *broadcast subnetwork*. The latter are the equivalent of broadcast shared links, and the former aggregate the remaining link types, including point-to-point and non-broadcast shared links. General topology subnetworks are represented as a collection of point-to-point links. Broadcast subnetworks are represented in the LSDB by the equivalent of a slink-NR.

OSPF link types OSPF considers five link types: *point-to-point links*, *virtual links*, *broadcast links* (the equivalent of broadcast shared links), and two types of non-broadcast shared links called *point-to-multipoint* and *Non-Broadcast Multi-Access* (NBMA). A point-to-multipoint link is represented through a collection of point-to-point links, as in IS-IS. An NBMA link is represented through the equivalent of a slink-NR, as in OSPF broadcast links and IS-IS broadcast subnetworks. The NBMA representation applies only to fully-meshed shared links. Since NBMA links have no broadcast capability, the control packets sent by one router must be transmitted individually to every other router attached to the link. OSPF includes several adaptations to minimize the amount of control traffic exchanged in this type of links. Virtual links are logical links used in hierarchical OSPF to overcome the topological restrictions of these types of networks. Because of their preponderance in current networks, we will concentrate on point-to-point and broadcast shared links, and leave aside the details of non-broadcast shared links and virtual links.

Except for virtual links, the OSPFv2 specification [36] refers to links as networks, where networks are understood as subnets and not as links. In fact, unlike OSPFv3, in OSPFv2 there is no distinction between links and subnets, and all the processing is per subnet. Recall from the discussion in Section 1.4.2 that multiple subnets can be assigned to a single link.

> *IS-IS abstracts the links between routers into two types, called general topology and broadcast subnetworks. OSPF abstracts the links into five types: point-to-point, broadcast, point-to-multipoint, Non-Broadcast Multiple Access (NBMA), and virtual links.*

5.4 General structure of the LSDB elements

The formats of OSPFv2 and OSPFv3 LSAs, and of IS-IS LSPs, are shown in Figures 5.9, 5.12, and 5.15, respectively. In the figures, we include all fields of LSAs and LSPs defined in the specifications, except unused fields. However, we will only describe the fields that are fundamental to the operation of LSR protocols. The figures also indicate the length of each field (in octets); a v indicates that the length is variable. Fields or groups of fields that can repeat several times are encapsulated in a darker rectangle. In this chapter, we will concentrate on the LSAs and LSPs needed to describe a single-area network. Like NRs, LSAs and LSPs are structured into two parts: a header and a body. The header contains the equivalent of the NR identifier (see Section 2.2.5) and NR freshness attributes (see Section 2.4.1); these fields are summarized in Figure 5.7.

5.4.1 LSA header

All LSA types share a common 20-octet header, which is almost identical in OSPFv2 and OSPFv3, as can be checked in Figures 5.9 and 5.15. LSAs are uniquely identified by three fields: the AR identifier (`Advertising Router`), the type of LSA (`LS Type`), and an additional field to distinguish among LSAs of the same type originated by the same router (`Link State ID`). We will call this set of fields the LSA ID. The LSA freshness is defined by the SN (`LS Sequence Number`), the age (`LS Age`), and the checksum (`LS Checksum`).

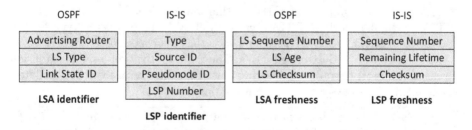

FIGURE 5.7: Fields that identify LSAs and LSPs and describe their freshness.

5.4.2   LSP header

Like LSAs, all LSP types share a common header (see Figure 5.12). LSPs can be of type L1 or L2, as indicated by the `Type` field. They are uniquely identified by the `Source ID`, which identifies the originating router, and the `Pseudonode ID`, which distinguishes among different LSPs originated by the same router. As referred above, the `Pseudonode ID` must be zero for LSPs describing routers and a non-zero value for LSPs describing transit shared links. In IP networks the `Source ID` field equals the SID. The freshness of LSPs is defined by the SN (`Sequence Number`), the age (`Remaining Lifetime`), and the checksum (`Checksum`).

The `LSP Number` field is used in the fragmentation process, and uniquely identifies the fragments of a given LSP. This is needed because IS-IS control packets are directly encapsulated in layer-2 packets, which have no support for fragmentation. In IS-IS, the fragmentation occurs only at the originating router, which must know a priori the minimum MTU among all network links to determine the maximum fragment size; there is no path MTU discovery process, as in IPv6. Moreover, in IS-IS, the LSP fragments are never reassembled and, therefore, they are listed individually at the LSDB; this is cumbersome. The LSP fragments are uniquely identified at the LSDB through the concatenation of `SID`, `Pseudonode ID`, and `LSP Number`, which is called LSP ID. OSPF packets are encapsulated in IP packets and, therefore, rely on the fragmentation and reassembly mechanism of IPv4 or IPv6.

5.4.3   IS-IS TLV records

The body of LSPs comprises a variable number of TLV records. The structure of IS-IS TLVs is shown in Figure 5.8.a. The `Type` field indicates the type of TLV, the `Length` field indicates the length of the `Value` field, and the `Value` field is the actual contents of the TLV, which varies according to its type. The contents of a TLV is limited to 255 octets.

Extended TLVs The structure of TLVs was later extended to enable the organization of the TLV contents in sub-TLVs (Figure 5.8.b) [29]. In this case, the contents of a TLV may include a variable number of sub-TLVs, allowing for a two-level organization of the routing information. The structure of sub-TLVs is similar to that of TLVs: each sub-TLV has a one octet `Type` field, a one octet `Length` field, and a `Value` field with zero or more octets. These TLVs are sometimes called extended TLVs. The main motivation for this extension was the support of traffic engineering functions. However, as will be seen later, extended TLVs have a more wide use today.

> Both LSAs and LSPs include a header that contains the fields required to uniquely identify them in the LSDB and to express their freshness. The body of LSPs is structured with Type-Length-Value records.

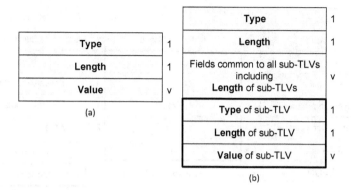

FIGURE 5.8: Structure of IS-IS TLVs; (a) regular TLVs, (b) TLVs organized in sub-TLVs (extended TLVs)

5.5 Topological and domain-internal addressing information in OSPFv2

The format of OSPFv2 LSAs is shown in Figure 5.9. In OSPFv2, the topological and addressing information is jointly provided by the Router-LSA and the Network-LSA. From the point of view of topological information, the Router-LSA can be considered the equivalent of the router-NR and the Network-LSA the equivalent of the slink-NR. However, the Router-LSA and Network-LSA also provide addressing information and can also be considered the equivalent of the aip-NR. This situation was not kept when evolving from OSPFv2 to OSPFv3, where Router-LSAs and Network-LSAs were restricted to provide topological information, as will be discussed in Section 5.7.

The identifiers of the network elements can end up being routable IPv4 addresses. As discussed in Section 5.2, the RID, if not explicitly configured, equals one the IPv4 addresses assigned to the router interfaces. Moreover, the shared link identifier always coincides with the IPv4 address assigned to DR interface attached to the link. Thus, OSPFv2 does not separate topological identifiers from routable addresses.

In Section 9.2.1.1 we show experiments that highlight the features of the OSPFv2 LSDB.

5.5.1 Router-LSA

The Router-LSA describes a router, its outgoing interfaces, and the IPv4 addresses assigned to itself and its interfaces.

Originating router The router that originates the LSA is identified in the Advertising Router field, using its RID.

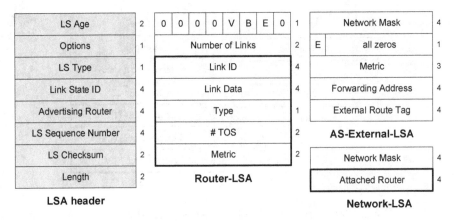

FIGURE 5.9: Format of OSPFv2 LSAs (single-area networks).

Link descriptions The router interfaces are characterized using the so-called *link descriptions*. More than one description may be needed to completely characterize an interface; as will be seen later, this is the case of point-to-point link interfaces. The number of link descriptions contained in a Router-LSA is indicated in the `Number of Links` field. The main fields of a link description are `Type`, `Link ID`, `Link Data`, and `Metric`. OSPFv2 considers four different types of link descriptions, discriminated by the `Type` field:

- type 1 - point-to-point link;

- type 2 - link to transit network;

- type 3 - link to stub network;

- type 4 - virtual link.

Type 1 and type 2 descriptions convey topological information, and type 3 descriptions convey addressing information. Type 4 descriptions are for virtual links and will be not be further discussed.

How are link descriptions used? Figure 5.10 shows the contents of the link description fields for each type of element being described. These elements are: (i) an IPv4 address assigned to a router, (ii) an interface to a point-to-point link, (iii) an interface to a transit shared link, and (iv) an interface to a stub shared link.

- **IPv4 address assigned to router** - An IPv4 address assigned to a router, i.e. a loopback address, is advertised through a type 3 link description, where `Link ID` is the IPv4 address, `Link Data` is a 32-bit subnet mask (255.255.255.255), and `Metric` has a value of zero.

- **Interface to point-to-point link** - An interface to a point-to-point link

Element	Type	Link Data	Link ID	Metric
IPv4 address assigned to router	3	IPv4 address	32-bit subnet mask	zero
Interface to point-to-point link	1	Neighbor RID	IPv4 address of interface	Interface cost
	3	Subnet IPv4 address	Subnet mask	Interface cost
Interface to transit shared link	2	IPv4 address of DR interface	IPv4 address of interface	Interface cost
Interface to stub shared link	3	Subnet IPv4 address	Subnet mask	Interface cost

FIGURE 5.10: Contents of link description fields (OSPFv2).

is characterized by two link description types: a type 1 description to convey topological information and a type 3 description to convey addressing information.

- The type 1 description identifies the link and the corresponding interface cost. It includes the RID of the neighbor (`Link ID`), the IPv4 address assigned to the router outgoing interface (`Link Data`), and the cost of the interface with the link (`Metric`). The `Link Data` field ensures uniqueness in the link identification when there are parallel links.

- The type 3 description identifies the prefix assigned to the link. It includes the subnet IPv4 address (`Link ID`) and the subnet mask (`Link Data`) that define the prefix assigned to the link, as well as the cost of the interface with the link (`Metric`). The latter is repeated information, already contained in the type 1 description.

The nomenclature around type 3 descriptions is confusing since, in the OSPFv2 specification this description is said to characterize a "link to stub network", whereas in this case it is used to characterize a point-to-point link.

Interestingly, OSPFv2 accommodates unnumbered point-to-point links, i.e. point-to-point links with no assigned prefix. In this case, the link is characterized through a single type 1 description, where `Link Data` is the number of the interface attaching to the link. The requirement for unnumbered point-to-point links comes from a time when subnets were classful and assigning a prefix to a point-to-point link implied wasting, at least, a class C address space (256 addresses). However, the existence of this feature is an implicit recognition that topological and addressing information can indeed be separated.

- **Interface to transit shared link -** An interface to a transit shared link is characterized through a type 2 description, which includes the shared link identifier, i.e. the IPv4 address of the DR interface attaching to the link (`Link ID`), the IPv4 address assigned to the router outgoing interface (`Link Data`), and the cost of the interface with the link (`Metric`).

- **Interface to stub shared link -** Finally, an interface to a stub shared link is characterized by a type 3 description, which includes the subnet IPv4 address (`Link ID`) and the subnet mask (`Link Data`) that define the prefix, and the cost of the router interface with the link (`Metric`). Stub shared links have no topological representation, since they are not identified in the network map.

Router-LSA identification The Router-LSAs are uniquely identified in the LSDB by two fields, the `Advertising Router` and the `LS Type`. In these LSAs, the `Link State ID` has no meaning and equals `Advertising Router`.

5.5.2   Network-LSA

The Network-LSA describes a transit shared link and the prefix assigned to it. The routers attached to the link are identified through their RIDs in the `Attached Router` field. The link identifier is placed in the `Link State ID` field. The IPv4 address contained in this field, together with the subnet mask contained in the `Network Mask` field, characterize the prefix assigned to the link. Thus, Network-LSAs convey both topological and addressing information.

Unlike Router-LSAs, Network-LSAs require the `Link State ID` field to be uniquely identified (see Section 2.2.5), since the same router can originate more than one Network-LSA.

5.5.3   Providing topological and addressing information

Figure 5.11 summarizes how the topological and addressing information is provided in OSPFv2. The figure includes the same information relative to OSPFv3, for comparison purposes. However, the OSPFv3 LSDB structure will only be discussed in Section 5.7.

Topological information The topological elements relevant for the construction of the network map are (i) the routers, (ii) the point-to-point links, and (iii) the transit shared links. The topological information concerns the way these elements are identified and connected to each other.

- A router is identified through its RID, contained in the Router-LSA it originates.

- A point-to-point link is described through two Router-LSAs, each originated by one link end router and describing one link direction. In particular, the point-to-point link interface is characterized by a type 1 link description,

Network element	OSPFv2	OSPFv3
router	Router-LSA (Advertising Router)	Router-LSA (Advertising Router)
point-to-point link	two Router-LSAs (Type 1 description)	two Router-LSAs (Type 1 description)
transit shared link	Network-LSA and Router-LSA (Type 2 description)	Network-LSA and Router-LSA (Type 2 description)

Topological information

Prefix assigned to	OSPFv2	OSPFv3
router	Router-LSA (Type 3 description)	Intra-Area-Prefix-LSA
point-to-point link	two router-LSAs (Type 3 description)	two Intra-Area-Prefix-LSAs
transit shared link	Network-LSA	Intra-Area-Prefix-LSA
stub shared link	Router-LSA (Type 3 description)	Intra-Area-Prefix-LSA

Addressing information

FIGURE 5.11: How topological and addressing information is provided in OSPFv2 and OSPFv3.

using the RID of the neighbor as the link identifier (concatenated with the IPv4 address of the router interface to distinguish between parallel lines).

- A transit shared link is characterized through a Network-LSA originated by the link DR, which includes the link identifier (i.e. the IPv4 address of the DR interface attaching to the link), and lists all routers attached to the link using their RIDs. Moreover, a router interface attached to a transit shared link is characterized through a type 2 link description included in the Router-LSA that describes the router, and using the link identifier as the reference to the link.

Addressing information The addressing information concerns the identification of the prefixes assigned to the network and their association with network elements. Prefixes may be assigned to four types of elements: (i) routers, (ii) point-to-point links, (iii) transit shared links, and (iv) stub shared links.

- **Addresses assigned to routers** - An IPv4 address assigned to a router is characterized through a type 3 link description included in the Router-LSA that describes the router. The association between address and router follows directly from the inclusion of the address in the Router-LSA.

- **Prefixes assigned to point-to-point links** - A prefix assigned to a point-to-point link is advertised by two Router-LSAs, each originated by one link end router, using type 3 link descriptions. The type 3 link descriptions are the same in both Router-LSAs, except for the interface cost. However, the interface cost is repeated information, already contained in the type 1 descriptions of the same link. The association between link and prefix is made by correlating the IPv4 addresses of the type 1 description (`Link Data`), with the prefix of the type 3 description (`Link Data` and `Link ID`): if the latter contains the former, this means that the prefix belongs to the link.

- **Prefixes assigned to transit shared links** - A prefix assigned to a transit shared link is advertised through a Network-LSA, and its association with the link follows directly from the fact that the Network-LSA also characterizes the link topologically.

- **Prefixes assigned to stub shared links** - A prefix assigned to a stub shared link is advertised through a Router-LSA originated by the single router attached to the link, using a type 3 link description. This type of link has no topological representation. A prefix assigned to a stub shared link is associated with the router it is attached to. The association between prefix and router follows directly from the inclusion of the prefix in the Router-LSA. Contrary to the cost information included in the type 3 descriptions of point-to-point links, which is useless, the cost information included in the type 3 descriptions of stub shared links is required to calculate the costs of paths leading to the link.

> *In OSPFv2, the topological and addressing information is provided jointly by the Router-LSA and the Network-LSA. From the perspective of topological information, the Router-LSA is the equivalent of the router-NR, and the Network-LSA is the equivalent of the slink-NR. Moreover, since these LSAs carry addressing information, they are also the equivalent of the aip-NR.*

5.6   Topological and domain-internal addressing information in IS-IS

L1 and L2 LSPs The format of IS-IS LSPs is shown in Figure 5.12. We start by noting that, in IS-IS, there are two types of LSDB required for the support of multi-area networking: the L1 LSDB and the L2 LSDB. However, the structure of these LSDBs is identical. Both LSDBs are formed by LSPs, and the LSPs are structured with the same set of fields irrespective of LSDB type. The distinction between LSPs of the L1 LSDB and LSPs of the L2 LSDB is made through the value of the IS Type field (2 bits), which is part of the LSP header; this field equals 1 in the first case and equals 3 in the second one. Note that TLVs are contained inside LSPs and, therefore, inherit its LSDB type. In what follows, we will make no distinction between L1 and L2 LSPs, since their structure is identical.

Nonpseudonode-LSPs and Pseudonode-LSPs An IS-IS router can only originate two types of LSPs: Nonpseudonode-LSPs and Pseudonode-LSPs. The Nonpseudonode-LSP describes a router, its interfaces, and the prefixes associated with it; the Pseudonode-LSP describes a transit shared link, but only topologically. A router can originate as many Pseudonode-LSPs as transit shared links where it is the DIS, but can only originate one Nonpseudonode-LSP. The LSP types are discriminated by the Pseudonode ID field, which equals zero in the first case and takes a non-zero value in the second.

TLV records LSPs include a variable number of TLVs to convey different types of routing information. The topological information is provided through the IS Neighbors TLV or the Extended IS Reachability TLV, which uses NSAP addresses as topological identifiers. From a routing information perspective, the topological TLVs are the equivalent of a router-NR when inserted in a Nonpseudonode-LSP, and the equivalent of a slink-NR when inserted in a Pseudonode-LSP. The topological TLVs keep the same structure in the IPv4 and IPv6 versions of IS-IS.

The addressing information is provided by the IP Reachability TLVs (that we may also refer to as the addressing TLVs). These TLVs differ in the IPv4 and IPv6 versions of IS-IS. The IPv4 version uses the IP Internal Reachability Information TLV or the Extended IP Reachability TLV, and the IPv6 version uses the IPv6 Reachability TLV. These TLVs cannot be included in Pseudonode-LSPs. When inserted in Nonpseudonode-LSPs, they are the equivalent of the aip-NR.

In Sections 9.2.1.4 and 9.2.1.5, we show experiments that highlight the features of the IPv4 and IPv6 IS-IS LSDBs.

5.6.1   IS Neighbors TLV and Extended IS Reachability TLV

There are currently two types of TLVs providing topological information (see Figure 5.12): the IS Neighbors TLV (type 2) and the Extended IS Reachability TLV (type 22). These TLVs are included in Nonpseudonode-LSPs and Pseudonode-LSPs to describe the neighbors of the network element they are associated with. When the TLV is inserted in a Nonpseudonode-LSP, the neighbors can be other routers or transit shared links (pseudonodes); when inserted in a Pseudonode-LSP, the neighbors can only be other routers.

How are neighbors described? Neighbors are described inside a TLV using

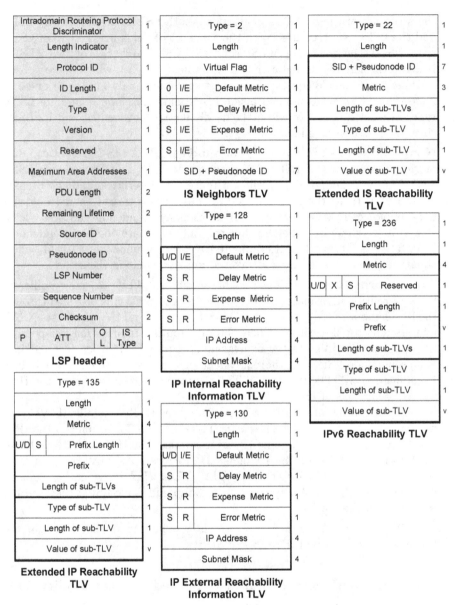

FIGURE 5.12: Format of IS-IS LSPs.

a structure formed by a set of fields (the neighbor structure), which repeats for each individual neighbor. Each neighbor is identified using a concatenation of the SID and the Pseudonode ID, which will be referred to as the Node ID. When the neighbor is a router, it is identified through the SID and a Pseudonode ID of zero; when the neighbor is a transit shared link, it is identified through the SID of the link DIS and a non-zero Pseudonode ID.

Interface cost information The neighbor structure also carries the cost of the outgoing interface leading to the neighbor in the `Default Metric` field (IS Neighbors TLV) or in the `Metric` field (Extended IS Reachability TLV). The length of `Default Metric` is 6 bits, allowing only for 64 different values. This small number was one of the motivations for the introduction of the Extended IS Reachability TLV, where the `Metric` field has a length of 3 octets. In Pseudonode-LSPs, the metric value is zero for all listed neighbors.

Sub-TLV records The Extended IS Reachability TLV may include a number of sub-TLVs in each neighbor structure. Examples of sub-TLVs are the IPv4 interface address (sub-TLV of type 6) and the IPv4 neighbor address (sub-TLV of type 8); the names of these sub-TLVs speak for themselves. Note that the inclusion of sub-TLVs is not mandatory.

The Extended IS Reachability TLV dropped most of the metric fields that exist in the IS Neighbors TLV, namely the `Delay Metric`, the `Expense Metric`, and the `Error Metric`. The philosophy behind the design of Extended IS Reachability TLV was to keep only the default metric and allow the inclusion of additional metrics only when required, using sub-TLVs.

Describing parallel point-to-point links A neighbor on a point-to-point link is identified through its SID and a Pseudonode ID of zero. Thus, IS-IS has no explicit way of differentiating between parallel point-to-point links (i.e. links leading to the same neighbor). Despite this, parallel links could be individually described in TLVs by using different neighbor structures with the same neighbor identifier. However, in actual implementations, only one link is described, actually the one with lowest interface cost. This lack of topological information is not problematic, since only the least cost interface is relevant for the computation of the shortest path costs.

5.6.2 IP Reachability TLVs

The IP Reachability TLVs provide the addressing information (see Figure 5.12). There are several different types: (i) the IP Internal Reachability Information TLV (type 128), (ii) the Extended IP Reachability TLV (type 135), and (iii) the IPv6 Reachability TLV (type 236). The first two describe IPv4 prefixes, and the last one IPv6 prefixes. The Extended IP Reachability TLV uses sub-TLVs and was introduced to widen the IP Internal Reachability Information TLV. The IPv6 Reachability TLV also uses sub-TLVs.

How are prefixes described? Prefixes are described inside a TLV using a structure formed by a set of fields (the prefix structure), which repeats for each individual prefix. The prefix is identified through the `IP Address` and `Subnet Mask` fields (in IP Internal Reachability Information TLVs) and through the `Prefix` and `Prefix Length` fields (in Extended IP Reachability TLVs and IPv6 Reachability TLVs).

The `Prefix` field in both the Extended IP Reachability TLV and IPv6 Reachability TLV is of variable length. It must contain an integer number of octets, the number of which depends on the `Prefix Length`. The number of octets packed in `Prefix` must equal the integer part of (`Prefix Length` + 7)/8.

The prefix structure also includes the cost from the originating router to the advertised prefix, in the `Default Metric` field (in IP Internal Reachability Information TLVs) or in the `Metric` field (in Extended IP Reachability TLVs and IPv6 Reachability TLVs). Note that `Default Metric` has a length of 6 bits and `Metric` has a length of 4 octets.

The association between prefixes and network elements In IS-IS, prefixes are always associated with routers, even if assigned to links, and are advertised through the IP Reachability TLVs included in Nonpseudonode-LSPs. A prefix assigned to a link is advertised by all routers attached to the link, even in the case of shared links. Pseudonode-LSPs are not allowed to include IP Reachability TLVs, i.e. they have no means to convey addressing information. This option unnecessarily penalizes the size of the LSDB.

The association between the prefixes and the network elements to which they are assigned is implicit in the fact that prefixes are always associated to routers and the Nonpseudonode-LSPs that represent them. In this case, the cost information included in the IP Reachability TLVs is always relevant. The cost of a path from a router to a prefix assigned to a link is always calculated in two steps: (i) determine the path cost from the router computing the path cost to the router advertising the prefix (e.g. using Dijkstra's algorithm); (ii) add the previous path cost to the cost included in the IP Reachability TLV that advertises the prefix (i.e. the cost of the interface with the link owning the prefix).

5.6.3   Providing topological and addressing information

Figure 5.13 summarizes how the topological and addressing information is provided in IS-IS (for simplicity, the extended TLVs are not included in the figure). It is interesting to compare these tables with the ones of Figure 5.11.

One shortcoming of the IS-IS LSDB structure is that the topological and the addressing informations cannot be conveyed separately, since the reachability and neighbor TLVs must be always flooded together in the same LSP.

Topological information As mentioned in Section 5.5.3, the topological elements relevant for the construction of the network map are (i) the routers, (ii) the point-to-point links, and (iii) the transit shared links.

- A router is identified through its SID (included in the `Source ID` field of its Nonpseudonode-LSP).

- A point-to-point link is characterized through two Nonpseudonode-LSPs, each describing one link direction and originated by one of the link end

Network element	IS-IS
router	Nonpseudonode-LSP (Source ID)
point-to-point link	Nonpseudonode-LSP (IS Neighbors TLV)
transit shared link	Nonpseudonode-LSP Pseudonode-LSP (IS Neighbors TLV)

Topological information

Prefix assigned to	IPv4 IS-IS	IPv6 IS-IS
router	Nonpseudonode-LSP (IP Internal Reachability Information TLV)	Nonpseudonode-LSP (IPv6 Reachability TLV)
point-to-point link	Nonpseudonode-LSP (IP Internal Reachability Information TLV)	Nonpseudonode-LSP (IPv6 Reachability TLV)
transit shared link	Nonpseudonode-LSP (IP Internal Reachability Information TLV)	Nonpseudonode-LSP (IPv6 Reachability TLV)
stub shared link	Nonpseudonode-LSP (IP Internal Reachability Information TLV)	Nonpseudonode-LSP (IPv6 Reachability TLV)

Addressing information

FIGURE 5.13: How topological and addressing information is provided in IS-IS.

routers. Point-to-point links are identified using the SID of the neighboring router concatenated with a zero Pseudonode ID.

- A transit shared link is characterized through a Pseudonode-LSP, originated by the link DIS, which includes the link identifier (SID of DIS concatenated with a non-zero Pseudonode ID). To allow cross-referencing, the Pseudonode-LSP includes the SIDs of the routers attached to the link, and the Nonpseudonode-LSPs of the routers attached to the link include the link identifier.

Addressing information The addressing information is provided by the IP Internal Reachability Information TLV or the Extended IP Reachability TLV (in IPv4 IS-IS) and the IPv6 Reachability TLV (in IPv6 IS-IS). A prefix, whether assigned to a router or to a link, is always associated with a router and its Nonpseudonode-LSP. Thus, the correspondence between a prefix and the network element to which it is assigned follows directly from having the IP Reachability TLVs inside Nonpseudonode-LSPs.

> *In IS-IS, there are two types of LSPs, one representing a router, called Nonpseudonode-LSP, and another representing a transit shared link, called Pseudonode-LSP. The topological information is provided by the IS Neighbors TLV or the Extended IS Reachability TLV, and is present in both LSP types. These TLVs are the equivalent of the router-NR when included in a Nonpseudonode-LSP, or the equivalent of the slink-NR when included in a Pseudonode-LSP. The addressing information is provided by the IP Internal Reachability Information TLV or the Extended IP Reachability TLV, in IPv4 IS-IS, and by the IPv6 Reachability TLV, in IPv6 IS-IS; these are the equivalent of the aip-NR. The Pseudonode-LSPs cannot include IP Reachability TLVs, which means that a prefix assigned to a shared link must be advertised by all nodes attached to the link, through their Nonpseudonode-LSPs. IS-IS forces the topological and addressing TLVs originated by a router to be flooded together in the same LSP.*

5.6.4   Additional IS-IS TLVs

Besides the TLVs mentioned so far, LSPs may include other TLVs, which will now be introduced (see Figure 5.14). Some of these TLVs may also be included in the control packets of IS-IS, which will be described in Section 6.1.

Area Addresses TLV The Area Addresses TLV lists all Area IDs assigned to the originating router, using the **Area Address** field. It is not included in Pseudonode-LSPs, but its presence is mandatory in Nonpseudonode-LSPs.

Protocols Supported TLV The Protocols Supported TLV identifies the network layer protocols supported by the IS-IS instance, using a 1-octet identifier called **Network Layer Protocol Identifier** (NLPID). The NLPID is 204 (0×cc) for IPv4 and 142 (0×8e) for IPv6. The use of this TLV is mandatory in Nonpseudonode-LSPs and in HELLO packets.

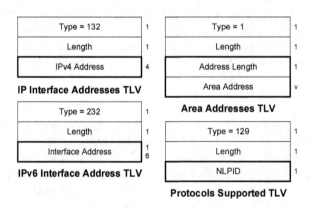

FIGURE 5.14: Format of additional IS-IS TLVs.

IP Interface Address TLV The IP Interface Address TLV lists one or more IPv4 addresses assigned to the IS-IS interfaces of the originating router. This TLV must be included in all Nonpseudonode-LSPs, when IPv4 routing is configured. This requirement is not easy to understand since, to begin with, a router may not need to disclose any IPv4 address and, if so, the IP Internal Reachability Information TLV provides a better solution for that purpose. Implementations usually include a single IPv4 address in this TLV. This TLV is also included in HELLO packets.

IPv6 Interface Address TLV The IPv6 Interface Address TLV has the same purpose as the IP Interface Address TLV, but for IPv6 IS-IS: it lists one or more IPv6 addresses assigned to the IS-IS interfaces of the originating router. This TLV is also used in HELLO packets. When used in HELLO packets the advertised addresses must be link-local addresses, but when used inside LSPs the advertised addresses cannot be link-local addresses.

Dynamic Hostname TLV An interesting feature of IS-IS is the inclusion of the Dynamic Hostname TLV, to allow the association of the SID with a name encoded in ASCII [48]. In this way, routers can easily exchange names with each other, and these names can be used, for example, when displaying the LSDB, instead of the harder hexadecimal notation. This is a very convenient feature for network managers, and is a good example of the flexibility introduced by IS-IS TLVs. OSPF does not have a similar feature.

From IPv4 IS-IS to IPv6 IS-IS Note that the extension of IS-IS to IPv6 was very simple, since the TLV based format of LSPs just required the creation of two new TLVs, namely the IPv6 Interface Address TLV and the IPv6 Reachability TLV. To signal IPv6 support, the Protocols Supported TLV must include an NLPID of 142 (0×8e).

5.7 Topological and domain-internal addressing information in OSPFv3

OSPFv3 was a major breakthrough regarding the separation between topological, addressing, and link information. Unlike OSPFv2, OSPFv3 does not use routable addresses as topological identifiers. The routable addresses of OSPFv3 are IPv6 addresses, but OSPFv3 kept using IPv4 addresses as topological identifiers. The format of OSPFv3 LSAs is shown in Figure 5.15.

The topological information is provided by Router-LSAs and Network-LSAs as in OSPFv2. The addressing information is provided by Intra-Area-Prefix-LSAs. The link information will be discussed in Section 5.9.

In Section 9.2.1.2, we show experiments that highlight the features of the OSPFv3 LSDB.

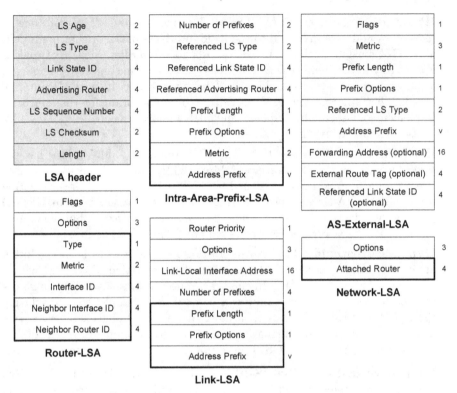

LS Age	2
LS Type	2
Link State ID	4
Advertising Router	4
LS Sequence Number	4
LS Checksum	2
Length	2

LSA header

Flags	1
Options	3
Type	1
Metric	2
Interface ID	4
Neighbor Interface ID	4
Neighbor Router ID	4

Router-LSA

Number of Prefixes	2
Referenced LS Type	2
Referenced Link State ID	4
Referenced Advertising Router	4
Prefix Length	1
Prefix Options	1
Metric	2
Address Prefix	v

Intra-Area-Prefix-LSA

Router Priority	1
Options	3
Link-Local Interface Address	16
Number of Prefixes	4
Prefix Length	1
Prefix Options	1
Address Prefix	v

Link-LSA

Flags	1
Metric	3
Prefix Length	1
Prefix Options	1
Referenced LS Type	2
Address Prefix	v
Forwarding Address (optional)	16
External Route Tag (optional)	4
Referenced Link State ID (optional)	4

AS-External-LSA

Options	3
Attached Router	4

Network-LSA

FIGURE 5.15: Format of OSPFv3 LSAs (single-area networks).

5.7.1 Router-LSA

The Router-LSA describes a router and its outgoing interfaces, and is the equivalent of the router-NR. Unlike OSPFv2, it does not include addressing information.

Originating router As in OSPFv2, the router that originates the LSA is identified in the Advertising Router field, using its RID.

Link descriptions The router interfaces are characterized through link descriptions, as in OSPFv2. However, the type 3 description was removed, since it provided addressing information and Router-LSAs are now restricted to provide topological information. The remaining link description types are identical to OSPFv2: type 1 (point-to-point links), type 2 (transit shared links), and type 4 (virtual links). As in OSPFv2, we do not address the type 4 descriptions.

The contents of the link descriptions is different from OSPFv2. All link descriptions include now (i) the interface cost (Metric), (ii) the interface identifier of the local interface (Interface ID), (iii) the interface identifier of the neighbor interface (Neighbor Interface ID), and (iv) the RID of the neigh-

Element	Type	Neighbor Router ID	Neighbor Interface ID	Interface ID	Metric
Interface to point-to-point link	1	RID of neighbor	Neighbor interface number	Local interface number	Interface cost
Interface to transit shared link	2	RID of DR	DR interface number	Local interface number	Interface cost

FIGURE 5.16: Contents of link description fields (OSPFv3).

bor (`Neighbor Router ID`). As discussed in Sections 2.2.1.3 and 2.2.1.4, the interface identifiers are required to ensure uniqueness in the link identification. What a neighbor is in link descriptions depends on the type of interface: in point-to-point link interfaces, the neighbor is the router on the other side of the link and, in transit shared link interfaces, the neighbor is the link DR. Thus, in transit shared link interfaces, `Neighbor Router ID` is the RID of the DR, and `Neighbor Interface ID` is the identifier of the DR interface attaching to the link. Note that the contents of these two fields coincide with the shared link identifier, as defined for OSPFv3 (see Section 5.2.2).

As in the case of OSPFv2, the `Link State ID` field is not needed to uniquely identify Router-LSAs; it is usually set to 0.

Figure 5.16 shows the contents of the link description fields, for each type of interface being advertised. Except for virtual links, there are now two possibilities only: (i) interface to a point-to-point link (type 1), and (ii) interface to a transit shared link (type 2). As in OSPFv2, stub shared links have no topological representation. The consequence in OSPFv3 is that Router-LSAs have no reference to this type of link.

5.7.2 Network-LSA

The OSPFv3 Network-LSA, like the Router-LSA, is restricted to provide topological information, and is the equivalent of the slink-NR. It describes transit shared links but not prefixes assigned to this type of links. Thus, in comparison with its OSPFv2 counterpart, the OSPFv3 Network-LSA no longer carries the subnet base address and the subnet mask. The routers attached to the shared link are identified by their RIDs, in the `Attached Router` field. The shared link is identified through the DR identifier (`Advertising Router`) and the number of the interface attaching the DR to the link (`Link State ID`).

5.7.3 Intra-Area-Prefix-LSA

OSPFv3 introduced the Intra-Area-Prefix-LSA to advertise IPv6 address prefixes and relate them to topological elements; this LSA is the equivalent of the aip-NR (see Section 2.2.2.2).

How are prefixes described? The IPv6 prefixes are described by IPv6 address (`Address Prefix`) and prefix length (`Prefix Length`). The LSA includes a `Metric` field to describe interface costs but, as discussed in Section 2.2.2.2, this information is only relevant when the LSA is advertising prefixes assigned to stub shared links.

The association between prefixes and network elements The association between prefixes and topological elements is done by relating each Intra-Area-Prefix-LSA with a topological LSA (Router-LSA or Network-LSA) using three fields: the `Referenced LS Type`, the `Referenced Advertising Router`, and the `Referenced Link State ID`. These three fields form a *pointer* linking the addressing and topological informations. The contents of `Referenced LS Type`, `Referenced Advertising Router`, and `Referenced Link State ID` equal the contents of `LS Type`, `Advertising Router`, and `Link State ID` of the referenced topological LSA, respectively. The `LS Type` of Router-LSAs and Network-LSAs are 0×2001 and 0×2002, respectively.

Prefixes can be assigned to four types of elements: (i) routers, (ii) point-to-point links, (iii) transit shared links, and (iv) stub shared links.

- **Addresses assigned to routers -** An IPv6 address assigned to a router is advertised through an Intra-Area-Prefix-LSA pointing to its Router-LSA and having an interface cost of zero.

- **Prefixes assigned to point-to-point links -** As in OSPFv2 and IS-IS, a prefix assigned to point-to-point link is advertised by the two link end routers. In OSPFv3, each router originates one Intra-Area-Prefix-LSA describing the prefix, pointing to the Router-LSA that describes the router, and including the cost of the interface with the link. The interface cost information is duplicated, since it is already present in the corresponding Router-LSAs.

- **Prefixes assigned to transit shared links -** A prefix assigned to a transit shared link is advertised through an Intra-Area-Prefix-LSA pointing to the Network-LSA that represents the link, and having an interface cost of zero. The Intra-Area-Prefix-LSA, as the Network-LSA, is originated by the link DR. Thus, unlike IS-IS, prefixes assigned to transit shared links need only be advertised by one router (and not all routers attached to the link).

- **Prefixes assigned to stub shared links -** A prefix assigned to a stub shared link is advertised through an Intra-Area-Prefix-LSA originated by the single router attached to the link, pointing to its Router-LSA, and containing the cost of interface with the link. The interface cost information is needed in this case, since stub shared links are not represented topologically (i.e. through Router-LSAs or Network-LSAs).

How many prefixes can an LSA advertise? One LSA can advertise more than one prefix, the number of which is defined by the `Number of Prefixes` field. However, these prefixes must refer to the same topological LSA, since the pointer fields (`Referenced Advertising Router`, Referenced Link State ID, and `Referenced LS Type` field) cannot be repeated on a prefix basis.

Intra-Area-Prefix-LSA identification One router may originate more than one Intra-Area-Prefix-LSA. Thus, this LSA requires the `Link State ID` field to be uniquely identified, besides the LSA type (`LS Type`) and the RID of the AR (`Advertising Router`); the `Link State ID` field carries a locally generated tag, unique for each LSA of this type that the router originates.

Comparison between IS-IS, OSPFv2, and OSPFv3 regarding the association between addressing and topological information In Figure 5.17, we compare how the topological and addressing informations are related in IS-IS, OSPFv2, and OSPFv3 (for simplicity, the extended TLVs of IS-IS are not included in the figure). In IS-IS, the association is implicit in the fact that a prefix, described by an IP Reachability TLV, is always associated with a router and included in the Nonpseudonode-LSP that describes the router. In both OSPFv2 and OSPFv3, prefixes can be associated with routers and links, and not only with routers as in IS-IS. In OSPFv2, the association is implicit in the fact that Router-LSAs and Network-LSAs provide both topological and addressing information. Prefixes are included in the topological LSA (Router-LSA or Network-LSA) that describes the network element (router or link) to which they are associated. In OSPFv3, the association is made through a pointer that comprises three fields, and relates the Intra-Area-Prefix-LSA that describes the prefix with the topological LSA that describes the network element to which they are associated.

5.7.4 Providing topological and addressing information

Figure 5.11 compares the way topological and addressing information is provided in OSPFv2 and OSPFv3. The topological information is provided in the same way in OSPFv2 and OSPFv3, with only slight differences in the contents of the LSAs and link descriptions. The main difference is in the way addressing information is provided. In OSPFv2 it is provided by Router-LSAs and Network-LSAs, and in OSPFv3 it is always provided by Intra-Area-Prefix-LSAs. Moreover, an IPv6 prefix is associated with the network element it was assigned to (router or link) by making a correspondence between the Intra-Area-Prefix-LSA that describes the prefix, and the topological LSA (Router-LSA or Network-LSA) that advertises the network element. Thus, prefixes assigned to routers, point-to-point links, and stub shared-links are associated with Router-LSAs, and prefixes assigned to transit shared links are associated with Network-LSAs.

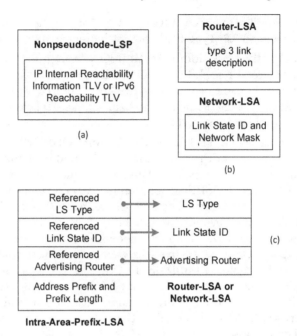

FIGURE 5.17: Association between addressing and topological information in (a) IS-IS, (b) OSPFv2, and (c) OSPFv3.

> *In OSPFv3, the topological information is provided by Router-LSAs and Network-LSAs and the addressing information is provided by Intra-Area-Prefix-LSAs. These LSAs are the equivalent of the router-NR, slink-NR, and aip-NR, respectively. The Intra-Area-Prefix-LSA describes one or more IPv6 prefixes and has a pointer to the topological LSA (Router-LSA or Network-LSA) that describes the network element to which the prefixes were assigned. The topological and addressing LSAs can all be disseminated separately, and this is a major difference regarding IS-IS TLVs.*

5.8   Domain-external addressing information

The domain-external addressing information is injected into a routing domain through its domain border routers. As discussed in Section 4.3, the domain border routers are called ASBRs in the case of OSPF and are simply called border routers in the Fcase of IS-IS. The domain border routers inject into the routing domain-external prefixes using specific LSAs and TLVs (included in Nonpseudonode-LSPs).

5.8.1 OSPF

AS-External-LSA In both OSPFv2 and OSPFv3, domain-external prefixes are advertised through the AS-External-LSA originated by an ASBR (see Figures 5.9 and 5.15). This LSA is the equivalent of the dep-NR introduced in Section 2.2.4.

How are prefixes described? In OSPFv2 the domain-external subnet address and subnet mask are carried in the `Link State ID` and `Network Mask` fields, and in OSPFv3 in the `Address Prefix` and `Prefix Length` fields.

How many prefixes can an LSA advertise? Each AS-External-LSA can only carry information relative to one address prefix.

Distinguishing between LSAs originated by the same ASBR In OSPFv3, the `Link State ID` is a locally generated tag that distinguishes among multiple AS-External-LSAs originated by the same ASBR. In OSPFv2, this distinction in implicit in the fact that `Link State ID` carries the IPv4 address being advertised.

Referring to other LSAs in OSPFv3 The `Referenced LS Type` and `Referenced Link State ID` of the OSPFv3 AS-External-LSAs provide the capability of referencing other LSAs. However, this capability is yet to be defined and, until then, `Referenced LS Type` should be set to zero and `Referenced Link State ID` should not be included.

Injecting default routes The AS-External-LSA can also be used to advertise default routes (0.0.0.0/0) within an AS, injected by the ASBR.

A comment on the OSPF designations As already mentioned in Section 4.3.1, the designations used by OSPF regarding domain-external addressing information are misleading. The domain border router is called ASBR even if it is interconnecting routing domains belonging to the same AS. Moreover, the designation "AS-External-LSA" hides the fact that this type of LSA may be used to advertise external prefixes that belong to the AS where the LSA is being disseminated.

In Sections 9.2.2, 9.2.3.1, and 9.2.5 we show experiments that highlight the features of the OSPFv2 LSDB structure with domain-external addressing information, coming or not from the same AS and including or not default routes.

5.8.2 IS-IS

IP External Reachability TLVs In IS-IS, IPv4 domain-external prefixes are advertised through the IP External Reachability Information TLV (type 130) or the Extended IP Reachability TLV (type 135), and IPv6 domain-external prefixes through the IPv6 Reachability TLV (type 236). These TLVs are the equivalent of the dep-NR introduced in Section 2.2.4, and are included in the Nonpseudonode-LSPs originated by the border routers. The two last

TLVs (type 135 and type 236) are also used to advertise domain-internal prefixes, as discussed in Section 5.6.2.

How are prefixes described? The prefix is identified through the IP Address and Subnet Mask fields (in IP External Reachability Information TLVs), and through the Prefix and Prefix Length fields (in Extended IP Reachability TLVs and IPv6 Reachability TLVs).

In the IPv6 Reachability TLV, the X flag indicates whether the advertised prefix is internal or external to the domain. The Extended IP Reachability TLV makes no distinction between these two types of prefixes.

Injecting default routes These TLVs can also be used to advertise default routes (0.0.0.0/0) within an AS, injected by the border router.

In Sections 9.2.2.1 and 9.2.3.1 we show experiments that highlight the features of the IS-IS LSDB structure with domain-external addressing information and with default routes.

5.8.3   OSPF E-type metrics

In OSPF, the Metric field of AS-External-LSAs carries an external cost associated with the advertised prefix (see Section 2.2.4). This cost can be interpreted as the cost of the external sub-path, from the ASBR that originates the LSA to the external prefix. OSPF considers two types of metric to determine the external path, called type E1 and type E2, and AS-External-LSAs include a flag, the E flag, to distinguish between them. The E flag defines how internal routers should calculate the path cost towards an external prefix. If the metric is of type E1 (E flag set to 0), the cost to the external prefix is obtained by adding the internal sub-path cost (the shortest path cost from the internal router to the ASBR that injected the LSA) and the external cost advertised by the LSA; the path calculated in this way is called an *E1 external route*. Otherwise, if the metric is of type E2 (E flag set to 1), the cost simply equals the external cost advertised by the LSA and the internal sub-path cost is ignored; the path calculated in this way is called an *E2 external route*. E2 routes always cross the ASBR that advertises the lowest external cost to an external prefix. The metric type and the external cost value are configured at the ASBRs.

We illustrate the use of the E-type metric of OSPF in the experiments of Section 9.2.4.

ASBRs inject domain-external addressing information into the routing domain using AS-External-LSAs (in case of OSPF), IP External Reachability Information TLVs or the Extended IP Reachability TLVs (in case of IPv4 IS-IS), or IPv6 Reachability TLVs (in case of IPv6 IS-IS); these are the equivalent of the dep-NR. In OSPF, ASBRs assign an external cost and metric type (E1 or E2) to each advertised external prefix; the metric type indicates whether or not the cost of the internal sub-path should be considered in the selection of the end-to-end path.

5.9 Link information

The link information is exchanged within the link to provide each router with the link layer addresses required to transport packets to the next-hop router. This issue was discussed in Section 2.2.3. Different technologies handle this problem differently.

OSPFv2 In OSPFv2, the Router-LSA advertises the IPv4 addresses of the router interfaces in the `Link ID` field of type 1 and type 2 link descriptions. Thus, a Router-LSA originated by one router provides to its neighboring routers the IPv4 addresses that should be used as layer-3 next hop addresses; the corresponding link layer addresses are obtained through the ARP protocol.

The Link-LSAs of OSPFv3 OSPFv3 includes the Link-LSA to help in this process, which is the equivalent of the la-NR introduced in Section 2.2.3. It is the only technology to do so. The Link-LSA is flooded only within the link, i.e. it has link flooding scope. The format of the Link-LSA is shown in Figure 5.15. Each LSA advertises the IPv6 link-local address assigned to the interface, in the `Link-Local Interface Address` field. Recall that, in IPv6, the layer-3 next-hop address is precisely the IPv6 link-local address, and the layer-2 address is obtained through the neighbor discovery process (see Chapter 4 of [17]). Unlike the la-NR introduced in Section 2.2.3, the Link-LSA does not include the identifier of the link where the addresses are to be used. Instead, it includes, in the `Link State ID` field, the identifier of the originating interface, which must be correlated with the interface identifier included in Router-LSAs (`Interface ID`) and Network-LSAs (`Link State ID`) to identify the link.

In Section 9.2.1.2 we show experiments that highlight the features of the OSPFv3 Link-LSAs.

Advertising configured IPv6 prefixes using Link-LSAs The Link-LSA also includes information on the IPv6 prefixes assigned to an interface. The prefixes are defined by (i) prefix address (`Address Prefix`) and (ii) prefix length (`Prefix Length`). This feature introduces a very powerful and elegant flexibility in OSPFv3, very much in the spirit of the separation between topological and addressing information: an IPv6 prefix assigned to a link need only be declared at one of the interfaces attached to the link, since the Link-LSA takes care of disseminating this prefix among neighboring interfaces. For example, an IPv6 prefix assigned to a shared link need not be configured at the link DR, but can be configured at any other interface attached to the link. This interface communicates the prefix information to the DR through the Link-LSA it disseminates on the link (and the DR then advertises the prefix to the remaining network through an Intra-Area-Prefix-LSA).

This feature will be illustrated in the experiment of Section 9.2.1.3.

IS-IS In IS-IS, the link information is conveyed through HELLO packets, using the IP Interface Address TLV (in IPv4 IS-IS) or the IPv6 Interface Address TLV (in IPv6 IS-IS). The IP Interface Address TLV carries the IPv4 address assigned to the interface that sends the HELLO, and the IPv6 Interface Address TLV carries the IPv6 link-local address. However, HELLO packets, despite having a link local scope, are not the best vehicle for this type of information. These packets are broadcast periodically, using short periods to allow a fine-grained maintenance of neighborhood relationships. They only need to include topological identifiers and not the heavier addressing information, which is thus broadcast much more frequently than necessary. This is precisely why OSPFv3 kept the use of IPv4 addresses (indeed their topological identifiers) in HELLO packets.

> *The link addresses are provided by the Link-LSA in OSPFv3, and this is the only technology to advertise this information in a separate NR; it is the equivalent of the la-NR. In OSPFv2, the link addresses are obtained from the Router-LSAs, and in IS-IS they are periodically broadcast by the neighboring routers in HELLO packets.*

6

OSPF and IS-IS Synchronization Mechanisms

The LSDB must be kept updated and synchronized across all routers of a routing domain. There are three synchronization mechanisms: (i) the Hello protocol, (ii) the flooding procedure, and (iii) initial LSDB synchronization process. These mechanisms are supported on the exchange of control packets among the routers. The Hello protocol is used to establish and maintain neighborhood relationships, to determine if two neighbors must synchronize their LSDBs, and to elect the designated router of shared links. The flooding procedure disseminates throughout the network the LSAs/LSPs originated by one router. This procedure may install, update, or delete LSAs/LSPs at the network. Finally, the initial LSDB synchronization is the process of exchanging routing information between two routers that have just established a neighborhood relationship, to ensure that their LSDBs get synchronized.

Chapter structure We start by introducing, in Section 6.1, the control packets of OSPF and IS-IS. The Hello protocol and the designated router election are discussed in sections 6.2 and 6.3. Then, Section 6.4 explains how routing information is disseminated, and Section 6.5 explains how it is updated. The initial LSDB synchronization process is covered in Section 6.6. Finally, Section 6.7 discusses when routing information must be originated and deleted.

6.1 Control packets

6.1.1 Packet types

OSPF and IS-IS use slightly different control packet types, which also do not coincide with the packet types of the generic protocol introduced in Chapter 2. The correspondence between the various control packets types is shown in Figure 6.1. In the IS-IS specification [1], the control packets are called *Protocol Data Units* (PDUs). Note that the designations of the control packet types are the same in the IPv4 and IPv6 versions of both protocols.

OSPF has five control packet types and IS-IS has four. The format of the control packets used in OSPFv2, OSPFv3, and IS-IS is shown in Figures 6.2,

generic	OSPF	IS-IS
HELLO	HELLO (type 1)	L1 LAN IS-IS HELLO (type 15)
		L2 LAN IS-IS HELLO (type 16)
		POINT-TO-POINT IS-IS HELLO (type 17)
UPDATE or DELETE	LS UPDATE (type 4)	L1 LINK STATE PDU (type 18)
		L2 LINK STATE PDU (type 20)
ACK (UPDATE) or ACK (DELETE)	LS ACK (type 5)	L1 PSNP (type 26)
		L2 PSNP (type 27)
PARTIAL LSDB REQUEST	LS REQUEST (type 3)	L1 PSNP (type 26)
		L2 PSNP (type 27)
LSDB SUMMARY	DB DESCRIPTION (type 2)	L1 CSNP (type 24)
		L2 CSNP (type 25)

FIGURE 6.1: Correspondence between the control packets of the generic protocol, OSPF, and IS-IS.

6.3, and 6.4, respectively. As in the case of LSAs and LSPs, fields or groups of fields that may repeat in a packet are encircled by a bold rectangle.

Hello packets Both OSPF and IS-IS use HELLO packets to support the Hello protocol (see Section 2.1). In OSPF, the packet is simply called HELLO. In IS-IS, there are two slightly different types of HELLO packets, one for point-to-point links, called POINT-TO-POINT IS-IS HELLO (or POINT-TO-POINT IIH), and another for shared links, called LAN IS-IS HELLO (or LAN IIH).

How are packet types identified? In IS-IS, except for one type of HELLO packet, there are separate packet types for packets transmitted/received by L1 and L2 interfaces. However, the L1 and L2 versions have exactly the same structure. The various packet types, including the distinction between the L1 and L2 versions, are identified through the PDU Type field (5 bits) of the packet header. In POINT-TO-POINT IS-IS HELLO packets this distinction is made in the Circuit Type field. In Figure 6.3 we include the TLVs exclusively used in IS-IS control packets. These TLVs are: (i) the IS Neighbors TLV of type 6, (ii) the LSP Entries TLV (type 9), and (iii) the Point-to-Point Three-Way Adjacency TLV (type 240). The TLVs used both in IS-IS control packets and in LSPs were introduced in Figure 5.12. In OSPFv2 and OSPFv3, the packet type is identified through the Type field of the packet header.

Dissemination of routing information The dissemination of routing information is performed, in the generic protocol, by UPDATE and DELETE packets (see Section 2.4.7). UPDATE packets transport the newest instances of LSAs/LSPs and DELETE packets transport the indications to delete them.

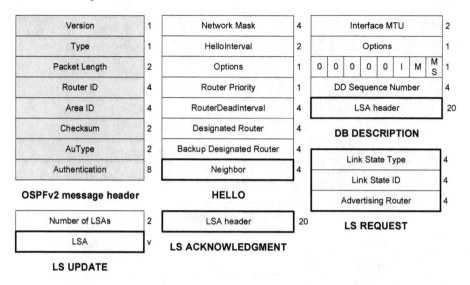

FIGURE 6.2: Format of OSPFv2 packets.

However, both OSPF and IS-IS use a single packet to perform these two functions, called LS UPDATE in OSPF and LINK STATE PDU in IS-IS.

Confusion between "LSP packets" and "LSP records" Unlike OSPF, where there is a clear distinction between the LSAs and the packets that transport them from router to router (i.e. the LS UPDATEs), in IS-IS there is no such distinction: they are both called Link State PDUs and abbreviated as LSPs. This is because one "LSP packet" can only carry one "LSP record", and the structure of "LSP packets" and "LSP records" is the same. To avoid confusion, we will refer to the "LSP packets" as LINK STATE PDUs (using capital letters), whenever necessary.

Acknowledgment packets In OSPF, the reception of an LSA is acknowledged through an LS ACKNOWLEDGMENT packet. In IS-IS, the reception of an LSP is acknowledged through a PARTIAL SEQUENCE NUMBER PDU (PSNP for short), but only in the case of point-to-point links. On shared links, the reception of LSPs is implicitly acknowledged through the periodic transmission of COMPLETE SEQUENCE NUMBER PDUs (CSNPs for short). This issue will be discussed in Section 6.4.2.

Advertising the LSDB summary Advertising the LSDB summary is performed by DB DESCRIPTION packets in OSPF and CSNPs in IS-IS. These packets carry summary information about *all* LSAs/LSPs contained in the LSDB. They are the equivalent of the LSDB SUMMARY of the generic protocol.

Requesting specific LSAs and LSPs Requesting the full contents of specific LSA and LSP instances is done through LS REQUEST packets in OSPF

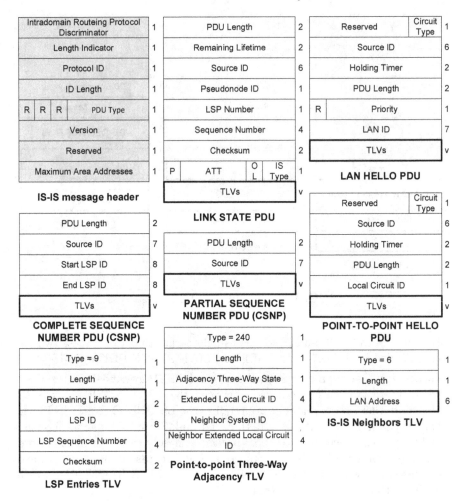

FIGURE 6.3: Format of IS-IS packets.

and through PSNPs in IS-IS. However, in IS-IS, PSNPs take this role only on shared links; as referred above, they are used as acknowledgments on point-to-point links. These packets are the equivalent of the PARTIAL LSDB RE-QUEST of the generic protocol.

Differences between OSPFv2 and OSPFv3 packet types The OSPFv2 and OSPFv3 control packets have a very similar structure. The main differences of OSPFv3 in relation to OSPFv2 are (i) the inclusion of the `Instance ID` and the removal of the authentication fields in the packet header, and (ii) the removal of the `Network Mask` in the HELLO packets.

Control packets that carry only summary information about LSAs and LSPs Except for the LS UPDATE of OSPF and the LINK STATE PDU

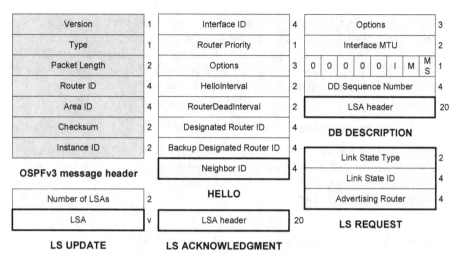

FIGURE 6.4: Format of OSPFv3 packets.

of IS-IS that carries a new LSP instance, all other control packets carry only summary information about LSAs and LSPs.

- In IS-IS, indications to delete LSPs (transported in LINK STATE PDUs) carry only the LSP header (the same is not true in OSPF).

- In OSPF, the LS ACKNOWLEDGMENT and DB DESCRIPTION packets carry only LSA headers; the LS REQUEST packets carries only the LSA ID, i.e. the LS Type, Link State ID, and Advertising Router fields.

- In IS-IS, the PSNPs and CSNPs carry only LSP summaries described through the LSP Entries TLV, which includes the LSP ID, Sequence Number, and Remaining Lifetime fields.

Dealing with large LSDB summaries A distinguishing feature of CSNPs is that they are prepared to handle LSDBs with large size, whose summary does not fit into a single CSNP. In this case, the LSDB summary is fragmented, and each fragment is transported on a different CSNP. The control of this process is done through the Start LSP ID and End LSP ID fields of CSNPs. Specifically, if several CSNPs are needed to carry the complete LSDB summary:

- the Start LSP ID of the first CSNP is set to 0000.0000.0000.00-00;

- the End LSP ID of the last CSNP is set to ffff.ffff.ffff.ff-ff;

- for all others CSNPs, the Start LSP ID and End LSP ID are set to the LSP ID of the first and last LSPs contained in the CSNP (and described in the corresponding LSP Entries TLV).

Note that if the LSDB fits into a single CSNP, the `Start LSP ID` and `End LSP ID` will be set to 0000.0000.0000.00-00 and `ffff.ffff.ffff.ff-ff`, respectively. In OSPF, this problem is handled by sending multiple DB DE-SCRIPTION packets, each with a fragment of the LSDB summary, during the initial LSDB synchronization process (see Section 6.6).

6.1.2   Encapsulation

OSPF packets are encapsulated in IP packets (IPv4 or IPv6) using the IP protocol number 89. IS-IS packets are encapsulated directly in layer-2 frames; for example, on Ethernet LANs, they are encapsulated in 802.3 frames with an LSAP value of 0×`fefe`.

6.1.3   Multicast addresses used on shared links

In IS-IS, the control packets transmitted on shared links use layer-2 multicast addresses. L1 type packets use the `AllL1ISs` address (01:80:c2:00:00:14), and L2 type packets use the `AllL2ISs` address (01:80:c2:00:00:15). In OSPF, the control packets transmitted on shared links can be either unicasted or use one of two multicast addresses: the `AllSPFRouters` address (224.0.0.5 for IPv4 and `ff02::05` for IPv6) or the `AllDRouters` address (224.0.0.6 for IPv4 and `ff02::6` for IPv6). The circumstances under which each type of address is used will become clear in Section 6.4.

6.1.4   Interface acceptance rules

In order for a packet to be accepted at an interface, a number of conditions must be satisfied, in addition to the ones that must be observed at lower layers. For example, regarding the acceptance at lower layers, an IPv4 packet can only be delivered to an OSPFv2 process if (i) the IPv4 destination address is either the IPv4 address of the receiving interface or one of the two IPv4 multicast addresses assigned to the protocol (`AllSPFRouters` or `AllDRouters`), (ii) the IP protocol number is 89, and (iii) the IP checksum is correct.

The acceptance rules depend on the specific parameters associated with an interface. For example, in OSPF each interface is assigned an area, and in IS-IS each interface is assigned a type (L1-only, L2-only, or L1/L2), and these configurations dictate if a packet can be accepted or not. The acceptance rules may be generic, i.e. may apply to any type of packet, or may only apply to specific packet types.

OSPFv2 In OSPFv2, a packet can only be accepted at an interface if (i) the protocol version number given by the `Version` field is 2, (ii) the `Area ID` field matches the Area ID configured for the interface, (iii) the IP destination multicast address is according to the interface state (non-DR/BDR routers cannot accept packets destined to the `AllDRouters` address), (iv) the OSPF checksum is correct, and (v) the packet is authenticated (in case authentication is required). The above rules are valid for all packets types. In addition, HELLO packets are only accepted if the `Network Mask`, `HelloInterval`, and `RouterDeadInterval` match the values configured for the interface. The `Network Mask` verification is not needed in the case of point-to-point links.

OSPFv3 The OSPFv3 acceptance conditions have some differences regarding OSPFv2. The version number (which must now be 3), the Area ID, the IP destination multicast address, and the checksum must also be verified. In addition, the `Instance ID` field must match one of the Instance IDs configured for the interface. The authentication is no longer done by the OSPF protocol; OSPFv3 relies on the authentication features of the IPv6 protocol, e.g. through the IP Authentication Header (see Chapter 5 of [17]). In addition to these conditions, HELLO packets are only accepted if the `HelloInterval` and `RouterDeadInterval` match the values configured for the interface, as in OSPFv2. Unlike OSPFv2, OSPFv3 HELLOs do not carry a network mask.

IS-IS In IS-IS, a packet can only be accepted at an interface if (i) the IS-IS checksum is correct and (ii) the packet is authenticated (in case authentication is required). The acceptance depends also on the interface type: L1-only interfaces only accept L1 packets and L2-only interfaces only accept L2 packets. The above rules are valid for all packet types. Additionally, PSNPs transmitted on shared links are only accepted by the link DIS and must be rejected by all non-DIS routers attached to the link. Note that unlike OSPF, IS-IS interfaces are not assigned to specific areas, since area identifiers are assigned to routers and not to interfaces. Thus, IS-IS routers do not constrain the acceptance of packets at interfaces based on the area information.

What next? When HELLO packets are accepted at an interface based on the rules described above, they must be delivered to the router process that deals with the Hello protocol (see Section 6.2). Other types of control packets are only further processed if the receiving interface is already adjacent to the interface that sent the packet. The conditions for two interfaces becoming adjacent will be explained in Section 6.2.3.

6.2 Hello protocol

In both OSPF and IS-IS, routers establish and maintain neighborhood relationships at specific interfaces using the Hello protocol, which was discussed in Section 2.1. This protocol allows (i) learning that a neighbor exits and is alive and (ii) determining the kind of relationship that can be established with the neighbor. The Hello protocol is also used to elect the DR, a process that will be discussed in Section 6.3.

6.2.1   Packet format

As referred in Section 6.1, OSPF uses one single type of HELLO packet, and
IS-IS uses two slightly different types, one for point-to-point links (POINT-
TO-POINT IS-IS HELLO) and another for LAN links (LAN IS-IS HELLO).
The format of these packets is shown in Figures 6.2, 6.3, and 6.4. In addition,
IS-IS HELLOs are differentiated according to the interface type (L1 or L2),
using the PDU Type field in the case of LAN IS-IS HELLOs and the Circuit
Type field in the case of POINT-TO-POINT IS-IS HELLOs. L1 LAN IS-IS
HELLOs have a PDU Type of 15 and L2 LAN IS-IS HELLOs have a PDU Type
of 16. In POINT-TO-POINT IS-IS HELLOs, the Circuit Type field indicates
the type of interface that transmits the packet: 1 means L1-only, 2 means L2-
only, and 3 means L1/L2.

Neighbor identification HELLO packets include the identifiers of the router
that sends the packet and of its neighbors. In OSPF, the sending router and
the neighbors are both identified through their RIDs, carried in the Router
ID and Neighbor fields, respectively.

In IS-IS, the sending router is identified by its SID, included in the packet
header. However, the way neighbors are identified differs in point-to-point and
shared links. In the LAN IS-IS HELLO packets (transmitted on shared links),
the neighbors are identified through the layer-2 addresses assigned to their
interfaces with the link, carried in an IS Neighbors TLV of type 6. Thus, IS-IS
does not use topological identifiers (i.e. NSAP addresses) to identify neighbors
in LAN IS-IS HELLO packets.

In the POINT-TO-POINT IS-IS HELLO packets (transmitted on point-to-
point links), the neighbors are not explicitly identified. To achieve the goal of
verifying the bidirectionality of communications, HELLO packets advertise the
state of the sending interface using the Point-to-Point Three-Way Adjacency
TLV. This issue will be further discussed in Section 6.2.6.

Area information HELLO packets also include area information, which is
important in determining whether adjacencies can be established (see Section
6.2.3). In OSPF, each interface is assigned a specific area, and the HELLO
packets sent by one interface indicate, in the Area ID field, the area the interface belongs to. In IS-IS, areas are assigned to routers and not to interfaces,
and more than one area can be assigned to a single router. IS-IS HELLO
packets indicate, in the Area Addresses TLV, the areas a router belongs to.

Mandatory IS-IS TLVs of HELLO packets The IS-IS HELLO packets
must also include the Protocols Supported TLV, and the IP Interface Address
TLV (IPv4) or IPv6 Interface Address TLV (IPv6), besides the TLVs mentioned above, i.e. the Area Addresses TLV, the IS Neighbors TLV of type 6
(shared links), and the Point-to-Point Three-Way Adjacency TLV (point-to-
point links).

OSPFv2 versus OSPFv3 HELLO packets Unlike OSPFv2, the OSPFv3
HELLO packets include the interface number of the sending interface in the
Interface ID field. The main use of this information is in the identification

of shared links. Recall that (i) one of the roles of the HELLO protocol is to allow that the DR advertises the shared link identifier, and that (ii) shared links are identified in OSPFv3 through the identifiers of the DR and of the DR interface attaching to the link (see Section 5.2.2). The OSPFv3 HELLOs carry the DR identifier in the `Designated Router ID` field and, as mentioned above, the interface identifier in the `Interface ID` field.

OSPFv3 HELLO packets also include more room for Options (three octets against one octet in OSPFv2). Finally, the `Network Mask` was removed, since no addressing information needs to be transported in these packets.

6.2.2   Liveness of relationships

Verifying the liveness of neighborhood relationships is done (i) by the periodic transmission of HELLO packets and (ii) by checking if the neighbor keeps transmitting HELLO packets.

The process of verifying the liveness of neighborhood relationships will be illustrated through experiments in Section 10.1.1.

Periodicity of HELLO transmissions In OSPF, the periodicity of HELLO transmissions is called `HelloInterval` and has a default value of 10 seconds. In IS-IS, the periodicity of HELLO transmissions can differ in the DIS and in other routers. The periodicity is called `dRISISHelloTimer` in the DIS and `iSISHelloTimer` in non-DIS routers; typical values are 3 and 10 seconds, respectively [12].

When is a neighbor considered dead? An interface considers a neighborhood relationship to be active as long as it keeps receiving HELLO packets from the neighbor. The time until a relationship is declared dead, upon ceasing to receive HELLO packets from a neighbor, is called `RouterDeadInterval` in OSPF and `Holding Timer` in IS-IS. The OSPF standard recommends the `RouterDeadInterval` to be 4 times the `HelloInterval`, and the IS-IS standard states that the `Holding Timer` must be 10 times the `iSISHelloTimer` value; however, IS-IS implementations use a value of 3. In IS-IS, different `Holding Timer` values can be used for the relationships with the DIS and with non-DIS routers; typical values are 10 and 30 seconds, respectively. Having a smaller `Holding Timer` value for the DIS relationships allows a faster recovery in the case of DIS failure.

The `RouterDeadInterval` (OSPF) and `Holding Timer` (IS-IS) values are both advertised in HELLO packets. The interpretation of this parameter when advertised by a router is "consider me dead if you do not receive HELLOs from me for more than this time". In OSPF, the `HelloInterval` is also included in HELLO packets. Both the `RouterDeadInterval` and `Holding Timer` are expressed in seconds and have a minimum value of 1 second. This sets a limit on how fast routers can recover from failures.

In OSPF, changing the `HelloInterval` and `RouterDeadInterval` values is a difficult task, due to the interface acceptance rules (see Section 6.1.4).

Adjacency type	Conditions
OSPFv2 adjacencies	Interfaces on same area
OSPFv3 adjacencies	Interfaces on same area and with same instance ID
IS-IS L1 adjacencies	L1 interfaces (L1-only or L1-L2) declaring same area
IS-IS L2 adjacencies	L2 interfaces (L2-only or L1-L2)

FIGURE 6.5: Conditions for becoming adjacent, related to the interface type and the area information.

Specifically, a HELLO packet received at an interface that does not match its configured `HelloInterval` and `RouterDeadInterval` values is immediately rejected. Thus, when changing these values, all routers attached to the same link must be reconfigured at the same time, to avoid service disruption. IS-IS is much more flexible in this respect.

6.2.3 Adjacencies

One of the roles of the Hello protocol is to determine if two neighbors can become *adjacent*. Only if two neighbors are adjacent can the connection between them be part of the network map. There are two basic conditions for becoming adjacent: (i) the Hello packets exchanged between the neighbors must comply with the interface acceptance rules described in Section 6.1.4; (ii) the communication between the neighbors must be bidirectional. However, additional conditions may be required. Moreover, both OSPF and IS-IS support different types of adjacencies. These issues will be discussed in the next sections. Figure 6.5 summarizes the conditions, related to the interface type and the area information, that neighbors must observe in order to become adjacent.

Adjacent versus fully adjacent The OSPF specification is not completely clear about what adjacent means. Most of the time adjacent neighbors are defined as neighbors that have already synchronized their LSDBs; other times the term *fully adjacent* is used for this purpose. Of course, *being able* to synchronize LSDBs or *having finished* to synchronize LSDBs are two different stages of a neighborhood relationship. In this book, the term adjacent will be used with the definition given above, i.e. adjacent neighbors are neighbors whose connection is part of the network map and that are ready for LSDB synchronization; the term fully adjacent will designate adjacent routers that have synchronized their LSDBs. Note that not all adjacent routers need to become fully adjacent, as will be discussed in Section 6.6. For example, on transit shared links, routers need only to synchronize their LSDBs with the link DR.

6.2.3.1 Bidirectionality of relationships

A router determines that it can communicate with a neighbor in both directions if its identifier is listed in the HELLO packets transmitted by the neighbor. As referred to above, in HELLO packets the neighbor identifier is included in the `Neighbor` field in the case of OSPF, and in the IS Neighbors TLV of type 6 in the case of IS-IS shared links. An interesting exception is the case of IS-IS point-to-point links, where bidirectionally is checked through the state of the neighboring interface, and not through its address; this issue will be discussed in Section 6.2.6.

6.2.3.2 OSPF adjacencies

OSPF neighbors can only become adjacent if they observe the interface acceptance rules (Section 6.1.4) and the bidirectionality conditions (Section 6.2.3.1).

Due to the interface acceptance rules, in both OSPFv2 and OSPFv3, two neighbors can only become adjacent if they belong to the same area, i.e. if they declare the same `Area ID`. Thus, an OSPF adjacency is characterized by one, and only one, area. Moreover, in OSPFv3, the neighbors must belong to the same instance, i.e. they must declare the same `Instance ID`.

OSPFv3 supports multiple OSPF instances (i.e. multiple adjacencies) running on the same link. This is similar to L1 and L2 adjacencies although, unlike IS-IS, the motivation for its introduction was not the organization of the network in multiple areas. As explained in Section 4.3.3, this feature can be used to overlay several logical networks on the same physical infrastructure, each with its own LSDB.

In Section 10.1.5 we illustrate through experiments the OSPF adjacency process.

6.2.3.3 IS-IS adjacencies

IS-IS neighbors can establish two types of adjacencies, called L1 and L2, for the exchange of routing information related to the L1 and L2 LSDBs. As mentioned in Section 6.1, each adjacency type uses specific control packets, which have the same structure but differ in type as being L1 or L2.

Recall from Section 4.3.3 that IS-IS interfaces can be configured as L1-only, L2-only, or L1/L2. L1-only interfaces only send and listen to L1 HELLO packets and can only establish L1 adjacencies. Likewise, L2-only interfaces only send and listen to L2 HELLO packets and can only establish L2 adjacencies. L1/L2 interfaces send and receive both types of HELLO packets and can establish both types of adjacencies.

Both adjacency types can only be formed if the interface acceptance rules

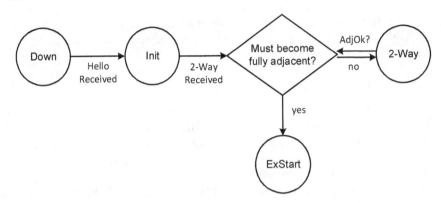

FIGURE 6.6: OSPF Hello state machine: evolution of states when successfully establishing a bidirectional relationship.

6.1.4) and the bidirectionality of communications (Section 6.2.3.1) are observed. However, an additional condition is imposed on L1 adjacencies.

In the case of L1 relationships, the areas configured at neighbors also determine if an L1 adjacency can be established. Recall that there can be more than one area assigned to a router, and that the area identifiers are advertised in HELLO packets through the Area Addresses TLV. Specifically, an L1 adjacency can only be established between two neighbors if they declare (in their L1 HELLO packets) at least one area in common. L2 adjacencies are not constrained by the declared areas, i.e. an L2 adjacency can be established between two routers placed in different areas. In fact, as will be seen in Chapter 7, this type of adjacency is used in IS-IS hierarchical networks to communicate routing information between areas.

Within a shared link, some adjacencies can be of type L1 and others of type L2. Suppose that two routers attach to a shared link with their interfaces configured as L1/L2. If the routers belong to the same area, they establish both L1 and L2 adjacencies. If they do not, they establish only L2 adjacencies.

In Section 10.1.6 we illustrate through experiments the IS-IS adjacency process.

6.2.4   OSPF Hello protocol state machine

In the OSPF specification [36], the process of establishing a bidirectional relationship is defined through a detailed state machine, corresponding to the Hello protocol part of the neighbor state machine, and described in Section 10 of the specification. Figure 6.6 shows the evolution of states when two routers successfully establish bidirectional communication over point-to-point or shared links; note that the figure does not include the complete state machine (only the part relative to the Hello protocol is included). A separate instance of the state machine is created for each neighbor on each interface.

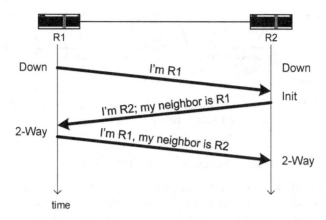

FIGURE 6.7: Establishing bidirectional communication in OSPF.

Evolution of states The state machine starts in the Down state. When the first HELLO packet is received from a neighbor (Hello Received event), the state machine transits to the Init state. At this point, the router already knows that the neighbor exists, but has not confirmed that bidirectional communication is possible. The state machine abandons the Init state when the router receives a HELLO packet from the neighbor with its own RID listed (2-Way Received event). The next state depends on whether the neighbors must start synchronizing their LSDBs (i.e. if they must become fully adjacent). If they must, the state machine goes to the ExStart state; otherwise it goes to the 2-Way state. For example, on shared links, routers only synchronize their LSDBs with the DR or BDR. The ExStart state is the initial state of the LSDB synchronization process, which will be detailed in Section 6.6.

Figure 6.7 illustrates the initial exchange of HELLO packets, where neighbors verify that bidirectional communications is possible (i.e. until they reach the 2-Way state).

Cold start on shared links During a cold start on a shared link all the interfaces move to the 2-Way state (and not to the ExStart state). This is because, when the interfaces abandon the Init state, the DR and BDR are yet to be elected. In fact, as will be seen in Section 6.3, the DR and BDR can only be elected among interfaces for which bidirectional communication has been verified.

The conditions to become fully adjacent must be reexamined whenever the DR or BDR change. Thus, when the DR and BDR get finally elected, the neighborhood relationships where at least one neighbor is DR or BDR move from the 2-Way state to the ExStart state, and start the LSDB synchronization process; the remaining neighborhood relationships, where neighbors are neither DR nor BDR, stay in the 2-Way state. The need for reexamining the adjacency conditions is referred to in the OSPF specification as the AdjOK? event.

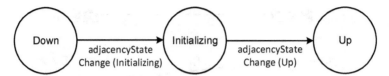

FIGURE 6.8: IS-IS Hello state machine: evolution of states when successfully establishing a bidirectional relationship.

6.2.5 IS-IS Hello protocol state machine

The IS-IS Hello protocol state machine for establishing bidirectional communications is similar to that of OSPF. There are also three states, with the same meaning but slightly different names: Down, Initializing (called Init in OSPF), and Up (called 2-Way in OSPF). Thus, an interface stays in the Down state until it receives the first HELLO from the neighbor. In the Initializing state, the interface already received a HELLO from the neighbor, but doesn't know if the neighbor received its own HELLO (i.e. bidirectional communication has not been verified). Finally, in the Up state the interface knows that the neighbor received its HELLO and, therefore, knows that bidirectional communication is possible. This state machine is shown in Figure 6.8. The initial LSDB synchronization process always starts from the Up state.

6.2.6 Bidirectionality of relationships over IS-IS point-to-point links

In the initial IS-IS specification, the POINT-TO-POINT IS-IS HELLO packets did not include the neighbor identifier or any other equivalent information, which precluded the verification of bidirectional communication. The assumption in this specification was that point-to-point links were fully reliable. This problem was corrected in RFC 3373 [24], later updated by RFC 5303 [25], which introduced the Point-to-Point Three-Way Adjacency TLV (type 240).

The Point-to-Point Three-Way Adjacency TLV The Point-to-Point Three-Way Adjacency TLV advertises the current state of the sending interface using the Adjacency Three-Way State field. In this field the Up state is coded as 0, the Initializing state as 1, and the Down state as 2. The TLV also includes the Neighbor System ID field to advertise the neighbor identifier, but its use is not mandatory. Thus, unlike the cases of OSPF and IS-IS shared links, a router does not verify bidirectional communication by checking if its identifier is included in the HELLO received from the neighbor; instead, it is verified when the received HELLO indicates that the neighbor is in the Initializing or Up states.

Comparing IS-IS point-to-point and shared links In Figure 6.9 we compare the verification of bidirectional communications in IS-IS point-to-point and shared links. R1 sends the first HELLO, declaring the Down state, in the

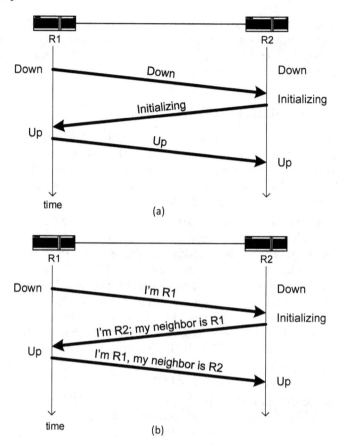

FIGURE 6.9: Establishing bidirectional communication in IS-IS (a) point-to-point links and (b) shared links.

case of point-to-point links, or the router identifier, in the case of shared links. When R2 receives this packet, it moves to the Initializing state. Then R2 replies with another HELLO packet where, in point-to-point links it declares the Initializing state, and in shared links it includes the neighbor identifier (the interface MAC address of R1). When R1 receives this packet, it understands that bidirectional communication with R2 is possible and moves to the Up state. Finally, in the next HELLO packet sent to R2, R1 declares the Up state in the case of point-to-point links or includes the neighbor identifier (the interface MAC address of R2) in the case of shared links. Again, when R2 receives this packet it understands that bidirectional communication with R1 is possible and moves to the Up state.

In Section 10.1.4 we illustrate through experiments the process of verifying bidirectional communications in IS-IS point-to-point links.

Comparing OSPF and IS-IS The process of verifying bidirectionality on IS-IS point-to-point links has the advantage, over that of OSPF, of declaring only the interface state, which requires a much smaller number of bits than declaring the neighbor identifier.

6.2.7   Immediate HELLO replies

Besides being transmitted periodically, HELLO packets can also be transmitted immediately, in response to specific events, without having to wait for the next time scheduled for a periodic HELLO. These packets will be called *immediate* HELLOs. The purpose of immediate HELLOs is to speed up convergence when there is a change in the network.

IS-IS Immediate HELLOs are part of the IS-IS specification [1]. For example, the specification states that, besides the periodic transmissions governed by the `iSISHelloTimer` or the `dRISISHelloTimer`, LAN IS-IS HELLOs must be transmitted when (i) the contents of the next HELLO would differ from the contents of the last transmitted one, or (ii) the interface determines that it will become the DIS or that it will resign from being the DIS. The transmissions are restricted such that there must be an interval of at least 1 second between consecutive HELLOs. The specification also states that the Hello timers can only be reset when a LAN IS-IS HELLO is transmitted as a result of (i) timer expiration (i.e. when it comes the time for a new periodic HELLO) or (ii) on becoming the DIS or resigning from DIS.

OSPF In OSPF, immediate HELLOs are not part of the specification, but are discussed in an Internet draft [26]. However, implementations send immediate HELLOs when a router receives a HELLO from a neighbor and the state of its relationship with the neighbor is less than 2-Way, i.e. the received HELLO still doesn't recognize the router as neighbor. These immediate HELLOs are unicasted to the neighbor and not multicasted as in periodic HELLOs. Moreover, unlike IS-IS, the Hello timer is only reset as a result of timer expiration. The Internet draft refers other circumstances where immediate HELLOs can be sent, but only the one described above seems to be implemented.

> *OSPF and IS-IS router interfaces become adjacent through the Hello protocol. The periodicity of HELLO packet transmissions is 10 seconds on LANs, in both OSPF and IS-IS. An adjacency between two interfaces can only be established if an interface accepts the HELLO packets sent by the other and if the bidirectionality of the communications is verified. However, additional conditions must be fulfilled, based on information provided by the HELLO protocol. IS-IS neighbors can establish two types of adjacencies, called L1 and L2, to exchange routing information related to the L1 LSDB or the L2 LSDB, respectively. OSPF and IS-IS adjacencies of type L1 can only be established between interfaces that declare the same area in their HELLO packets.*

6.3 Designated router election

The designated routers, called DR in OSPF and DIS in IS-IS, are both elected through the Hello protocol, but the election processes are slightly different. OSPF also elects a second router as part of this process, called BDR, with the role of replacing the DR in case of failure. In both protocols, the election process is influenced by a configurable priority called `Router Priority` in OSPF and `Priority` in IS-IS; a higher priority means that the router will have more chances of being elected, and a priority of zero means that the router will not participate in the election process.

6.3.1 Packet support

The designated router election is supported on the periodic transmission of HELLO packets. To this end, HELLO packets include (i) the identifiers of the active neighbors on the link, (ii) the identifier(s) of the elected designated router(s), and (iii) the priority values.

How are neighbors listed in HELLO packets? In OSPF, the neighbors are listed in the `Neighbor` field and, in IS-IS, in the IS Neighbors TLV of type 6. This TLV is different from the IS Neighbors TLV included in LSPs, which is of type 2. In the type 6 IS Neighbors TLV, a neighbor is an interface identified through its MAC address. The priority value is carried in the `Router Priority` field in OSPF and in the `Priority` field in IS-IS.

How are designated routers identified in HELLO packets? In OSPF, the DR and BDR are identified in HELLO packets through the `Designated Router` and `Backup Designated Router` fields. In OSPFv2, these fields carry the IPv4 addresses assigned to the DR and BDR interfaces attaching to the link. In OSPFv3, these fields carry the RID of the DR and the BDR. In IS-IS, the DIS is identified in HELLO packets through the `LAN ID` field, which carries the SID of the DIS.

Dissemination of shared link identifiers through HELLO packets Recall that the identifier of a shared link is based on the identifier of its designated router. In OSPFv2, the identifiers of the shared link and of its DR coincide. In OSPFv3 and IS-IS, the link is identified using two elements: the RID of the DR (in the case of OSPFv3) or the SID of the DIS (in the case of IS-IS), concatenated with a local tag generated by these routers to ensure uniqueness in the link identification. One of the roles of the Hello protocol is to communicate the (complete) link identifier from the DR or DIS to the neighboring routers. As mentioned above, in OSPFv3, the RID is included in `Designated Router` field and, in IS-IS, the SID is included in the `LAN ID` field. Moreover, in OSPFv3, the tag is included in the `Interface ID` field and, in IS-IS, the tag is the Pseudonode ID and is included in the `LAN ID` field.

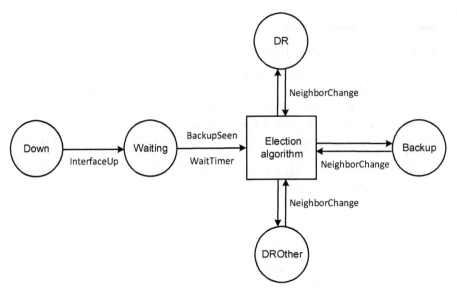

FIGURE 6.10: Interface state machine concerned with shared link interfaces.

These tags play no role in the designated router election, but are generated and communicated to all neighbors as soon as a router becomes designated.

6.3.2   OSPF election process

In OSPF, the DR and the BDR are elected such that once designated, no other router can take over their role unless one or both abandon the link, because of either a management action or a failure. The objective is to ensure that a joining router always gets the most up-to-date LSDB on a shared link during the initial synchronization process.

6.3.2.1   The interface state machine

The OSPF election process is defined through the interface state machine, described in Section 9 of the OSPF specification [36]. Figure 6.10 shows the part of the interface state machine concerned with the election of the DR and BDR. This state machine defines the state of an interface as being DR (if the interface is the link DR), Backup (if the interface is the link BDR), or DR other (if the interface is neither the link DR nor the link BDR). This state must be determined initially, as soon as the interface is switched on, or when a change in the link occurs (e.g. due to a failure of the current DR or BDR).

An interface that is switched off is in the Down state. When the interface is switched on (InterfaceUp event), it is placed in the Waiting state, where it tries to determine who the DR and BDR are. In this state, the interface listens to HELLO packets, but cannot elect a BDR or a DR; moreover, it must advertise DR and BDR addresses of 0.0.0.0 in the HELLO packets it sends. The interface abandons this state under two events: (i) after a timeout period, called `Wait Timer` (`WaitTimer` event), or (ii) when, after establishing bidirectional communication with a neighbor, that neighbor declares itself as being the DR or the BDR (`BackupSeen` event). The `Wait Timer` equals `RouterDeadInterval` and, therefore, has a default value of 40 seconds. When the interface abandons the `Waiting` state, it must run the election algorithm.

The election algorithm must also be run when a change in the link occurs, triggered by the `NeighborChange` event, which happens when: (i) a new neighbor attaches to the link, (ii) an existing neighbor abandons the link, (iii) a neighbor newly declares itself as DR or BDR, (iv) a neighbor no longer declares itself as DR and BDR, or (v) a neighbor changes its `Router Priority`.

6.3.2.2 Election algorithm

The election algorithm of the DR and BDR is described in Section 9.4 of the OSPFv2 specification [36]. It considers the possibility that several neighbors declare themselves as DR or BDR, and has the following steps (also summarized in Figure 6.11):

1. List all neighbors for which bidirectional communication has been established; the router running the algorithm must also be placed in the list.

2. Determine the new BDR from the list of step 1, but excluding the routers declaring themselves as DR. The new BDR is the router ranked first according to the following criteria. First, divide the routers in two groups: (i) the routers that declared themselves as BDR (group 1) and (ii) the remaining routers (group 2). The algorithm considers first the routers of group 1, and only if this group is empty (i.e. no routers declared themselves as BDR) considers the routers of group 2. Within each group, the routers having higher `Router Priority` are ranked first and, in case of a tie, the ones with higher RID are ranked first.

3. Determine the new DR, considering first the routers that declared themselves as DR. As in the case of the BDR election, the router ranked first is the one having the highest `Router Priority` and, in case of a tie, the one with the highest RID. If no router declared itself as DR, the new DR is the BDR determined in step 2.

4. This step evaluates a condition. If there has been a change in the DR or BDR status of the router running the algorithm, i.e. if the router is the new DR or BDR, or is no longer the DR or BDR, and if steps 2 and 3 were run for the first time, then these two steps must be run a second time. Otherwise, the algorithm proceeds to step 5.

5. Finally, according to the calculation above, the router must be

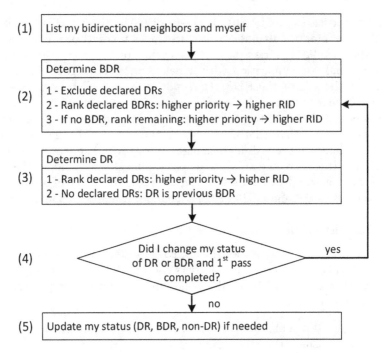

FIGURE 6.11: DR and BDR election algorithm.

placed in the DR state (if it was elected DR), in the `Backup` state (if it was elected BDR), or in the `DR Other` state (otherwise).

An important feature of this election algorithm, clear from steps 2 and 3, is that it favors already declared DRs and BDRs. Another feature is that it may elect the same router both as DR and BDR. However, the router that is twice elected cannot be the one running the algorithm.

6.3.2.3   Special cases

The algorithm seems complex, so it is important to highlight its main characteristics. We discuss several cases below.

Cold start When all interface are switched on at the same time (i.e. a cold start scenario), the interface with the highest `Router Priority`, or highest RID in case of a tie, becomes the link DR, and the one with the second highest `Router Priority`, or highest RID in case of a tie, becomes the link BDR.

In this case, the interfaces enter the `Waiting` state at approximately the same time and stay there for 40 seconds (`Wait Timer` default value). During this period, they keep receiving HELLO packets from their neighbors and learn about the `Router Priority` and RID of each one. The interfaces abandon the `Waiting` state at approximately the same time, and immediately run the election algorithm. When this happens, no interface has yet declared itself as

DR or BDR and, therefore, they all elect the same DR and BDR. However, the behavior is different in the DR, BDR, and non-DR/BDR routers.

The DR elects itself as both DR and BDR in the first pass of the algorithm and is forced to do a second pass since its role has changed to DR. In the second pass, in step 2 it is excluded from the computation and elects the correct BDR and, in step 3, it confirms itself as DR. Thus, the DR arrives at the correct conclusion when running the election algorithm for the first time.

The BDR elects the correct DR both as DR and BDR in the first pass of the algorithm, and then stops the algorithm since its role did not change. The computation is corrected when it receives a HELLO from the DR, indicating the correct DR and BDR. In this case, the correct DR and BDR are computed in the first pass of the algorithm, but a second pass is necessary since the role of the router changed to BDR. The behavior of a non-DR/BDR router is similar to that of the BDR. The only exception is that, when it receives the correct information from the DR, it only performs one pass over the algorithm, since its role did not change.

The experiments of Sections 10.2.1.1 and 10.2.1.2 illustrate the cold start case.

Router joining link when DR and BDR already elected A router joining a link that already has a DR and a BDR produces no changes regarding these roles. When its interface is switched on at the link, it is placed in the Waiting state, but will abandon it as soon as bidirectional communication is established with either the DR or the BDR (through the BackupSeen event). Thus, unlike the cold start case, the interface will not stay in the Waiting state for Wait Timer seconds. When the Waiting state is abandoned, the interface runs the election algorithm. Suppose that, when the algorithm is started, both the DR and BDR have established bidirectional communication with the joining router. Since in step 2 the algorithm considers first the routers that declared themselves as BDR, and only the current BDR did, that router is elected BDR. Likewise, since in step 3 the algorithm considers first the routers that declared themselves as DR, the current DR is also elected DR, and the algorithm stops.

The experiment of Section 10.2.1.3 illustrates this case.

Only one router started at link If only one router attaches to a shared link, that router will be elected DR. The router will enter the Waiting state and will abandon it through the WaitTimer event. It will then run the election algorithm. In the first pass of the algorithm, the router elects itself as both DR and BDR. However, it is forced to do a second pass, since its role changed to DR. In step 2 of this pass, the router is excluded from the calculation, since it is now the DR, and a BDR of 0.0.0.0 (i.e. no BDR) is determined; step 3 confirms that the router is the DR.

The experiment of Section 10.2.1.4 illustrates this case.

DR failure When the DR fails, the old BDR is promoted to DR and the new

BDR is the router that, among the remaining ones, has the highest `Router Priority`, or highest RID in case of a tie.

When the old BDR detects the failure, it runs the election algorithm. In step 2, it confirms itself as BDR, and in step 3, it elects itself as DR, since the DR failed. Now, because its role changed from BDR to DR, it must perform a second pass over the algorithm. In step 2, the router must remove itself from the calculation, since it is now the DR. Thus, only the remaining routers are considered in this step, and the one with highest `Router Priority` is elected BDR. In step 3, the router is confirmed as DR. Thus, the old BDR arrives at the correct conclusion when it runs the algorithm for the first time following the DR failure.

The remaining routers, upon detecting the DR failure, will initially declare the old BDR, as DR and BDR simultaneously. They will have to wait for a HELLO packet sent by the new DR, declaring itself as DR, to run the election algorithm again and arrive at the correct conclusion.

The experiment of Section 10.2.1.5 illustrates the DR failure case.

6.3.2.4 Procedure for defining the DR and BDR

The election algorithm suggests a procedure that network managers can follow in order to define what routers are the DR and BDR, if the running solution is unsatisfactory: (i) switch off all interfaces attached to the link; (ii) switch on the interface that one wishes to become the DR and wait 40 seconds (`Wait Timer`); (iii) repeat the previous step for the interface that one wishes to become the BDR; (iv) finally, switch on all other interfaces. Unfortunately, this procedure disrupts service, but there is no alternative.

6.3.2.5 Events triggered by the DR election

When a new DR is elected, the shared link identifier changes. Thus, the DR must originate a new Network-LSA and the routers attached to the link must originate new instances of their Router-LSAs, containing the new link identifier. In OSPFv2, the link identifier is included in the `Link State ID` field of Network-LSAs and in the `Link' ID` field of the type 2 link descriptions of Router-LSAs. In OSPFv3, the link identifier is included in the `Advertising Router` and `Link State ID` fields of Network-LSAs, and in the `Neighbor Router ID` and `Neighbor Interface` ID fields of the type 2 link descriptions of Router-LSAs.

6.3.3 IS-IS election process

The process of electing a designated router is much simpler in IS-IS than in OSPF. In IS-IS, the interface with highest priority always takes the role of DIS and, in case of a tie, the interface with highest MAC address becomes the DIS. The priority is advertised in the `Priority` field of HELLO packets. On shared links, a distinct DIS may exist for L1 and L2 adjacencies.

6.3.3.1 Election algorithm

Under stable conditions, the election process is run at an interface every time a HELLO packet is received. The election candidates are (i) the interface performing the election and (ii) its adjacent neighbors. As mentioned above, among these candidates the one with highest priority wins, and if there is a tie, the one with highest MAC address wins.

The waiting time When an interface is switched on, the election process can only start after a waiting period of 2 × iSISHelloTimer (20 seconds default). However, the IS-IS implementations that we have analyzed use a shorter period. The waiting period is the equivalent of the Waiting state of OSPF.

The LAN ID field The current DIS is identified in the LAN ID field of HELLO packets through its SID. The LAN ID also includes the Pseudonode ID, i.e. the local tag generated by the DIS to uniquely identify the link. During the waiting period, the LAN ID field carries the SID of the sending router and a local tag generated by the router (which will become the Pseudonode ID in case this router is elected DIS). After the DIS election, the LAN ID field of non-DIS routers is copied from the one transmitted by the DIS. At least some implementations use immediate HELLOs during the waiting period; however, as in the case of OSPF, this feature is not part of the standard.

Changing the DIS Configuring a new DIS is a simple task in IS-IS: an interface can become the DIS at a shared link by increasing its priority to a value higher than the priorities of the remaining interfaces. When this configuration is done, the interface immediately elects itself as DIS and transmits a HELLO packet (i) indicating in the LAN ID field that is now the DIS and (ii) advertising the new priority value.

Detecting a DIS failure When the DIS fails, the remaining routers elect a new DIS as soon as they detect the failure. The DIS typically uses a smaller Holding Timer than the remaining routers. Thus, detecting a DIS failure is faster than detecting the failure of any other router.

Sections 10.2.2.1, 10.2.2.2, and 10.2.2.3 illustrate, through experiments, the behavior of the DIS election process during cold start, when a new DIS is configured, and when the DIS fails.

6.3.3.2 Events triggered by the DIS election

As in the case of OSPF, when a new DIS is elected the link identifier changes. Thus, the DIS must originate a new Pseudonode-LSP, and the routers attached to the link must originate new instances of their Nonpseudonode-LSPs. The shared link identifier is the SID of the DIS concatenated with the Pseudonode ID assigned by the DIS. It is included in the topological TLVs that describe the neighbors of a router (the IS Neighbors TLV or the Extended IS Reachability TLV).

Moreover, if the new DIS contains in its LSDB the Pseudonode-LSP orig-

inated by the previous DIS, it must delete it. We explain the deletion process
of LSPs in Section 6.5.3. This behavior is not allowed in OSPF, since only the
originating router can delete its self-originated LSAs. Moreover, when a DIS
understands that it must resign (e.g. because a neighbor advertised a higher
Priority), it must also initiate the deletion of its Pseudonode-LSP.

> *The designated router (called DR in OSPF and DIS in IS-IS) is elected
> through the Hello protocol. OSPF also elects a backup designated router
> (called BDR) in this way, to replace the DR in case of failure. In
> OSPF, the DR and BDR are elected such that once elected no other
> router can takeover their role. In IS-IS, the DIS is elected according
> to a configurable priority.*

6.4 Flooding procedure

OSPF and IS-IS use the controlled and reliable flooding procedure described in
Section 2.4 to disseminate routing information, with a few exceptions. Recall
from the discussion in Section 5.1 that the indivisible flooding elements of
OSPF and IS-IS are the LSAs and the LSPs.

The reliable flooding procedure works in the following way:

- When a router originates an LSA/LSP to be flooded, it transmits the
  LSA/LSP on all its interfaces.

- When a router receives on one interface a new LSA/LSP (i.e. not already
  contained in its LSDB) or a fresher (i.e. more recent) instance of a stored
  LSA/LSP, it retransmits the LSA/LSP on all other interfaces; it does not
  retransmit the LSA/LSP on the incoming interface.

- When a router receives an LSA/LSP for which there is a stored instance
  with equal or higher freshness, the LSA/LSP is discarded.

- A router that receives an LSA/LSP on one interface acknowledges its reception to the router that transmitted it. The acknowledgment must be done
  even if the LSA/LSP is to be discarded.

The notion of freshness in OSPF and IS-IS will be discussed in Sections
6.5.1 and 6.5.2. These sections present the freshness attributes and explain
how these attributes are used to rank the freshness of LSAs/LSPs.

Besides the actions enumerated above, the arrival of LSAs/LSPs at routers
trigger additional events, which will be discussed in Section 6.5.4. We will
concentrate now on the flooding procedure.

OSPF and IS-IS deviate slightly from the procedure described above regarding flooding on shared links. The specific features of OSPF and IS-IS will
be detailed in the next sections.

In Section 10.3, we illustrate the flooding procedure of OSPF and IS-IS through experiments.

6.4.1 Flooding in OSPF

OSPF uses the controlled and reliable flooding procedure described above, with some optimizations in the case of shared links. The reliable flooding uses ACK protection in both point-to-point and shared links.

Packet support The LSAs and their acknowledgments are transported in LS UPDATE and LS ACKNOWLEDGMENT packets, respectively. The format of these packets is shown in Figures 6.2 (OSPFv2) and 6.4 (OSPFv3). Both packet types can transport information on more than one LSA. The LS ACKNOWLEDGMENT packets carry only the headers of the acknowledged LSAs, and not their full contents. We highlight that what needs to be acknowledged are the individual LSAs, and not the LS UPDATE packets, which are just LSA containers. When an LSA is transmitted, it is placed in a list, called `Link state retransmission list` in the OSPF specification [36], where its stays until the reception of the LSA is finally acknowledged. The retransmission timeout for LSAs is a parameter configurable per interface, called `RxmtInterval`, which is typically 5 seconds for LANs; the maximum number of retransmission attempts has not been specified.

Flooding on shared links On shared links, the DR acts as an intermediary in the flooding process. An LSA that is flooded on a shared link is sent first to the DR, which then retransmits it on the link to all other routers. Subsequently, each of these routers confirms the reception of the LSA sent by the DR. This is achieved by using different IP multicast addresses (see Section 6.1): the DR and BDR transmit packets using the `AllSPFRouters` multicast address, which is read by all routers attached to the link; non-DR/BDR routers transmit packets using the `AllDRouters` multicast address, which is only read by the DR and BDR. Thus, when a non-DR router needs to flood an LSA on a shared link, it sends the LSA to the DR encapsulated in an LS UPDATE packet with multicast address `AllDRouters`; the DR then retransmits the LSA to all other routers encapsulated in another LS UPDATE packet, but now with multicast address `AllSPFRouters`. The BDR also receives the initial LSA, but ignores it and waits for the LSA transmitted by the DR; if for some reason the DR fails, the BDR takes care of the retransmission itself.

Minimizing the number of ACK transmissions In principle, an LSA broadcast on a shared link must be acknowledged by all other routers attached to it. However, OSPF includes several optimizations to minimize the number of acknowledgment transmissions. First, the transmission of LS ACKNOWLEDGMENT packets can be delayed in the hope of grouping several LSA acknowledgments in a single packet; this feature is called *delayed acknowledgment*. Second, the acknowledgment can be implicit. If a router sends an LSA through an interface and, while waiting for the corresponding ac-

	received by DR	received by BDR	received by non-DR/BDR
sent by DR (AllSPFRouters)	NA	send ACK	send ACK
sent by BDR (AllSPFRouters)	send ACK	NA	send ACK
sent by non-DR/BDR (AllDRouters)	retransmit LSA	do nothing	not heard

* some ACK transmissions may be suppressed

FIGURE 6.12: Summary of OSPF flooding on shared links.

knowledgment, receives the same LSA on that interface, it considers that the acknowledgment was effectively done; this is called an *implied acknowledgment*. Moreover, the router doesn't need to acknowledge the reception of this LSA. The exact procedure for the transmission of LS ACKNOWLEDGMENT packets is summarized in Table 19 of the OSPF specification [36].

Summary of flooding on shared links Figure 6.12 summarizes the flooding procedure on shared links, when a router receives an LSA and determines that it must be flooded (e.g. because the LSA is missing from the LSDB or is a fresher instance of a stored LSA).

- If the LSA is sent by the DR or by the BDR, all other routers hear it (because it was transmitted using the `AllSPFRouters` address) and acknowledge its reception.

- If the LSA is sent by a non-DR/BDR router, only the DR and BDR hear it, since it was sent to the `AllDRouters` address. In this case, the DR retransmits the LSA on the link. The reaction to this transmission is according to the previous case, except that the non-DR/BDR router that originated the LSA doesn't need to acknowledge the packet.

Three examples In Figure 6.13, we give three examples of flooding on shared links. In Figure 6.13.a, the LSA to be flooded arrives at a non-DR/BDR router. In this case, the router encapsulates the LSA into an LS UPDATE packet and transmits it on the link using the `AllDRouters` multicast address (step 1). This packet is only read by the DR and BDR, but these routers do not acknowledge the LSA reception. The DR retransmits the LSA using the `AllSPFRouters` address so that it reaches all other routers (step 2), and each of these routers, except the router that initially sent the LSA, acknowledge the reception using an LS ACKNOWLEDGMENT (step 3); the initial router considers the LSA sent by the DR as an implied acknowledgment. In Figure 6.13.b, the LSA arrives at the BDR, and the BDR sends it immediately to all other routers using the `AllSPFRouters` multicast address (step 1); each of these routers,

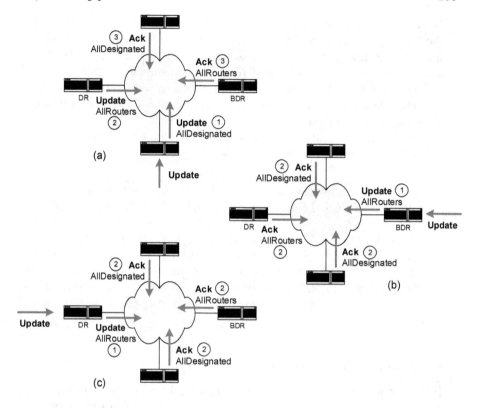

FIGURE 6.13: Example of OSPF flooding on shared links.

including the DR, acknowledge the reception of the LSA (step 2). A similar behavior is obtained when the LSA arrives at the DR (Figure 6.13.c).

LSA retransmissions If an LSA needs to be retransmitted on a shared link, it will only be sent to the neighbors from which an acknowledgment was not received. Thus, the LSA will not be retransmitted using an IP multicast address but, instead, the IP unicast addresses of the unresponsive neighbors. This option relieves the neighbors that already received and acknowledged the LSA from having to read it again. However, one LSA (and, therefore, one LS UPDATE packet) must be sent to each unresponsive neighbor. The acknowledgments of unicasted LSAs must also be unicasted to the router that transmitted the LSA.

Motivation for using DR as flooding intermediary The motivation for using the DR as an intermediary in the flooding process is often said to be scalability. This deserves some discussion. In fact, the total number of transmissions required to flood an LSA on a shared link is the same with or without an intermediary. If there are N routers on a shared link, flooding without an intermediary requires the transmission of an LSA and $N-1$ acknowledgments, i.e. a total of N packet transmissions. If the shared link has an intermediary, the flooding requires one LSA transmission to the intermediary, another one from the intermediary, and $N - 2$ acknowledgments (due to the implied acknowledgment rule), i.e. a total of N packet transmissions again.

When using the DR as a flooding intermediary, some gains in the number of packet transmissions may occur when the same LSA arrives simultaneously at several routers attached to the shared link. However, these gains are not very high due to the implied acknowledgment rule. The advantage of having a DR is more obvious in the initial LSDB synchronization process, as will become clear in Section 6.6.

> In OSPF, reliable flooding uses ACK protection in both point-to-point and shared links. LSAs are encapsulated in LS UPDATE packets and are acknowledged using only the LSA headers encapsulated in LS AC-KNOWLEDGMENT packets. On shared links, the DR acts as an intermediary in the flooding process: an LSA is first transmitted to the DR, and the DR retransmits the LSA to all other routers, which then acknowledge the packet reception to the DR. OSPF includes several optimizations to minimize the number of acknowledgment transmissions on shared links.

6.4.2   Flooding in IS-IS

IS-IS uses a different flooding mechanism on point-to-point and shared links.

Packet support In IS-IS, the LSPs are carried in LINK STATE PDUs; these packets carry only one LSP (Nonpseudonode-LSP or Pseudonode-LSP). The packets used to acknowledge the correct reception of LSPs differ in point-to-point and shared links. In point-to-point links, this role is played by the PSNPs and, in shared links, by the CSNPs that are periodically transmitted by the DIS. Both the PSNPs and CSNPs can acknowledge the reception of several LSPs simultaneously. The PSNPs are also used in the flooding process of shared links, but not as acknowledgments; they are used to request the retransmission of lost LSPs. The PSNPs and CSNPs carry only LSP summaries, and not complete LSPs. The LSP summary is provided by the LSP Entries TLV (see Figure 6.3), which comprise the `LSP ID`, the `LSP Sequence Number`, and the `Remaining Lifetime` fields.

Point-to-point links IS-IS uses different mechanisms on point-to-point and shared links to protect the transmission of LSPs. On point-to-point links, the mechanism is similar to OSPF: it uses ACK protection where the acknowledgments are transported in PSNPs. The retransmission timeout is called `minimumLSPTransmissionInterval` and has a default value of 5 seconds. IS-IS includes an optimization relative to OSPF: if a router receives an LSP for which it has a fresher instance, the router replies immediately with that LSP instance (instead of the PSNP) and this LSP implicitly acknowledges the reception of the initial one.

Shared links On shared links, the protection of LSP transmissions is not performed using explicit acknowledgments (as in OSPF and IS-IS point-to-point links). Instead, it relies on the periodic transmission of LSP summaries encapsulated in CSNPs, which is carried out by the DIS. The periodicity of CSNP transmissions is called `completeSNPInterval` and is typically 10 seconds. The periodic transmission of CSNPs is also used in the initial LSDB synchronization process over shared links (see Section 6.6.2.2).

The flooding process on shared links has two cases, depending on whether the LSP is flooded by a non-DIS router or by the DIS. In the first case, the router transmits the LSP on the link and waits for the next CSNP sent by the DIS, to check if the summary of its LSP is already listed there. If not, it retransmits the LSP on the link and repeats this process until the LSP summary gets listed. Thus, in this case, the periodic transmission of CSNPs acts as an implicit acknowledgment of the reception of flooded LSPs.

In the second case, the DIS transmits the LSP on the link and, afterwards, adds its summary to the CSNPs transmitted on the link. The DIS has no mechanism to confirm the correct reception of this LSP. However, in case of loss, non-DIS routers will notice that they are missing the LSP, or that they have an outdated instance of the LSP, since the subsequent CSNPs transmitted by the DIS include the summary of the newest (and previously transmitted) LSP instance. In this case, the non-DIS routers request the transmission of the LSP by sending a PSNP. The DIS replies by resending the LSP, and this process is repeated until the LSP is effectively received. Note that in this case, there can be a storm of PSNP transmissions, since all non-DIS routers will request the LSP, and IS-IS has no mechanism to suppress the transmission of scheduled PSNPs. Figure 6.14 illustrates this case. The DIS broadcasts an LSP but the LSP gets lost (step 1). In the next CSNP, the DIS includes the summary of the lost LSP instance (step 2). Non-DIS routers, upon receiving the CSNP, notice that they do not have the LSP (or that they have an outdated instance of the LSP), and request its transmission using a PSNP (step 3); finally, the DIS replies by retransmitting the LSP (step 4).

> In IS-IS, LSPs are transported in LINK STATE PDUs. The acknowledgments are performed through the LSP Entries TLV, but included in different control packets on point-to-point and shared links. On point-to-point links, the acknowledgments are included in PSNPs and, on shared links, in the CSNPs periodically transmitted by the DIS.

6.4.3 Guard-time between LSA and LSP transmissions

In both OSPF and IS-IS, the interval between two consecutive transmissions of the same LSA or LSP instance cannot be less than 5 seconds, called `MinLSInterval*` in OSPF and `minimumLSPTransmissionInterval` in IS-IS. These intervals, besides limiting the rate of LSA and LSP transmissions, im-

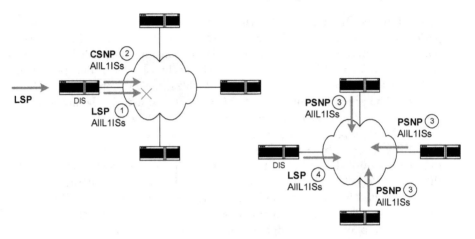

FIGURE 6.14: Flooding on shared links (IS-IS).

pose a guard-time required for the correctness of the process of deleting LSAs and LSPs, as discussed in Section 2.4.6.

6.5   Updating procedure

The routing packets that circulate in the network are associated with one of the following actions upon their reception at a router: (i) insertion in the LSDB of a new LSA/LSP (being disseminated for the first time), (ii) update of a stored LSA/LSP by a more recent instance, and (iii) deletion of an existing LSA/LSP.

In both OSPF and IS-IS, the delete indications are provided through special types of LSAs/LSPs, rather than through explicit DELETE control packets (as suggested in Section 2.4.6). Thus, the routing information that is disseminated in OSPF and IS-IS networks consists only of LSAs/LSPs.

LSAs/LSPs possess both a freshness attribute and a validity period, which are used to determine the actions performed at a router upon their reception or their validity expiration. The actions also depend on whether the router performing the action is the LSA/LSP originator.

These issues will be discussed in the next sections.

6.5.1   Freshness attributes

Both OSPF and IS-IS use three types of attributes to express the freshness of an LSA/LSP: (i) the SN, (ii) the checksum, and (iii) the age. However, the way these attributes are used differ in OSPF and in IS-IS.

	SN	**checksum**	**age**
LSA (OSPF)	LS Sequence Number	LS Checksum	LS Age
LSP (IS-IS)	Sequence Number	Checksum	Remaining Lifetime

FIGURE 6.15: Freshness attributes of LSAs (OSPF) and LSPs (IS-IS), and corresponding header fields.

The freshness attributes are included in the headers of LSAs/LSPs. The fields that describe the freshness of an LSA are the LS Sequence Number, the LS Checksum, and the LS Age; the corresponding fields of the LSP header are the Sequence Number, the Checksum, and the Remaining Lifetime (see Figure 6.15).

The SN In both OSPF and IS-IS, the SN is taken from the linear space of 32-bit numbers. In OSPF, the SN is a signed value, growing from 0×80000001 (initial value) to 0×7fffffff (final value); in IS-IS, it is an unsigned value, growing from 0×00000001 (initial value) to 0×ffffffff (final value). Using a linear space of numbers poses the problem of what to do when the final value is reached. This problem will be addressed in Section 6.5.6.

The age The age expresses how old an LSA/LSP instance is, in relation to its creation time. The age has a maximum value, called MaxAge* in both OSPF and IS-IS. The maximum value is the lifetime of the LSA/LSP. The age counts up in OSPF, from zero to MaxAge*, but counts down in IS-IS, from MaxAge* to zero. The MaxAge* is 1 hour (3600 seconds) in OSPF and 20 minutes (1200 seconds) in IS-IS. The age fields of LSAs/LSPs have a length of two octets, and count in seconds. The need for using the age as a freshness attribute is debatable. Because LSAs/LSPs are aged, their age needs to be refreshed before the maximum value is reached. This issue will be addressed in Section 6.5.5.

The age of LSAs/LSPs is updated while they stay at routers, using the routers' local clocks, but also while they are being flooded, to account for processing and transmission delays. The update value can be interface specific since different interface speeds lead to different transmission delays. IS-IS specification says that the age of an LSP must be decremented by at least 1 second whenever the LSP is transmitted through an interface. OSPF just says it must be incremented by a value greater than zero.

The checksum In both OSPF and IS-IS, the checksum is computed over the complete contents of the LSA/LSP, with the exception of the age fields (LS Age in OSPF and Remaining Lifetime in IS-IS). The checksum is essentially used to verify if two LSA/LSP instances with the same SN have indeed the same contents. However, in OSPF the checksum is also used to rank the freshness.

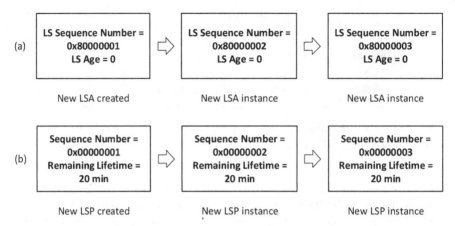

FIGURE 6.16: Generation of successive instances of (a) LSAs or (b) LSPs.

Generation of successive instances of LSAs or LSPs When a router
needs to originate a new LSA/LSP instance (e.g. because it has newer routing
information that needs to be disseminated throughout the network), it increments the SN by one in relation to the previous instance, and sets the age to
the initial value, which is zero in OSPF and 20 minutes in IS-IS. This process
is illustrated in Figure 6.16.

> *The freshness of LSA/LSP instances is determined by three attributes:
> the SN, the age, and the checksum. The SN is taken from the linear
> space of 32-bit numbers. The age expresses how old an instance is; it
> increases from zero to 1 hour in OSPF, and decreases from 20 minutes
> to zero in IS-IS. The checksum is mainly used to determine if two
> instances with the same SN have the same contents.*

6.5.2 Determining which LSA and LSP is fresher

The rules for determining which LSA/LSP is fresher are summarized in Figure
6.17. They depend on the freshness attributes introduced in Section 6.5.1: SN,
age, and checksum.

Role of the final age value Both OSPF and IS-IS assign a special meaning
to the final age value, regarding the freshness of LSAs/LSPs: in a circum-
stance of equality of other factors defining freshness, an LSA/LSP with an
age equal to the final value is always considered the freshest. As discussed in
Section 6.5.3, this feature is used in both OSPF and IS-IS for the deletion of
LSAs/LSPs.

OSPF In OSPF, an LSA instance with a higher SN is always considered
fresher. The checksum is used as a tiebreaker when two LSA instances have
the same SN, and the highest checksum wins. The age is used as a tiebreaker

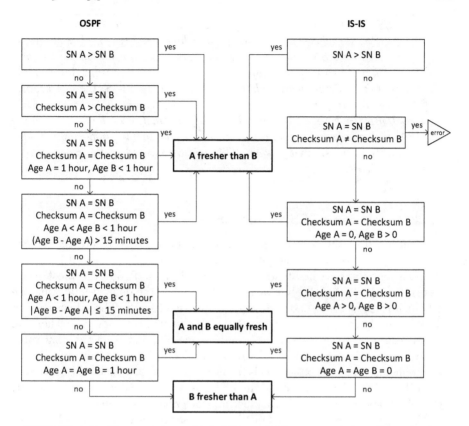

FIGURE 6.17: Determining which LSA or LSP instance, A or B, is fresher.

when two LSA instances have the same SN and checksum. In this case, if only one instance has an age of 1 hour (MaxAge*), that instance is considered fresher. Otherwise, if the ages differ by more than 15 minutes (MaxAgeDiff*), the LSA with lower age is considered fresher. Finally, if the ages differ by 15 minutes or less, the instances are considered equally fresh.

IS-IS In IS-IS, the freshness of an LSP is mostly determined by the SN and the age (Remaining Lifetime). The checksum is used to verify if two LSPs with the same SN have the same contents and, unlike OSPF, is not used to rank the freshness. An instance with a higher SN is always considered fresher. If two instances have the same SN, IS-IS uses the checksum to verify if their contents are the same. If the checksums are the same and one of the instances has zero age, that instance is considered fresher. Otherwise, if the two instances have the same SN, same checksum, and both have a non-zero age or both have zero age, they are considered equally fresh. Finally, when two instances have the same SN but different checksums, an error condition is generated; the actions taken in this case will be discussed is Section 6.5.4.

> *An LSA/LSP instance with higher SN is always considered fresher. In OSPF, the checksum is used as a tiebreaker if two LSA instances have the same SN (higher checksum wins), and the age is used as a tiebreaker if two instances have the same SN and checksum (lower age wins, except if the age equals the final value). In IS-IS, the checksum is used to verify if two LSP instances with the same SN have the same contents. In both OSPF and IS-IS, if two instances have the same SN and checksum but one has an age equal to the final value, that instance is considered fresher. This feature is used in the deletion of LSAs/LSPs.*

6.5.3 Deletion of LSAs and LSPs

As mentioned in the previous section, the indications to delete LSAs/LSPs are provided through the age attribute. Specifically, both in OSPF and IS-IS, when a router wishes to delete an LSA/LSP, it floods the most recent instance of the LSA/LSP, but with an age equal to the defined final value, i.e. 1 hour in case of LSAs and zero seconds in case of LSPs. These instances are disseminated throughout the network because they are considered fresher than stored instances with the same SN and an age different from the final value. When they arrive at a router, the router replaces the stored instance by the incoming one since the latter one is fresher and, next, deletes the instance from the LSDB since the instance has an age equal to the final value. Interestingly, the OSPF specification calls this mechanism *premature aging*.

The delete indications in OSPF unnecessarily carry the complete LSA, whereas in IS-IS they carry only the LSP header.

In Section 10.4 we illustrate through experiments the process of deleting LSAs and LSPs.

Guard-time between delete indication and next LSA or LSP instance In Section 2.4.6 we highlighted the need for a guard-time between the indication to delete an LSA/LSP and the flooding of the next LSA/LSP instance, having the same NR identifier and an SN equal to the initial value. In OSPF and IS-IS this guard-time is implemented through the restriction imposed on the minimum time between successive LSA/LSP transmissions, which is 5 seconds in both technologies (see Section 6.4.3).

What router can delete an LSA or LSP? In OSPF, the indication to delete an LSA can only be issued by the router that originated the LSA. In IS-IS, the indication to delete an LSP can be issued, in some special situations, by routers that did not originated it. One important case is when the DIS changes. Here, the new DIS, besides flooding the new Pseudonode-LSP, has also to delete the Pseudonode-LSP originated by the previous DIS. The advantage of this procedure is that the old Pseudonode-LSP does not remain in the LSDB until it expires, unnecessarily occupying memory resources.

		Action	OSPF	IS-IS
Incoming (I) not self-originated	1	freshness of I > freshness of S or I is new	replace or install and flood	
	2	freshness of I = freshness of S	discard	discard
	3	freshness of I < freshness of S		discard, transmit S on receiving interface
	4	SN of I = SN of S and checksum of I ≠ checksum of S	NA	set age of I to final value and flood
Incoming (I) self-originated	5	freshness of I > freshness of S and router interested	increment SN by one (relative to I) and flood	
	6	SN of I = SN of S and checksum of I ≠ checksum of S	NA	increment SN by one (relative to I) and flood
	7	freshness of I ≤ freshness of S and router interested	discard	
	8	router not interested or I is new	set age of I to final value and flood	

FIGURE 6.18: Actions on the reception of LSAs and LSPs.

> *When a router wants to delete an LSA/LSP, it floods an instance of the LSA/LSP with the current SN and an age equal to the final value, i.e. 1 hour in OSPF and zero seconds in IS-IS. This instance is considered the freshest one and, therefore, replaces the stored one when arriving at a router, but gets immediately deleted since its age equals the final value.*

6.5.4 Actions on the reception of LSAs and LSPs

When an LSA/LSP is received at a router, the router performs one or more actions that depend mainly on the relative freshness of the incoming and stored LSAs/LSPs, and on whether the incoming LSA/LSP is self-originated, i.e. is being received by the LSA/LSP originator. The behavior of OSPF and IS-IS differ slightly in this respect. In Figure 6.18, we summarize the actions taken by a router upon the reception of LSAs/LSPs. In the figure the incoming LSA/LSP is identified by the letter "I" and the stored LSA/LSP by the letter "S". We first divide the actions in two groups, differing on whether the incoming LSA/LSP is self-originated or not: actions 1 to 4 belong to the first group and actions 5 to 8 to the second.

6.5.4.1 Incoming LSA/LSP is not self-originated

Action 1 If the freshness of the incoming LSA/LSP is higher than the stored one or if the incoming LSA/LSP is new (nonexistent at the LSDB), then the router must replace the stored instance by the incoming one or install the new one and flood the LSA/LSP. This behavior is the same in OSPF and IS-IS. Note that the incoming LSA/LSP may be an indication to delete the stored one but, even in this case, the action is to replace and flood; the indication to delete has an age equal to the final value and, because of this, it gets immediately deleted after being stored.

Actions 2 and 3 When the freshness of the incoming LSA/LSP is equal to or lower than the stored one, then the incoming LSA/LSP must be discarded, since it is not new and, therefore, its flooding must be stopped. IS-IS takes an additional action when the freshness of the incoming LSA/LSP is lower than the stored one: it transmits the stored LSP through the interface where the incoming LSP was received, to ensure that the sending neighbor gets properly updated.

Action 4 As mentioned in Section 6.5.2, in IS-IS, when two LSPs have the same SN but different checksums, an error condition is generated, since the LSPs have the same SN but different contents. When this happens, the router sets the age of the LSP to zero, causing it to expire, and floods the expired instance (i.e. a delete indication) to remove the LSP from the routing domain. OSPF does not take a specific action in this case, simply using the checksum to rank the LSA freshness.

6.5.4.2 Incoming LSA/LSP is self-originated

Receiving self-originated LSAs/LSPs is a relatively common occurrence, e.g. during the flooding process. What is not so frequent is to receive self-originated LSAs/LSPs fresher (e.g. with higher SN) than the last instance generated by the router. This may happen when a router is switched off and then switched on before the expiration of its self-originated LSAs/LSPs at neighboring routers.

Note that a router receiving a self-originated LSA/LSP may no longer be interested in it. For example, this happens in OSPF when a router receives a self-originated Network-LSA and is no longer the DR of the corresponding link. This Network-LSA, if still stored at the router, is aging and waiting to be purged.

Actions 5 and 6 If the freshness of the incoming LSA/LSP is higher than the stored one, and the router is still interested in the LSA/LSP, it must flood a new instance with the SN incremented by one in relation to the received SN. This behavior is the same in OSPF and IS-IS, and ensures that previous incarnations of self-originated LSAs/LSPs are replaced by the most up-to-date instances of these LSAs/LSPs. IS-IS also performs the same action when the incoming and stored LSPs have the same SNs but different checksums (i.e. different contents).

Action 7 If the freshness of the incoming LSA/LSP is equal to or lower than the stored one, and the router is still interested in the LSA/LSP, no action should be taken and the LSA/LSP must be discarded, since the incoming LSA/LSP is not new. This behavior is the same in OSPF and IS-IS, and is also similar to actions 3 and 4.

Action 8 It may also happen that the receiving router is no longer interested in the incoming LSA/LSP or that the incoming LSA/LSP is no longer stored at the router. The action to be taken is to set the age of the incoming LSA/LSP to the final value and flood the expired instance (i.e. a delete indication) to remove the LSA/LSP from the network.

> *The actions to be taken when an LSA/LSP arrives at a router depend mainly on the relative freshness of the incoming and of the stored LSA/LSP instance (if it exists), and whether the incoming LSA/LSP is self-originated. When the incoming LSA/LSP is not self-originated, if its freshness is higher than the stored instance one, or if there is no stored LSA/LSP, then the incoming LSA/LSP replaces the stored one or is installed for the first time, and is subsequently flooded; otherwise, it is discarded. When the incoming LSA/LSP is self-originated, it may happen that its freshness is higher than the stored one. This means that the incoming LSA/LSP is outdated, and the router must flood a new instance of the LSA/LSP with the SN incremented by one in relation to the incoming one. Besides this general behavior, there are other events leading to specific actions in both OSPF and IS-IS.*

6.5.5 Unsolicited generation of LSAs and LSPs

Since LSAs/LSPs have an age attribute and get aged, the age needs to be refreshed (i.e. reset to the initial value) before reaching its final value. Otherwise, under stable conditions (i.e. if there are no changes in the network), the LSAs/LSPs would (smoothly) reach their lifetime and be purged from the LSDB, causing network malfunction. The age refreshment is carried out by generating, at some point in time, new LSA/LSP instances with the SN incremented by one, and the age is reset to the initial value. The refresh interval needs to be smaller than the lifetime of LSA/LSP instances. In OSPF, the refresh interval is 30 minutes (called `LSRefreshTime*`), and in IS-IS it is 15 minutes (called `maximumLSPGenerationInterval*`).

> *To avoid the premature expiration of LSAs/LSPs, new LSA/LSP instances with the SN incremented by one are generated when the age of the previous instance reaches 30 minutes, in OSPF, or 15 minutes, in IS-IS.*

6.5.6   Wrap around of SNs

OSPF and IS-IS deal differently with the wraparound problem, which was discussed in Section 2.4.8. OSPF explicitly deletes the LSA instance with the highest SN when there is a request for flooding a new LSA instance, whereas IS-IS freezes the LSA instance until it expires. Thus, wraparound is faster in OSPF. In particular, in OSPF, when the originating router needs to flood a new LSA instance and the current one has SN equal to the final value (0×7ffffffff), it first deletes the current instance using the premature aging mechanism described in Section 6.5.3, and then it floods the new LSA instance with the initial SN value (0×80000001). In IS-IS, when the SN reaches the final value ($0 \times$ffffffff), the update of the current LSP instance freezes for 21 minutes (`MaxAge* + ZeroAgeLifetime*`), to make sure it expires prior to transmitting a new LSP instance with the initial SN value (0×00000001).

Given the 5 seconds interval between successive LSA/LSP transmissions (see Section 6.4.3) and the SN space of 32 bits, it would take more than 600 years to go from the lowest to the highest SN value! This clearly shows that the SN wraparound is a rare event.

> *In OSPF, when there is an attempt to increase the SN of an LSA beyond the final value (the wraparound problem), the originating router first deletes the LSA, and only then floods the new LSA instance with an SN equal to the initial value. In IS-IS, when the SN of an LSP reaches the final value, the originator freezes its update for 21 minutes, to make sure that the LSP expires at the LSDB of all routers, before flooding a new LSP instance with an SN equal to the initial value.*

6.5.7   Age expiration of LSAs and LSPs

When an LSA/LSP instance stored at an LSDB expires, i.e. when its age reaches the final value (its lifetime), some action needs to be performed. The action depends on whether the router is the originator of the LSA/LSP. Figure 6.19 summarizes these actions.

OSPF In OSPF, when an LSA expires at its originating router, the router floods a delete indication to remove the LSA from the LSBD of all other routers and, subsequently, purges the LSA from its own LSDB. When the LSA expires at a non-originating router, the router simply purges the LSA from its LSDB.

IS-IS The process in IS-IS is similar with two exceptions. First, when an LSP expires, it remains one additional minute at the LSDB before being purged (this parameter is called `ZeroAgeLifetime*`). Second, any router, and not just the LSP originator, must flood expired LSP instances. At first sight this procedure seems to lead to a flooding storm of LSPs. However, LSPs tend to expire at approximately the same time. Thus, it is highly probable that an expired LSP being flooded finds recently stored instances with equal freshness

	OSPF	**IS-IS**
non-LSA/LSP originator	purge from LSDB	flood delete indication, wait 1 minute, and purge from LSDB
LSA/LSP originator	flood delete indication and purge from LSDB	flood delete indication, wait 1 minute, and purge from LSDB

FIGURE 6.19: Actions on the expiration of LSAs and LSPs.

(same SN and zero age), in which case it will be discarded, decreasing the chances of a flooding storm. Nevertheless, these additional features used by IS-IS seem unnecessary.

> *When the age of an LSA/LSP expires at the originating router, the router must flood a delete indication to remove the LSA/LSP from the network and purge the LSA/LSP from its LSDB. Other routers just purge the LSA/LSP from the LSDB, except in IS-IS, where the routers also flood a delete indication.*

6.5.8 Protection against corruption of LSAs and LSPs

Both OSPF and IS-IS take measures to protect against the corruption of LSAs/LSPs. The protection is based on the checksum calculated over most of their fields. The checksum is not computed over the age field (`LS Age`, in case of OSPF, and `Remaining Lifetime`, in case of IS-IS), to allow updating the age without recalculating the checksum. Corruption of LSAs/LSPs can happen in two situations: during transmission between two routers, or while residing at a router's memory. The first situation is much more frequent than the second one, due to the lack of reliability of some communications media. To handle the first situation, the checksum is verified whenever an LSA/LSP instance arrives at a router and, if the checksum is invalid, the arriving instance is discarded. The second situation is handled through periodic verification of the checksum of all LSAs/LSPs stored at a router. In OSPF, this happens whenever the age of stored instances reaches a multiple of 5 minutes (`CheckAge*`) and in IS-IS every 15 minutes (`maximumLSPGenerationInterval`). If a checksum failure is detected, the routing process has to be restarted.

6.6 Initial LSDB synchronization

Two routers that become adjacent may need to synchronize their LSDBs. In both OSPF and IS-IS, a router must synchronize its LSDB with all its

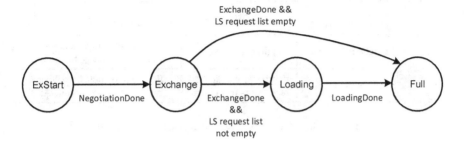

FIGURE 6.20: Evolution of states during a successful initial LSDB synchronization in OSPF.

neighbors on point-to-point links and with selected neighbors on shared links; in case of IS-IS, it synchronizes with the DIS and, in case of OSPF, with both the DR and the BDR. The initial LSDB synchronization process starts as soon the two routers become adjacent via the Hello protocol, and has two phases, the first to advertise the LSDB summary and the second to retrieve specific LSAs/LSPs. OSPF and IS-IS use slightly different methods for initial LSDB synchronization, neither of them adhering exactly to the protocol of Section 2.5. We will explain them in the next sections.

> OSPF and IS-IS routers synchronize their LSDBS with all their neighbors on point-to-point links; on shared links, the synchronization is with the DR and the BDR, in case of OSPF, and with the DIS, in case of IS-IS.

6.6.1 OSPF

In OSPF, the LSDB synchronization process is defined through a detailed state machine, which is part of the neighbor state machine, and is described in Section 10 of the OSPF specification [36]. The evolution of states during a successful synchronization process is shown in Figure 6.20. A router maintains a separate state machine for each synchronization instance it establishes. The synchronizations are per neighbor interface, and not just per neighbor. For example, two routers connected by two parallel point-to-point links must synchronize on both links.

Packet support The LSDB synchronization process of OSPF is supported on four types of control packets: (i) DB DESCRIPTION packets advertise the LSDB summary, (ii) LS REQUEST packets request specific LSAs, (iii) LS UPDATE packets transport complete LSAs, and (iv) LS ACKNOWLEDGMENT packets acknowledge the reception of complete LSAs.

6.6.1.1 Database description process

Two routers start an LSDB synchronization process by establishing a master/slave relationship and exchanging LSA headers using DB DESCRIPTION packets. The neighbor with the highest RID becomes the master. The exchange of DB DESCRIPTION packets is protected through a Stop-and-Wait (SW) protocol (see Section 1.8.1). This process is called *database description* in the OSPF specification.

States of database description process The database description process is further divided into two states: the first one to elect the master (called ExStart) and the second one to exchange the LSA headers (called Exchange). The process is controlled by three flags: the I-bit, set while the router is in the ExStart states; the M-bit, set if the router has additional LSA headers to send; and the MS-bit, set whenever the router believes it is the master.

Protection of DB DESCRIPTION transmissions As mentioned above, the exchange of DB DESCRIPTION packets is protected by a SW protocol. The SW protocol is controlled by the master, and the corresponding sequence numbers are contained in the DD Sequence Number field of DB DESCRIPTION packets (see Figures 6.2 and 6.4); we will refer to these sequence numbers as ddSN, to avoid confusion with the sequence numbers of LSA instances.

ExStart state In the beginning of the ExStart state, each router considers itself as master. The initial packet can be sent by either router; it will have an arbitrary ddSN defined by the sending router and the three control flags set. A router that receives this packet determines whether it is master or slave by comparing its RID with the RID of the neighbor, contained in the received packet (Router ID field). From then on only the master takes the initiative of sending DB DESCRIPTION packets, and the slave just replies to these packets using the ddSN of the master. The packets sent by the slave serve as acknowledgments to the packets sent by the master, and vice-versa. The master increments the ddSN by one in every new packet it sends, except in the case of retransmissions. For the initial packet, which may be sent by the master or the slave, and for any subsequent packets sent by the master, if the response is not received within a predefined time interval (called RxmtInterval), the packet is retransmitted using the same ddSN.

Exchange state The transition from the ExStart to the Exchange state occurs either when a router learns it is the slave (i.e. receives a packet with the three control bits set and the neighbor RID is higher than its own) or when it gets the confirmation that it is the master (i.e. receives a packet with the I-bit and the MS-bit cleared and the neighbor RID is lower than its own). When a router enters the Exchange state, it places all its LSA headers in a list called Database summary list, and transmits these headers to the neighbor in DB DESCRIPTION packets. Note that several packets may be needed to transmit all LSA headers contained in the Database summary list. The LSA headers are removed from the list when their reception is acknowledged by the neighbor.

Use of the M-bit The OSPF standard [36] is not very clear on how a router abandons the `Exchange` state and, in particular, on the handling of the `M-bit`. The standard seems to imply that the `M-bit` is cleared as soon as the last LSA header is transmitted for the first time (this is also the interpretation given in Figure 6.7 of [11]). Specifically, in its Section A.3.3 (page 196), the specification says in relation to the `M-bit`: "When set to 1, it indicates that more Database Description packets are to follow". However, analyzing the behavior of existing implementations (see Section 10.5.1) we conclude the following:

- The `M-bit` is only cleared when the `Database summary list` becomes empty, i.e. when the reception of all LSA headers has been acknowledged;

- Both routers have to signal their neighbor that the `Database summary list` has emptied, i.e. a router has to send at least one DB DESCRIPTION packet with the `M-bit` cleared;

- The slave abandons the `Exchange` state when it receives a DB DESCRIP-TION packet from the master with the `M-bit` cleared and its `Database summary list` is empty;

- Likewise, the master abandons the `Exchange` state when it receives a DB DE-SCRIPTION packet from the slave with the `M-bit` cleared and its `Database summary list` is empty.

Note that the master is always the last one to abandon the `Exchange` state. When this happens, the database description process ends, and no further DB DESCRIPTION packets are sent.

6.6.1.2   Loading process

Requesting LSAs When a router receives LSA headers (during the `Exchange` state), it compares the received information with the one stored in its LSDB. If the router verifies that the neighbor has fresher instances of existing LSAs or LSAs the router is missing, it places the corresponding LSA IDs in a list called `Link state request list`, and transmits these IDs to the neighbor in LS REQUEST packets. Note that, unlike DB DESCRIPTION packets, LS REQUEST packets do not carry complete LSA headers, but only LSA IDs, i.e. the three fields that identify an LSA. Upon receiving an LS REQUEST, the neighbor replies by flooding the complete LSAs in an LS UPDATE packet. The LSA IDs are removed from the `Link state request list` when the corresponding LSAs have been received (in LS UPDATE packets). A router can start sending LS REQUEST packets even before the database description process has ended.

Protection of LSA requests The transmission of LSA headers contained in LS REQUEST packets is ACK protected using `RxmtInterval` as the timeout period: if a requested LSA is not received within this interval, the request is retransmitted. Moreover, there can be only one LS REQUEST packet outstanding at a time.

Transmission of LS REQUESTs and LS UPDATEs on shared links
On shared links, an LS REQUEST is directed to the neighbor that contains the requested LSA (this neighbor can only be the DR or the BDR); thus, the packet is sent to the unicast address of the neighbor, and not to an OSPF multicast address. Likewise, LS UPDATE packets sent in response to LS REQUEST packets are unicasted to the requesting neighbor.

Reaching the Full state When the database description process ends, the router can move to the `Loading` state or to the `Full` state. It will move immediately to `Full` state if the `Link state request list` is empty, meaning that all requests sent to the neighbor (if any) have been satisfied. Otherwise, the router moves to the `Loading` state. In this state, it will keep sending LS REQUEST packets to the neighbor until they are satisfied by LS UPDATE packets and the `Link state request list` becomes empty, at which point it moves to the `Full` state. When the `Full` state is reached, the LSDB synchronization process ends. In this state, the router is said to be *fully adjacent* to its neighbor.

Comparison with the generic protocol The protocol used by OSPF is not significantly different from that of Section 2.5, the main difference being that OSPF does not use a packet to request the LSDB summary. This is not required since OSPF starts by electing a master, which subsequently controls the exchange of LSDB summaries. Besides that, the remaining three packets of Section 2.5 all have an equivalent in OSPF: LSDB SUMMARY is the equivalent of DB DESCRIPTION, PARTIAL LSDB REQUEST the equivalent of LS REQUEST, and UPDATE the equivalent of LS UPDATE.

6.6.1.3 Dealing with outdated self-originated LSAs

An important scenario, discussed in Section 2.5.3, occurs when a router is switched off and switched on again, and finds at a neighbor previous incarnations of self-originated LSAs, i.e. outdated instances of these LSAs with higher or equal freshness.

Recall that neighbors exchange summary information about the LSAs stored at their LSDBs during the database description phase (using DB DESCRIPTION packets). Thus, during this phase, a router learns if its neighbor has LSAs originated by itself with higher freshness than the ones it just created. If this is the case, the router requests from the neighbor the transmission of the complete LSAs (using LS REQUEST packets); this behavior is common to all LSA instances with higher freshness found at the neighbor, and is not specific of self-originated LSAs.

However, when the router receives a self-originated LSA from a neighbor with freshness higher than the stored one, it performs action 5 of Section 6.5.4; it floods a new instance of the LSA with the SN incremented by one in relation to the SN received from the neighbor. This ensures that all other routers receive the most up-to-date LSA instance.

The case of equal SNs A scenario that is worth considering is when the old and new instances have the same SN (recall that the freshness of LSAs is defined by SN, checksum, and age). This scenario may happen if a router is switched on, off, and then on again, without waiting much time between these actions. In this case, and according to the OSPF freshness rules, the checksum is the tiebreaker. There are three possibilities when comparing the freshness of the new instance (just created by the router) and the old instance (left at the neighbor):

- **The new instance has a higher checksum than the old instance -** In this case, the new instance is considered fresher. The neighbor will then request the instance as part of the LSDB synchronization process (using an LS REQUEST packet) and will flood it throughout the network. Thus, the network LSDB will end up being updated with the most recent information.

- **The new instance has a lower checksum than the old instance -** In this case, the old instance is considered fresher. Thus, according to the rule stated above (action 5 of Section 6.5.4), the router will flood a new instance of the LSA with the SN incremented by one. Again, the network LSDB will end up being updated with the most recent information.

- **The new and old instances have equal checksum -** In this case nothing will happen, and the outdated information remains in the network until the next instance of the LSA is generated, which can take as long as 30 minutes (`LSRefreshTime*`). However, the probability of having the same checksum but different contents is very low.

6.6.1.4 Example

Consider the example of Figure 6.21, illustrating the initial LSDB synchronization process between routers R1 and R2, with RIDs 1.1.1.1 and 2.2.2.2, respectively. R1 was switched off and switched on again before expiration of its self-originated LSAs at the LSDB of R2. Suppose that when R1 is switched on, R2 contains in its LSDB Router-LSAs relative to routers R1, R2, and R3. The figure shows the exchange of packets following the establishment of the neighborhood relationship, i.e. after the routers enter the `ExStart` state. The exchange of packets follows these steps:

1. Suppose that R1 is the first to send a DB DESCRIPTION packet. It chooses a ddSN of 634, sets the `MS-bit` since it believes in being the master, sets the `I-bit` since it is in the `ExStart` state, and sets the `M-bit` since it has LSA headers to send.

2. When R2 receives the packet it verifies that its RID is higher than the neighbor's. Thus, it replies with a DB DESCRIPTION packet

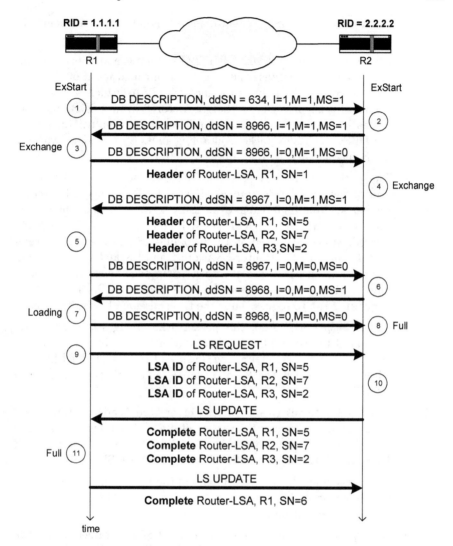

FIGURE 6.21: Initial LSDB synchronization example (OSPF).

having a new ddSN (ddSN = 8966) and the MS-bit set, since it is
now certain of being the master.

3. Upon receiving the packet from R2, R1 moves to the Exchange state
   and sends, in another DB DESCRIPTION packet, the header of its
   single and just created LSA, a Router-LSA with SN=1. In this
   packet, R1 clears the I-bit since it has abandoned the ExStart
   state. It also clears the MS-bit and uses the ddSN of the previously

received packet (ddSN = 8966), since it is now aware of being the slave.

4. R2 receives this DB DESCRIPTION packet and moves to the Exchange state, because the neighbor confirmed it is not the master. Moreover, R2 analyzes the incoming LSA header, verifies that is has a fresher instance of that LSA stored in its LSDB, and concludes it is not interested in its full contents. Finally, R2 sends a DB DESCRIPTION packet containing all its LSA headers, with the ddSN incremented by one (ddSN = 8967).

5. The packet received at this step acknowledges the correct reception of the LSA header previously sent by R1. Thus, R1 removes the header from its Database summary list and, since the list became empty, it sends a DB DESCRIPTION packet with the M-bit cleared. Moreover, based on the incoming LSA headers, R1 verifies that it is missing the Router-LSAs of R2 and R3, and that the neighbor has a fresher instance of its self-originated Router-LSA. Therefore, it places the corresponding LSA headers in its Link state request list to request later the transmission of its full contents.

6. As in the previous step, this packet acknowledges the correct reception of the three LSA headers previously sent by R2. Thus, R2 empties its Database summary list and sends a DB DESCRIPTION packet with the M-bit cleared, and with the same ddSN (8967).

7. When R1 receives this packet, it enters the Loading state, since its Link state request list is not empty, meaning that R1 still needs to receive complete LSAs from R2. R1 also replies with a DB DESCRIPTION packet with the same ddSN (ddSN = 8968).

8. Upon receiving this packet, R2 enters the Full state since its Link state request list is empty. At this point, R2 completes the database description process.

9. R1 requests the complete contents of the LSAs contained in its Link state request list using an LS REQUEST packet unicasted to R2. This step could have occurred before, as soon as R1 determined it had to request complete LSAs from the neighbor.

10. Upon receiving the LS REQUEST packets, R2 replies sending the complete contents of the requested LSAs in an LS UPDATE packet, unicasted to R1.

11. When R1 receives the LS UPDATE packet, it enters the Full state since it has completed receiving the requested LSAs. Now, because R1 received an instance of a self-originated Router-LSA with a higher SN (with SN=5), it creates a new instance of this LSA (with SN=6) and floods it throughout the network.

The experiments of Section 10.5.1 illustrate the initial LSDB synchronization of OSPF.

> *In OSPF, the LSDB synchronization process between two neighboring interfaces evolves according to four states. Upon becoming adjacent through the Hello protocol, the routers start exchanging DB DESCRIPTION packets, to elect a master for the control of the subsequent communication phase (ExStart state) and to exchange LSA headers (Exchange state). Based on the received LSA headers, both routers check if they are missing any LSA instances and, if so, request them from the neighbor using an LS REQUEST packet; the neighbor replies with an LS UPDATE packet containing the full contents of the requested LSAs (Loading state). Neighbors become fully adjacent when they receive all requested LSAs (Full state).*

6.6.2 IS-IS

IS-IS uses a process slightly different from OSPF for initial LSDB synchronization. Routers do not engage into a master-slave relationship prior to exchanging LSP headers. Moreover, the synchronization process differs substantially in point-to-point and shared links.

Packet support The initial LSDB synchronization is supported in three types of control packets:

- CSNPs provide the LSDB summary of the sending router. They are used both in point-to-point and shared links and carry only LSP summaries (through LSP Entries TLVs). The CSNPs are the equivalent of the DB DESCRIPTION packets of OSPF.

- PSNPs have different roles in point-to-point and shared links. In point-to-point links, they are used to acknowledge the reception of LSPs (in this role, they are the equivalent of the LS ACKNOWLEDGMENT packet of OSPF). In shared links, they are used to request the transmission of LSPs from a neighbor (in this role, they are the equivalent of the LS REQUEST packet of OSPF). PSNPs carry only LSP summaries (through LSP Entries TLVs).

- LINK STATE PDUs transport, between neighbors, the complete Nonpseudonode-LSPs and Pseudonode-LSPs required to synchronize their LSDBs. They are the equivalent of the LS UPDATE packets of OSPF. However, LINK STATE PDUs can only carry one LSP, whereas LS UPDATEs can carry several LSAs.

6.6.2.1 Point-to-point links

When two routers become adjacent on a point-to-point link, they immediately send to each other their self-originated Nonpseudonode-LSPs. Each router also sends to the neighbor a CSNP containing its LSDB summary, to give summary information about the LSPs currently stored in its LSDB. If, based on this information, the neighbor detects that the router is missing some LSPs or has outdated instances of existing LSPs, the neighbor sends to the router the complete LSPs. Note that, unlike the case of shared links, each router transmits only one CSNP on point-to-point links.

The CSNP is not ACK protected and, therefore, may not be received by the neighbor. However, initially, the router also schedules the transmission of all its complete LSPs for a later time. This delay, if sufficient, allows receiving the CSNP from the neighbor and canceling the transmission of the complete LSPs for which the neighbor already has the freshest instance.

The transmission of complete LSPs is ACK protected using the PSNPs as acknowledgments. Thus, both neighbors get initially synchronized even if the CSNPs are lost. We may say that during the initial LSDB synchronization on point-to-point links, IS-IS delays the transmission of all its complete LSPs to a neighbor, hoping to receive sooner, via a CSNP, summary information on the LSPs the neighbor really needs to receive.

6.6.2.2   Shared links

The initial LSDB synchronization on shared links is similar to the reliable flooding mechanism explained in Section 6.4.2. It also relies on the periodic transmission of the LSDB summary by the DIS. Moreover, similarly to point-to-point links, when a router establishes a new adjacency on a shared link, it transmits its self-originated LSPs on the link.

A non-DIS router that has just joined a shared link waits to receive the first CSNP from the DIS. If the router, based on the LSDB summary included in the CSNP, verifies that the DIS has more recent information, i.e. unknown LSPs or fresher instances of known LSPs, then it sends a PSNP to the DIS asking for these LSPs. Thus, in the case of shared links, PSNPs take a role similar to the LS REQUEST packets of OSPF. The router keeps sending PSNPs until it receives the requested LSP.

Moreover, a non-DIS router also takes the initiative of sending an LSP to the DIS if it verifies, based on the received CSNPs, that the LSP is unknown to the DIS or that the DIS has an older instance of the LSP. The correct reception of LSPs is not explicitly acknowledged using PSNPs, as is the case of point-to-point links. Instead, the periodic transmission of CSNPs acts as an implicit acknowledgment; if a non-DIS router does not see the transmitted LSP listed in the next CSNP transmitted by the DIS, it retransmits that LSP on the link. Note that IS-IS packets are transmitted on shared links using layer-2 multicast addresses targeting all IS-IS routers. Thus, when a non-DIS router sends an LSP and there are several other neighbors interested in this LSP, they will all receive it (if the LSP is not lost) and, therefore, no longer need to request it to the DIS.

6.6.2.3 Dealing with outdated self-originated LSPs

During the initial LSDB synchronization process, a router may find at a neighbor previous incarnations of its self-originated LSPs. These LSPs may have an SN equal to or higher than the LSP just created by the router, but with outdated information. This issue was discussed in Section 2.5.3.

As explained in previous sections, when a router establishes a new adjacency, it sends to the neighbor its self-originated LSPs. Upon receiving these LSPs, the neighbor verifies if it has stored instances of these LSPs. If so, it compares their freshness and, in case they have the same SNs, compares their checksums. There are two cases where the instance stored at the neighbor is found outdated:

- **The freshness of the instance stored at the neighbor is higher than the incoming one** - In this case, according to action 3 of Section 6.5.4, the neighbor sends back to the LSP originator information on this outdated LSP instance and its SN. This information can be provided in two ways: using a CSNP or by sending the complete LSP to the router. The CSNP advertises only summary information, but this summary includes the SN of LSPs, which is all that is needed in this case. Providing this information through CSNPs is always possible on point-to-point links (both routers can do it), but on shared links only the DIS can do it (since only the DIS is allowed to send CSNPs on shared links). Finally, when the LSP originator receives information that a neighbor has a previous incarnation of a self-originated LSP with higher freshness, it performs action 5 of Section 6.5.4: it floods a new instance of the LSP, with the SN incremented by one in relation to the SN held by the neighbor.

- **The SNs of the instance stored at the neighbor and of the incoming instances are the same but the checksums are different** - In this case, the router performs action 4 of Section 6.5.4: it sets the age of the LSP to zero seconds (`Remaining Lifetime`) and floods this delete indication. Finally, when the LSP originator receives this delete indication, it performs action 5 of Section 6.5.4, since the delete indication is considered fresher than the stored one: it floods a new instance of the LSP with the SN incremented by one in relation to the SN held by the neighbor.

In Section 10.5.2.3, we illustrate through an experiment the role of the checksum in the IS-IS initial LSDB synchronization process.

6.6.2.4 Examples

Point-to-point links Consider the example of Figure 6.22. Similarly to the OSPF example, R1 is switched off and on again, before expiration of its self-originated Nonpseudonode-LSP at R2. Moreover, R2 contains a Nonpseudonode-LSP of router R3. In the figure, the LSPs are identified through their LSP IDs. Initially, R1 and R2 send to each other their complete

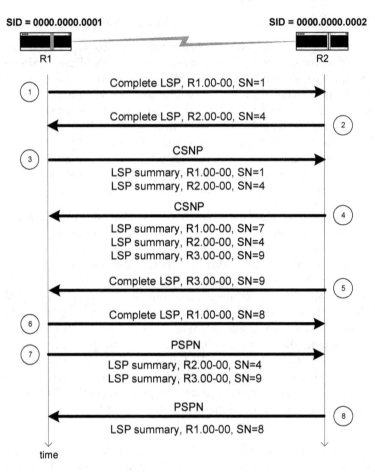

FIGURE 6.22: Initial LSDB synchronization example (IS-IS point-to-point links).

self-originated Nonpseudonode-LSPs (steps 1 and 2). The Nonpseudonode-LSP of R1 has SN=1, and the one of R2 has SN=4. The routers also send to each other CSNPs containing their LSDB summaries (steps 3 and 4). R1 advertises its Nonpseudonode-LSP (R1.00-00 with SN=1), and the Nonpseudonode-LSP just received from R2 (R2.00-00 with SN=4). R2 advertises its own Nonpseudonode-LSP (R2.00-00 with SN=4), the one of R3 (R3.00-00 with SN=9), and one Nonpseudonode-LSP of R1 (R1.00-00 with SN=7) with an SN higher than the one just created by R1. Note that R2 already received the newest Nonpseudonode-LSP of R1 (R1.00-00 with SN=1, created after R1 was switched on), but discarded it since it had an SN lower than the stored one. Based on the CSNP sent by R1 (at step 3), R2 learns that R1 is missing the Nonpseudonode-LSP of R3, and decides to send it (step 5). Moreover, based on the CSNP sent by R2 (at step 4), R1 learns that R2 has an instance of its Nonpseudonode-LSP, and decides to flood a new instance of this LSP with the SN incremented by one (step 6). The transmissions of LSPs is acknowledged using PSNPs. In step 7, R1 acknowledges the reception of R2.00-00 (with SN=4) and R3.00-00 (with SN=9); in step 8, R2 acknowledges the reception of R1.00-00 (with SN=8).

In Section 10.5.2.2 we illustrate through an experiment the initial LSDB synchronization process of IS-IS over point-to-point links.

Shared links Consider now the example of Figure 6.23. Assume that R2 is the DIS and, as in previous examples, R1 is switched on and switched off after a while. Upon becoming adjacent, R1 and R2 transmit on the link their self-originated LSPs. R1 transmits its Nonpseudonode-LSP (R1.00-00 with SN=1) (step 1) and R2 transmits its Nonpseudonode-LSP (R2.00-00 with SN=8) (step 2) and also the Pseudonode-LSP (R2.01-00 with SN=1) that represents the transit shared link where R2 is the DIS (step 3). R2 also transmits the previous incarnation of the Nonpseudonode-LSP of R1 (R1.00-00 with SN=4) (step 4), since R1 transmitted (in step 1) the same LSP but with lower SN. When R1 receives this LSP, it transmits on the link a new instance of the LSP, with the SN incremented by one in relation to the received SN (R1.00-00 with SN=5) (step 5). The next packet seen in the example (step 6) is the CSNP transmitted by R2 (since it is the link DIS). In this packet, R2 advertises R1.00-00 with SN=5, R2.00-00 with SN=8, R2.01-00 with SN=1, and R3.00-00 with SN=7. Based on this information, R1 learns that it is still missing the Nonpseudonode-LSP of R3. It requests this LSP using a PSNP (step 7), and R2 replies with the complete LSP (step 8).

In Section 10.5.2.1 we illustrate through an experiment the initial LSDB synchronization process of IS-IS over shared links.

> *In the IS-IS initial LSDB synchronization process, routers start by transmitting on the link their self-originated LSPs. The subsequent steps are different in point-to-point and shared links. On point-to-point links, each router sends to its neighbor a CSNP containing its LSDB summary, and schedules the transmission of all its complete LSPs for a later time. The transmission of an LSP is canceled if the router finds, through the received CSNPs, that the neighbor already has the freshest instance of that LSP. On shared links, the initial LSDB synchronization process is based on the periodic transmission of CSNPs, carried out by the DIS. A non-DIS router joining the link waits for the next CSNP transmitted on the link and, based on its contents, requests from the DIS the LSPs it is missing or for which it has older instances, using a PSNP.*

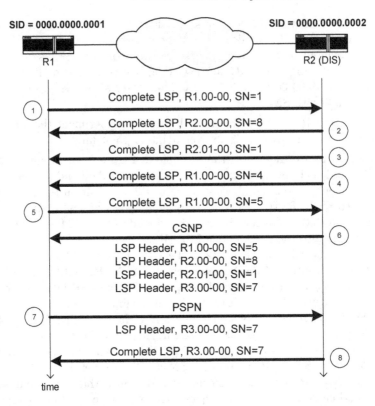

FIGURE 6.23: Initial LSDB synchronization example (IS-IS shared links).

6.7   Origination and deletion of routing information

We now give a more detailed explanation about when LSAs and LSPs are actually originated. We left this subject to the last section, since the origination of LSAs/LSPs depends on the protocol state machines and actions described in the previous sections. Some of the issues covered in this section were already introduced in previous ones.

New LSA/LSP instances must be created or deleted when the routing information that they convey changes. This may happen in many situations, e.g. when (i) the cost of a router interface changes, (ii) a neighbor fails, (iii) a prefix is assigned to (or removed from) a link, or (iv) a DR is elected (or dismissed) at a shared link.

When a new instance of an LSA/LSP is originated, the SN is incremented by one and the age is set to the initial value (see Figure 6.16).

Unsolicited generation of LSAs and LSPs One mechanism that is common to OSPF and IS-IS is the unsolicited generation of LSAs/LSPs.

As explained in Section 6.5.5, new LSA/LSP instances must be created before their expiration (i.e. before their age reaches the lifetime). Specifically, a new LSA instance must be created when the age of the previous one reaches 30 minutes (`LSRefreshTime*`) and a new LSP instance must be created when the age of the previous instance reaches 15 minutes (`maximumLSPGenerationInterval*`).

Expiration of LSAs and LSPs Another mechanism that is common to OSPF and IS-IS is the expiration of LSAs/LSPs. As explained in Section 6.5.7, when an LSA/LSP expires at its originating router, the router must originate a delete indication to have the LSA/LSP removed from the LSDB of other routers. In IS-IS, this behavior is also followed by non-originating routers.

In the next sections, we detail the procedures used by OSPF and IS-IS to originate new LSA/LSP instances (and possibly delete them).

6.7.1 OSPF

In both OSPFv2 and OSPFv3, routers originate Router-LSAs, Network-LSAs, and AS-External-LSAs. OSPFv3 routers also originate Link-LSAs and Intra-Area-Prefix-LSAs. This is true for single area networks. As will be seen in Chapter 7, in hierarchical OSPF networks routers originate other types of LSAs.

According to the OSPF specification [36], in general, the contents of an LSA may change when (i) the state of an interface changes, (ii) the DR changes, or (iii) the relationship with a neighboring router changes to/from the `Full` state. We highlight that, when the neighbor state machine is involved, it is the entry into or exit from the `Full` state that dictates the origination of LSAs/LSPs.

Router-LSAs A Router-LSA describes one router and its interfaces. It is created at a router immediately upon the activation of an OSPF process, and must be deleted when the OSPF process is deactivated. The decision on when to add or remove an interface to a Router-LSA depends on the neighbor and interface state machines, and various configuration actions.

In the case of a point-to-point link, a router adds/removes a type 1 link description (point-to-point link) when it reaches/abandons the `Full` state with the corresponding neighbor (and only at this point). This is true for OSPFv2 and OSPFv3. Moreover, in the case of OSPFv2, it adds/removes a type 3 link description (link to stub network) as soon as the interface is switched on/off, and regardless of the neighbor state. Recall from Section 5.7.1 that type 3 link descriptions are not included in OSPFv3 Router-LSAs.

In the case of a shared link (OSPFv2 and OSPFv3), a type 3 link description is added when the interface is switched on and enters the `Waiting` state. In the case of a stub shared link, the interface remains in this state. Otherwise, if the interface is attached to a transit shared link, the router adds a type 2 link description (link to transit network) to its Router-LSA and removes the link 3 description when (i) it is the link DR and becomes fully adjacent to at least one other router or when (ii) it is not the link DR and becomes fully adjacent to the DR. The router replaces the link 2 description by the link 3 description if these conditions cease to hold, and removes the link description if the interface is switched off. It is possible that a router connected to a transit shared link advertises first a Router-LSA with a type 3 link description characterizing the interface and, when the DR gets elected, replaces this description by a type 2 description (originating then a new Router-LSA instance); recall that, in a cold start scenario, the election of the DR takes approximately 40 seconds.

Network-LSAs A Network-LSA describes a transit shared link and its attached routers. It is created by the link DR when it establishes its first full adjacency on the shared link; the DR later updates the LSA when it becomes fully adjacent to new neighbors on the link or when existing adjacencies are terminated. The Network-LSA must be deleted when the originating router ceases having at least one fully adjacent neighbor or when the router loses its DR role. This behavior is the same in OSPFv2 and OSPFv3.

AS-External-LSA AS-External-LSAs describe prefixes external to the OSPF routing domain and are originated by ASBRs. A new AS-External-LSA is originated whenever the ASBR learns a new domain-external prefix (e.g. through the inter-domain routing protocol), and must be removed when the ASBR loses the prefix (e.g. because the inter-domain routing protocol indicated that the prefix no longer exists).

Link-LSAs (OSPFv3) A router originates one Link-LSA for each link it is attached to, if there is at least one other router attached to the link. This LSA advertises the IPv6 link-local address and the IPv6 address prefixes assigned to a link interface (see Section 5.9). Thus, when one of these addresses changes, a new Link-LSA instance is originated. The Link-LSA must be deleted when the router ceases having neighbors at the link.

Intra-Area-Prefix-LSAs (OSPFv3) An Intra-Area-Prefix-LSA describes the IPv6 prefixes associated with a router. These prefixes can be assigned to the router itself or to its directly attached links.

A prefix assigned to a transit shared link is advertised through an Intra-Area-Prefix-LSA originated by the link DR (the LSA is referenced to the Network-LSA that describes the link topologically). However, as pointed out in Section 5.9, these prefixes need not be configured at the DR; they may be configured at other routers attached to the same link and communicated to the DR through Link-LSAs. Thus, each DR examines the Link-LSAs associated with the interface where it is the DR to determine which prefixes must be added/removed to/from its Intra-Area-Prefix-LSA. Specifically, the DR advertises only the prefixes announced in Link-LSAs originated by fully adjacent routers; it also advertises its own prefixes associated with the interface.

Prefixes assigned to the router itself or to other types of links (i.e. point-to-point and stub shared links) are advertised through Intra-Area-Prefix-LSAs referenced to the Router-LSA that represents the router. These prefixes are added to the LSA as soon as they get configured at the router and are deleted when removed from the router.

As pointed out in the OSPFv3 specification [8], a router may need to move a prefix from one Intra-Area-Prefix-LSA to another. For example, this happens when a router is attached to a stub shared link (with an assigned prefix) and the link changes to a transit shared link (because a second router attached to it). In this case, the prefix assigned to the shared link was initially advertised by an Intra-Area-Prefix-LSA referenced to a Router-LSA, and has to be transferred to an Intra-Area-Prefix-LSA referenced to the Network-LSA that represents (topologically) the newly created transit shared link.

Changes in interface costs Note that new Router-LSAs and Intra-Area-Prefix-LSAs (OSPFv3) have to be originated when the cost of an interface is reconfigured at a router.

Changes in the LSA identifier Moreover, any configuration action that involves changing the LSA identifier (e.g. the router RID), causes the deletion of all LSAs previously originated by the router, and the origination of new ones (with the new LSA ID).

6.7.2 IS-IS

IS-IS routers originate only two types of LSPs: Nonpseudonode-LSPs and Pseudonode-LSPs. The IS-IS specification [1] is not so detailed as that of OSPF concerning the origination of LSPs. IS-IS does not have a state equivalent to the `Full` state of OSPF. In origination events that involve neighborhood relationships, what dictates the origination of an LSP is creation or destruction of an adjacency.

According to the IS-IS specification, the events that trigger the origination of an LSP are: (i) an adjacency going up/down, (ii) change of circuit status (L1-only, L2-only, L1/L2), (iii) change of DIS, (iv) change of SID, (v) change of interface cost, and (vi) insertion/removal of an address.

> *The origination of LSAs and LSPs is triggered by various events and actions, such as an adjacency coming up or down, the insertion or removal of a prefix, changing an interface cost, the age expiration of an LSA or LSP, and a change of designated router.*

7

OSPF and IS-IS Hierarchical Networks

The multi-area structures of both OSPF and IS-IS are restricted to a two-level hierarchy, with only one area allowed in the upper level. The lower level areas must attach directly to the upper one, such that all traffic between the lower level areas is forced to cross the upper one. This is illustrated in Figure 7.1. The structure of hierarchical OSPF and IS-IS networks was already introduced in the overview chapter (see Section 4.3 and Figure 4.1). Both OSPF and IS-IS use the distance vector routing (DVR) approach for inter-area routing discussed in Section 3.2.3.

The areas are identified through an *area identifier*, referred to as the Area ID or AID. In both OSPFv2 and OSPFv3, the Area ID is a 32-bit number, unique within the routing domain, and expressed in the dotted-decimal notation used for IPv4 addresses. The Area ID of the backbone is always 0.0.0.0, and that is why the backbone is also called area 0. In IS-IS, the Area ID is part of the complete router identifier, i.e. the NET, as discussed in Section 5.2.1. Unlike OSPF, there is no specific Area ID reserved for the backbone.

As explained in Section 4.3, in OSPF each router interface must be assigned an Area ID, and all interfaces attached to the same link must belong

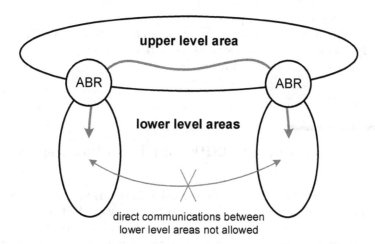

FIGURE 7.1: Communication between lower level areas in OSPF and IS-IS hierarchical networks.

to the same area. ABRs have one interface connected to the backbone, with an assigned Area ID of 0.0.0.0, and one or more interfaces connected to lower level areas, each assigned a non 0.0.0.0 Area ID. Recall from the discussion in Section 6.2.3.2 that, in OSPF, the router interfaces are prevented from establishing neighborhood relationships if they do not belong to the same area.

In IS-IS, the Area IDs are assigned to the whole router and not to individual interfaces. However, more than one Area ID can be configured at a router. To support hierarchical routing, IS-IS uses two types of LSDB (the L1 and L2 LSDBs) and two types of adjacencies (the L1 and L2 adjacencies). L1 and L2 adjacencies only exchange routing information relative to the L1 and L2 LSDBs, respectively. Routers are classified according to the type of adjacencies they support: L1 routers support only L1 adjacencies, L2 routers support only L2 adjacencies, and L1/L2 routers support both types of adjacencies. Moreover, an L1 adjacency can only be established with two routers if they have at least one area in common; L2 adjacencies are not restricted by the configured areas. Accordingly, a hierarchical IS-IS network can be configured as follows (see Figure 4.1):

- The area border routers are L1/L2 routers, since they must support both L1 and L2 adjacencies. Moreover, since routers must be assigned an Area ID, L1/L2 routers must be configured with a lower level Area ID;

- The backbone is implemented through L2 adjacencies;

- The lower level areas are implemented through L1 adjacencies.

Chapter structure In this chapter, we start by comparing, in Section 7.1, the hierarchical structures of OSPF and IS-IS. Then, Section 7.2 shows how distance vector routing applies to the case of hierarchical networks. Section 7.3 introduces the LSDB structure and distance vectors of OSPF and IS-IS. Finally, Section 7.4 discusses several restrictions that may lead to non-optimal path selection in hierarchical networks.

7.1   Comparison of OSPF and IS-IS hierarchical structures

The structures of OSPF and IS-IS hierarchical networks are usually considered very different. This idea seems to be rooted on the way OSPF and IS-IS routers handle the Area IDs: in OSPF the Area IDs are per interface and in IS-IS they are per router. However, as will be seen, this is not a significant difference.

In OSPF, each interface is assigned one Area ID, and ABRs maintain one LSDB for each area they are directly attached to, i.e. for each distinct Area ID.

IS-IS distinguishes between interfaces attached to the backbone and interfaces attached to lower level areas, not through the Area ID as in OSPF, but through the adjacency type: connections to the backbone are made through L2 adjacencies, and connections to lower level areas are made through L1 adjacencies. However, there is still one question remaining: it seems that an L1/L2 router is only capable of attaching to a single lower level area, since it must be assigned an Area ID of a lower level area. This IS-IS restriction was present in the first implementations, but vendors already offer solutions that overcome it, exploring the two features that we describe below.

The first feature comes from the initial IS-IS specification [1], which already supported the possibility of assigning different Area IDs to a single router. The various Area IDs are advertised through the Area Addresses TLV, which we introduced in Section 5.6.4. According to Section 7.1.5 of the specification, this feature was thought of for facilitating the reconfiguration of areas within a domain without disrupting the network operation (see also Section 7.4.8 of [11]). However, its consequences are broader. What this effectively means is that a router can have more than one NET assigned to it, as long as they differ only on the Area ID. Thus, although IS-IS routers are identified by the NET and the NET incorporates the Area ID, leading us to think that a router must belong to a single area, the fact is that routers can be assigned different NETs, one for each area they directly attach to and, therefore, can be seen as belonging to all these areas. Unfortunately, IS-IS literature insists in emphasizing that IS-IS border routers belong to a single area and that this is a marked difference in relation to OSPF. It is not!

The second feature, this one already added by vendors, is the possibility of associating the Area IDs configured at a router with specific interfaces. This feature is sometimes called IS-IS Multiarea Support [2]. In this case, an L1/L2 router maintains a distinct L1 LSDB for each lower level area, i.e. for each distinct Area ID, just as in OSPF. Note that the addition of this feature does not create any interoperability concerns; it is just a matter of local support at routers and does not require the addition of new TLVs. Such a feature seems not in the spirit of the initial IS-IS specification, but the specification certainly does not disallow it.

We may then conclude that the structure of hierarchical OSPF and IS-IS networks is indeed very similar, the main difference being essentially the nomenclature. The area border routers are called ABRs in OSPF and L1/L2 routers in IS-IS. But both store one LSDB per area they directly attach to, and both exchange addressing information between areas. IS-IS names differently the LSDB of the backbone (called L2 LSDB) and that of the lower level areas (called L1 LSDBs), but this is again a naming issue. In both cases, the interfaces of area border routers must be associated with specific areas. In IS-IS, the interface with the backbone is the L2 interface, and in OSPF it is the interface associated with area 0.

> *Both OSPF and IS-IS restrict their multi-area networks to a two-level hierarchy, where the lower level areas can only communicate through the upper one. Moreover, they both use a DVR approach in their inter-area routing protocol. The area border routers are called ABRs in OSPF and L1/L2 routers in IS-IS. In both OSPF and IS-IS, each area border router interface is associated with an area and the border routers maintain as many LSDBs as areas they directly attach to.*

7.2   DVR in hierarchical networks

In this section, we give an example of DVR in the context of hierarchical networks, which applies to both OSPF and IS-IS. This issue was already addressed in Section 3.2.3, in the more general setting of multi-area networks.

Consider the network of Figure 7.2. It includes two lower level areas (areas 1 and 2) and the backbone, with a total of four ABRs: R2 and R3 interconnect the backbone with area 1, and R4 and R5 interconnect the backbone with area 2. Consider the problem of determining how router R1 learns about prefix ap1 and computes the shortest path to this prefix; actually, R1 will only be able to determine the sub-path of area 1, from itself to the outgoing ABR. The figure includes near each interface the interface costs that are relevant for this discussion.

Collection of information in area 2 First, routers R4 and R5 collect information on prefix ap1 using the LSDB of area 2. Based on this LSDB, the ABRs determine the (intra-area) shortest paths from themselves to the prefix. The shortest path cost is 10 (5+5) from R4 to ap1 and is 30 (25+5) from R5 to ap1.

Dissemination and computation at the backbone Each router then injects a distance vector into the backbone advertising the prefix and its shortest path cost to it: R4 advertises (ap1, 10) and R5 advertises (ap1, 30). The DVs are flooded within the backbone, eventually reaching R2 and R3.

Routers R2 and R3 then determine the shortest path costs from each one to ap1 by combining two pieces of information: (i) the path costs included in the distance vectors received from R4 and R5, and (ii) the shortest path costs from each one to R4 and R5. The latter information is obtained from the LSDB of the backbone. The shortest path from R2 to R4 is R2 → R7 → R6 → R4 with a cost of 50, and the one from R2 to R5 is R2 → R7 → R5 with a cost of 20. Using now the information contained in the DVs, R2 determines that the cost to ap1 via R4 is 50+10=60 and the cost via R5 is 30+20=50, and concludes that the shortest path cost it can provide is 50 (via R5). It then injects this information into area 1, using the distance vector (ap1, 50). Router R3 does the same type of computation. The shortest path from R3 to R4 is R3 → R6 → R7 with a cost of 20, and the one from R3 to R5 is R3 →

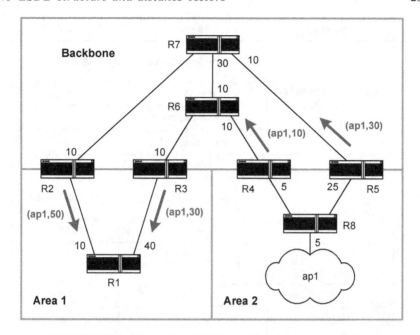

FIGURE 7.2: Example of DVR in hierarchical networks.

R6 → R7 → R5 with a cost of 30. Thus, R3 concludes that shortest path cost it can provide to ap1 is 30 (via R4) and injects this information into area 1, using the distance vector (ap1, 30).

Computation at router R1 Finally, R1 decides on the path to ap1 based on the DVs injected by R2 and R3 and on the information contained in the LSDB of area 1. The path from R1 to ap1 via R2 has cost 10+50=60, and the path via R3 has cost 40+30=70. Thus, R1 determines that its shortest path to ap1 is via R2 with a cost of 60, and inserts this information in its forwarding table.

7.3 LSDB structure and distance vectors

The support of hierarchical routing in OSPF and IS-IS may require the addition of new elements to the LSDB. Moreover, one needs to understand how distance vectors are actually implemented in the two technologies. Figure 7.3 summarizes the correspondence between the LSDB elements of the generic protocol required to disseminate routing information across areas (introduced in Section 3.1.3), and the ones of OSPF and IS-IS.

generic	OSPFv2	OSPFv3
area-external-prefix-NR (aep-NR)	Network-Summary-LSA	Inter-Area-Prefix-LSA
	IPv4 IS-IS	**IPv6 IS-IS**
	IP Internal Reachability Information TLV or Extended IP Reachability TLV	IPv6 Reachability TLV
domain-external-NR (dep-NR)	**OSPFv2**	**OSPFv3**
	AS-External-LSA	AS-External-LSA
	IPv4 IS-IS	**IPv6 IS-IS**
	IP Internal Reachability Information TLV or Extended IP Reachability TLV	IPv6 Reachability TLV
domain-border-router-NR (dbr-NR)	**OSPFv2**	**OSPFv3**
	ASBR-Summary-LSA	Inter-Area-Router-LSA

FIGURE 7.3: Correspondence between the LSDB elements required to disseminate routing information across areas of the generic protocol, OSPF and IS-IS, from a routing information perspective.

7.3.1 OSPF

Area-internal routing information OSPF keeps the LSAs of single-area networks to describe the routing information internal to an area. In OSPFv2, the topological and the area-internal addressing information is provided by Router-LSAs and Network-LSAs. In OSPFv3, the topological information is provided by Router-LSAs and Network-LSAs and the area-internal addressing information by Intra-Area-Prefix-LSAs. Given that the topological information is restricted to describing a single area, the ABRs originate as many Router-LSAs as areas they are directly attached to, but a Router-LSA injected in one area only describes the interfaces belonging to that area. The Router-LSAs also include information on whether the originating router is an ABR or an ASBR, using the B-bit and E-bit flags: the B-bit is set when the originating router is an ABR, and the E-Bit is set when the originating router is an ASBR.

Area-external addressing information In OSPF, the distance vectors that distribute the area-external prefixes across areas (prefixes external to the area but internal to the routing domain), i.e. the equivalent of the aep-NR (see Section 3.1.3), are the Network-Summary-LSA in OSPFv2, and the Inter-

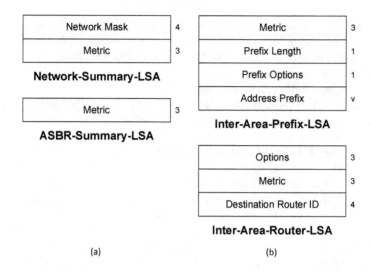

FIGURE 7.4: LSAs specific of hierarchical (a) OSPFv2 and (b) OSPFv3.

Area-Prefix-LSA in OSPFv3 (see Figure 7.4). These LSAs are flooded with area scope and, as mentioned in Section 3.2.3, achieve two goals simultaneously: they communicate the area-external prefixes among ABRs and, at the same time, disseminate the prefixes inside areas, so that they become known by internal routers.

In the Network-Summary-LSAs (OSPFv2), the advertised prefixes are defined by the subnet IPv4 address and subnet mask, carried in the `Link State ID` and `Network Mask` fields, respectively (recall that the `Link State ID` field belongs to the LSA header common to all LSA types). In the Inter-Area-Prefix-LSAs (OSPFv3), the corresponding information is carried in the `Address Prefix` and `Prefix Length` fields, respectively. The `Metric` field of both LSAs is the shortest path cost from the originating ABR to the prefix it advertises. Each LSA can only advertise one prefix. The `Link State ID` field distinguishes among LSAs advertising prefixes originated by the same router. In Network-Summary-LSAs this field carries an IPv4 address, as referred above; in Inter-Area-Prefix-LSAs, it carries a locally generated tag unique for each advertised prefix.

> *In OSPF, the distance vectors that distribute area-external addressing across areas, i.e. the equivalent to the aep-NR, are the Network-Summary-LSA in OSPFv2 and the Inter-Area-Prefix-LSA in OSPFv3. These LSAs are flooded with area scope and are also used to disseminate the routing information inside areas.*

Domain-external addressing information OSPFv2 and OSPFv3 use the mixed approach with dep-NRs and dbr-NRs to advertise domain-external addressing information across areas (see Section 3.4). The equivalent of the dep-

NR is the AS-External-LSA, and the equivalent of the dbr-NR are the ASBR-Summary-LSA in OSPFv2, and the Inter-Area-Router-LSA in OSPFv3.

AS-External-LSAs were already introduced in the context of single-area networks (see Section 5.8.1): they describe domain-external prefixes and are originated by ASBRs. In hierarchical OSPF, these LSAs are flooded with domain scope, i.e. bypassing the ABR overlay, and, as such, bring no information on how to route towards the originating ASBRs.

This problem is solved through the ASBR-Summary-LSAs, in OSPFv2, and the Inter-Area-Router-LSAs, in OSPFv3. These LSAs are originated by ABRs to describe the ASBRs of the areas they directly attach to, and are disseminated as distance vectors through the inter-area routing protocol. Contrarily to AS-External-LSAs, ASBR-Summary-LSAs and Inter-Area-Router-LSAs are flooded with area scope. The AS-External-LSA references the originating ASBR through its RID, included in the `Advertising Router` field; the ASBR-Summary-LSA carries the RID of the originating ASBR in the `Link State ID` field, and the Inter-Area-Router-LSA carries it in the `Destination Router ID` field. The difference in the fields used to carry the RID of the originating ASBR reflects the flooding scope of the messages: AS-External-LSAs use the `Advertising Router` field since they are flooded unmodified throughout the domain; ASBR-Summary-LSAs and Inter-Area-Router-LSAs use another field (the `Link State ID`), since the `Advertising Router` field must be reserved to identify the ABRs that originate these LSAs while they are being disseminated in the ABR overlay. In this solution, ABRs need to know if there are ASBRs connected to the areas they directly attach to. This information is provided by the E-bit of Router-LSAs.

> *In OSPF, the AS-External-LSA is flooded using domain scope, and the information on how to route towards the originating ASBR is provided by the ASBR-Summary-LSA in OSPFv2 and the Inter-Area-Router-LSA in OSPFv3. The last two LSAs are the equivalent of the dbr-NR and are disseminated as distance vectors among the ABRs.*

Origination and deletion of LSAs The distance vectors (i.e. the Summary LSAs of OSPFv2 and the Inter-Area LSAs of OSPFv3) are originated by the ABRs whenever new destinations (prefixes or ASBRs) are added, or the cost information regarding an already advertised destination changes. The advertisements must respect the restrictions described in Section 7.4. Moreover, when a previously advertised destination becomes unreachable, or when its advertisement becomes forbidden due to the restrictions of Section 7.4, the corresponding LSA must be deleted.

In Section 11.1.2 we illustrate the LSDB structure of hierarchical OSPF networks through experiments.

7.3.2   IS-IS

IS-IS does not introduce new LSDBs elements to support hierarchical routing.

Area-internal routing information As in OSPF, IS-IS keeps the TLVs of single-area networks to describe the routing information internal to an area. The topological information is provided by the IS Neighbors TLV or the Extended IS Reachability TLV of Nonpseudonode-LSPs and by the Pseudonode-LSPs. The area-internal addressing information is provided by the IP Internal Reachability Information TLV or the Extended IP Reachability TLV in IPv4 IS-IS and the IPv6 Reachability TLV in IPv6 IS-IS.

The distinction between areas is made through the LSP type: the LSPs that describe lower level areas are of L1 type, and the ones that describe the upper level area are of L2 type. The L1/L2 routers originate LSPs of both types. As in the case of Router-LSAs, the IS Neighbors TLV or Extended IS Reachability TLV of one area only describes the neighbors belonging to that area.

Area-external addressing information IS-IS uses the IP Reachability TLVs of single-area networks as distance vectors, to distribute the area-external prefixes among L1/L2 routers. As mentioned above, these TLVs are the IP Internal Reachability Information TLV or the Extended IP Reachability TLV in IPv4 IS-IS and the IPv6 Reachability TLV in IPv6 IS-IS; in their role of distance vectors, they are the equivalent of the aep-NR introduced in Section 3.1.3. The Metric field of these TLVs is the shortest path cost from the originating L1/L2 router to the advertised prefix. The prefix is defined by the IP Address and Subnet Mask fields in the IP Internal Reachability Information TLV and the Prefix and Prefix Length fields in the Extended IP Reachability and IPv6 Reachability TLVs.

As in the case of OSPF, the IP Reachability TLVs fulfill two roles: they communicate area-external prefixes between L1/L2 routers, and they also disseminate these prefixes inside areas.

The initial specification of IS-IS did not allow the injection of L2 addressing information into the L1 LSDB. Thus, routers from lower level areas had no access to external prefixes, and communication with other areas was only possible through default routes pointing to the L1/L2 routers. If a lower level area had more than one L1/L2 router, then internal routers would use the nearest one in the path cost sense. For that purpose, L1/L2 routers signal their presence by setting the ATT bit in its L1 Nonpseudonode-LSP (see Figure 5.12). This type of routing was blind to the actual location of destinations and, therefore, globally optimal routing might not be achieved. To see this, consider the example of Figure 7.5, where we include the router interface costs. The shortest path from R1 to R4 is via R3, with a cost of 30. However, if area-external information is not injected into the L1 area and R1 uses the closest L1/L2 router, then the selected path is the path via R2, with a cost of 40.

Apparently the reason for not allowing the injection of L2 addressing information into the L1 LSDB was to avoid the possibility of reinjecting this information into the backbone, as there was no way to distinguish among area-internal and area-external prefixes in IP Internal Reachability Information TLVs (see discussion in Section 7.4.6 of [11]). The problem was solved

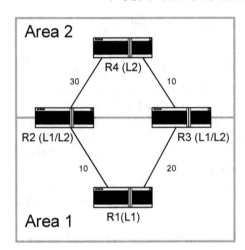

FIGURE 7.5: Consequences of not injecting area-external prefixes in L1 areas
- Example network.

through RFC 2966 [28], which specified that the eight bit of the Default
Metric field in both IP Internal and IP External Reachability Information
TLVs, unused until then, should be used to distinguish among the two types
of prefixes. The bit is called the Up/Down bit (U/D) and is set when the corresponding prefix is injected by an L1/L2 router into a lower level area, i.e.
when it is an area-external prefix. This bit was also included in the IPv6
Reachability TLV. The injection of L2 addressing information into L1 areas
is sometimes referred to as *route leaking*.

> In IS-IS, the distance vectors that distribute the area-external prefixes
> across areas, i.e. the equivalent of the aep-NR, are the IP Internal
> Reachability Information TLV or the Extended IP Reachability TLV
> in IPv4 IS-IS, and the IPv6 Reachability TLV in IPv6 IS-IS. The
> Up/Down bit distinguishes between area-internal or area-external pre-
> fixes inside an L1 area (lower level area).

Domain-external addressing information IS-IS uses the dep-NR only
approach to advertise domain-external addressing information across areas
(see Section 3.4). As in the case of area-external addressing information, IS-IS
uses the TLVs of single-area networks as distance vectors to advertise domain-
external prefixes across areas. Thus, it uses the IP External Reachability Information TLV or the Extended IP Reachability TLV in IPv4 IS-IS and the IPv6
Reachability TLV in IPv6 IS-IS, which are the equivalent of the dep-NR. These
TLVs are originated by DBRs and ABRs and disseminated as distance vectors between areas. The relevant information is included in the IP Address,
Subnet Mask, and Default Metric fields (IP External Reachability Information TLV) or in the Prefix, Prefix Length, and Metric fields (Extended

IP Reachability TLV and IPv6 Reachability TLV). The IPv6 Reachability TLV is also used to advertise domain-internal prefixes; the X flag makes the distinction between domain-internal and domain-external prefixes. Note that the initial specification of IS-IS did not support the presence of ASBRs in L1 areas, since the IP External Reachability Information TLV was not defined for L1 LSPs; this situation would be corrected by RFC 2966 [28].

> *In IS-IS, the domain-external prefixes are advertised across areas through the IP External Reachability Information TLV or the Extended IP Reachability TLV, in IPv4 IS-IS, and the IPv6 Reachability TLV, in IPv6 IS-IS. These TLVs are originated by DBRs and ABRs and disseminated as distance vectors between the L1/L2 routers.*

Origination and deletion of TLVs IS-IS follows rules similar to OSPF for the origination and deletion of distance vectors, which in this case are the IP Reachability TLVs. Thus, a new IP Reachability TLV must be originated when a new prefix is added or the path cost to an existing prefix is changed, subject to the restrictions described in Section 7.4. Moreover, an IP Reachability TLV must be deleted when the corresponding prefix is removed, or when its advertisement becomes forbidden due to the restrictions of Section 7.4. Recall that the origination and deletion of TLVs implies the origination of new instances of the LSPs they belong to.

In Section 11.1.3 we illustrate the LSDB structure of hierarchical IS-IS networks through experiments.

7.4 Restrictions on shortest path selection

Both OSPF and IS-IS impose two restrictions on the way the DVs are advertised that may prevent globally optimal routing. This issue was discussed in Section 3.3. Specifically, the DVs cannot advertise inside an area:

- Paths to area-internal destinations;

- Paths to area-external destinations that cross that area.

Taking the network of Figure 7.6 as an example, the first restriction mentioned above forbids R3 from advertising in area 1 the cost of the path R3 → R4 → R2 → R1 → ap1 even if the intra-area paths R3 → R1 → ap1 or R3 → R2 → R1 → ap1 have higher cost. The second restriction forbids R3 from advertising in area 1 the cost of the path R3 → R2 → R4 → ap2 even if the path R3 → R4 → ap2 has higher cost. As discussed in Section 3.3, the first restriction implies that the path selected between a router and a destination located in the same area might not be the shortest one.

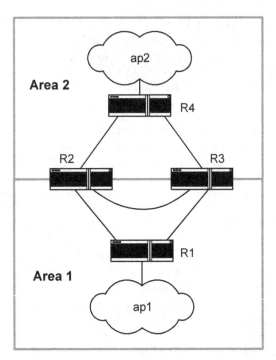

FIGURE 7.6: Restrictions on shortest path selection - Example network.

Another feature that restricts the routing selection process is the preference given to intra-area paths over inter-area paths when building the forwarding table. Taking again the network of Figure 7.6 as an example, this rule forces R3 into selecting the intra-area path R3 → R4 → ap2 even if the inter-area path R3 → R2 → R4 → ap2 has lower cost.

In Section 11.2, we present experiments that illustrate the restrictions on the shortest path selection process of hierarchical OSPF and IS-IS networks.

> *OSPF and IS-IS forbid the injection of distance vectors into an area when they advertise paths to area-internal prefixes or paths to area-external prefixes that cross the area. Moreover, when computing its forwarding table, a router must give preference to intra-area paths over inter-area paths. These restrictions may prevent globally optimal routing from being achieved.*

Part IV

Case studies

8

Tools and Configurations

Experimenting is an important step in acquiring effective skills in computer networking. You cannot become a computer networking specialist just by reading books! Well, this part of the book proposes and discusses a large number of experiments that illustrate the main features of OSPF and IS-IS. The experiments were designed so that they can be easily reproduced by the reader (and some of them are really fun!). They require IP routers implementing single-area and hierarchical OSPF and IS-IS for both IPv4 and IPv6 networks, and a protocol analyzer. However, they can be performed in an *emulated* environment, i.e. using only the operating system of routers and not the real equipment. In our experiments we have used GNS3 as the network emulator, a Cisco IOS image as the router operating system, and Wireshark as the protocol analyzer.

Structure of this part of the book This chapter presents the main features of GNS3, Cisco IOS, and Wireshark, and explains the basic router configurations. The next chapters discuss several types of experiments. Chapter 9 addresses the structure of the forwarding tables, the route selection process, and the LSDB structure of single-area networks. Chapter 10 discusses the synchronization mechanisms. Finally, Chapter 11 addresses the extensions required to support hierarchical networks.

Two words of caution There are no perfect specifications and no perfect implementations of LSR protocols, nor are the tools used to emulate and analyze them perfect (as anything in our lives...). Thus, when running experiments with LSR protocols, you may be faced with unexpected behavior due to unclear specifications, implementation bugs, undocumented vendor specific features, intrinsic hardware and software limitations, and so on. However, most of the time the unexpected behavior arises because you failed to understand that specific bit of the protocol... So, be prepared for the surprises!

   LSR protocols are *asynchronous* protocols. One feature of this type of protocol is that the sequence of events leading to a given outcome may vary. For example, the sequence of LS UPDATE packets transmitted on a shared link in reaction to some router failure depends (i) on the relative timing of the failure detection by the neighbors of the failed router, and (ii) on the load of the routers and links in the various routes between the neighbors and the shared link. The beauty of asynchronous protocols—if they are correct—is that the outcome is always the same irrespective of the sequence of events.

Thus, always be prepared for a different sequence of events when you try to reproduce the experiments of this part of the book.

What is a converged network? We will refer many times to converged networks. A *converged network* is a network in a stable state. In the context of routing protocols, this means that the routing information used to build the forwarding tables is not changing and is consistent across all routers. Following an event such as a router failure or a cold start, there is usually a transient period when the network is adapting to the new conditions and the routing information changes; however, if the protocol is correct, the routing information becomes stable after a while. Thus, network convergence is always the endpoint of LSR experiments that disturb the network stability.

One way to determine if a network has converged in LSR experiments is to observe the control packets transmitted by the routers, e.g. using Wireshark. The conditions differ slightly in OSPF and IS-IS. In OSPF, a converged network is a network where only HELLO packets with correct information are exchanged between routers. The information transmitted by an HELLO packet is correct if it lists all neighbors of the sending router and, on shared links, if it also lists the correct DR and BDR. The IS-IS conditions are the same, except on shared links, where, besides the HELLO packets transmitted by all routers, the DIS also transmits periodic CSNPs. In this case, network convergence is achieved when the routers transmit only correct HELLO packets, and the DISes, besides correct HELLO packets, also transmit CSNPs advertising a fully updated LSDB summary.

8.1   Getting used to GNS3

GNS3 is a network emulator that allows testing network configurations using the actual operating system of networking equipment (e.g. Cisco IOS or Juniper JUNOS) and using Wireshark for packet analysis. In our experiments we have used version 2.1.3 of GNS3.

To perform experiments with Cisco devices you must install first a Cisco IOS image. Other elements such as the links, the Ethernet switches, and the PCs are provided by GNS3.

Central pane Figure 8.1 shows the main window of GNS3. The central pane displays the network you have configured. In this case, the network has five routers (R1 to R5) connected through Fast Ethernet links, except the connection between R4 and R5, which is a serial link. Moreover, router R5 is connected to an Ethernet switch. The interfaces are labeled with their names.

Left pane The network devices can be drag-and-dropped from the left pane, where they are organized in categories. For example, in the first button (Browse Routers) you find the routers, in the second one (Browse Switches)

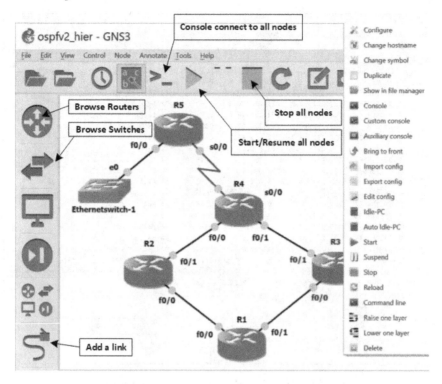

FIGURE 8.1: GNS3 main window.

the switches, and in the last one (**Add a link**) the links. These are the network elements that you need for almost all experiments of this book.

Upper pane The upper pane includes buttons for the main actions performed on a network. For example, to start all devices at the same time click **Start/Resume all nodes**, and to stop them all click **Stop all nodes**. The button **Console connect to all nodes** is also very useful and opens the consoles of all devices at the same time.

Performing actions on individual devices There are many actions that you can perform on individual devices. For that, you must right click the device. Figure 8.1 shows the window that opens when right clicking router R3. You can, for example, **Start**, **Stop**, or **Reload** just this router, you can open its console (**Console** button), configure the router (**Configure** button), or edit its startup-config file (**Edit config** button). In addition, to run a Wireshark capture at a link, right click on the link and select **Start capture**.

These are just the main features of GNS3; try to explore other features on your own. Have fun!

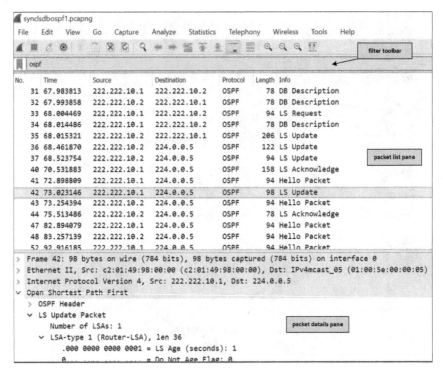

FIGURE 8.2: Wireshark main window.

8.2 Getting used to Wireshark

Wireshark is a protocol analysis tool. It can capture all packets that cross a network interface (in both directions) and decode their complete contents, e.g. it identifies the protocols of the various layers and the source and destination addresses. It is a very useful tool for computer networking specialists. Moreover, Wireshark is fully integrated with GNS3. When emulating a network using GNS3, several Wireshark captures can be run simultaneously, to observe different points of the network. We could not have a better environment to study computer networks! Moreover, Wireshark is easy to install and to work with. In our experiments, we have used version 2.4.4 of Wireshark.

Figure 8.2 shows the main window of Wireshark. In this case, the window shows information on the packets exchanged during an initial LSDB synchronization in OSPF. We highlight three main areas: (i) the filter toolbar, (ii) the packet list pane, and (iii) the packet details pane.

Filter toolbar A filter lets you choose what you want to see. In Wireshark, the filters are specified in the filter toolbar, and there are many predefined filters. For example, the **ospf** keyword, shown in Figure 8.2, lets us see only

OSPF packets. In our experiments, we will use **ospf** or **isis** filters. Wireshark allows us to be more specific regarding a protocol. For example, to see only OSPF HELLO packets you can use the **ospf.hello** filter.

Packet list pane Immediately below the filter toolbar, you find the packet list pane. This pane has one line per observed packet and displays the most important information regarding each packet. It has seven columns: (i) `No.` is the packet number, (ii) `Time` is the capture time, (iii) `Source` is the source address, (iv) `Destination` is the destination address, (v) `Protocol` is the highest-layer protocol decoded by Wireshark, (vi) `Length` is the packet length, and (vii) `Info` includes additional information, such as the packet type.

The packet number and the packet capture time are relative to the first observed packet. The first observed packet is given a number of 1 and a capture time of 0. The source and destination addresses are the IP addresses, unless the packet has no IP layer (which is the case of the IS-IS packets); in this case, the source and destination MAC addresses (if they exist) are displayed.

Packet details pane Below the packet list pane, you find the packet details pane. This pane shows all fields of the packet selected in packet list pane. In the figure, the packet selected is packet 42 (an LS UPDATE packet) and the packet details pane shows its Ethernet and IPv4 headers (not expanded), and shows that the OSPF packet contains a Router-LSA with an LS Age of 1 second (most of the contents is not shown due to lack of space).

How Wireshark captures will be shown in this book In this book we will show the results of many Wireshark captures. However, we will not present them as in Figure 8.2, i.e. as Wireshark snapshots, because we want to save space and add important information not displayed in the packet list pane. What we do is (i) to export each capture to an Excel file, (ii) remove the unwanted columns (we will always remove the `Length` column), and (iii) edit the `Info` column such that it includes the most relevant information for the explanation of the experiment, which we extract from the packet contents.

8.3 Getting used to Cisco IOS

8.3.1 Some IOS background

A router runs an *operating system* to manage its hardware and software resources and provide its services. The functions are similar to the operating systems of our personal computers. The operating system of Cisco routers is called *IOS* (Internetwork Operating System). Cisco IOS can be configured through *commands* to perform different tasks. In our experiments, we have used version 12.4(25d) of a Cisco IOS c3725 image.

Cisco IOS access levels There are two levels of access to Cisco IOS commands (also called *user modes*): the *user EXEC level* and the *privileged EXEC* *level*. In the user EXEC level, you only have access to some commands, the innocuous ones (e.g. **connect**, **login**, **ping**, and **show**). In the privileged EXEC level, you have access to all commands, including the more powerful and potentially destructive ones (e.g. **configure**, **debug**, **erase**, and **setup**). A network administrator will generally work in the privileged EXEC level. The prompt ">"" indicates that you are configuring in user EXEC level and the prompt "#" indicates that you are configuring in the privileged EXEC level. To go from user EXEC level to privileged EXEC level, you must enter the **enable** command, and a password may be requested. In GNS3, routers are started in the privileged EXEC level.

Configuration methods The configuration of a router may be (i) delivered via network download, (ii) copied from an image stored in the router's memory, or (iii) typed in from the terminal. The third method will be used in our experiments, and the command **configure terminal** indicates to Cisco IOS that this method will be used.

Configuration modes Cisco IOS has several *configuration modes*, which indicate the router element (e.g. interface or protocol) that is being configured. The *global configuration* mode is the entry level for the router configurations; the commands entered in this mode affect the entire router. Important configuration modes in the context of the experiments of this book are the *interface mode*, where the interfaces are configured, and the *router mode*, where the IP routing protocols are configured.

Configuration files Every router has two configuration files: the running-config file and the startup-config file. The *startup-config* file contains the commands that are loaded when a router is switched on. The *running-config* file is in RAM and contains the commands that are entered during a configuration session. The **write** command saves the running-config contents into the startup-config file. Use **write** frequently if you don't want to lose your work.

8.3.2   Configuration of interfaces and IP addresses

The first thing to configure at a router is its interfaces and IP addresses. You need to switch on the interfaces and assign each one an IP address and a subnet mask. This may not be strictly required in the case of IPv6 interfaces, which may be configured to obtain an IPv6 link-local address automatically. We will illustrate these configurations on a single router with two interfaces: a Fast Ethernet interface and a serial interface. This setup is shown in Figure 8.3, together with the IPv4 and IPv6 addresses that will be assigned to each interface.

Router and interface names The routers and the interfaces have names under which they are known for configuration purposes. In this case, the router name is **R1** and the interface names are **FastEthernet0/0** and **Serial0/0**. Fortunately, the interface names can be abbreviated, in this case to f0/0 and s0/0, respectively. The name of a router can be changed using the **hostname**

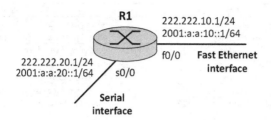

FIGURE 8.3: One router, two of its interfaces, and the IP addresses assigned to them.

command in the global configuration mode, but you will not need to do that in the experiments of this book.

IPv4 addresses and interfaces Figure 8.4.a shows the sequence of Cisco IOS commands that must be entered when configuring the two interfaces. The **configure terminal** command puts Cisco IOS in the global configuration mode; a router configuration session always starts with this command. The **interface f0/0** command places IOS in the interface mode, to configure interface f0/0. The **ip address** command configures the IPv4 address (222.222.10.1) and subnet mask (255.255.0.0) of the interface, and the **no shutdown** command switches on the interface. Commands 5 to 7 repeat the actions of commands 2 to 4 but for interface s0/0. Note that there is no difference in the basic interface configurations of Fast Ethernet and serial interfaces. The **end** command forces IOS to exit the global configuration mode, and the **write** command saves the configuration into the startup-config file.

IPv6 addresses and interfaces The configuration of the IPv6 addresses is shown in Figure 8.4.b. It is exactly the same as the IPv4 addresses, with one exception: the command that configures IPv6 addresses is **ipv6 address**, and the prefix length must be entered in slash notation. We will see later that, in some cases, there is no need to configure IPv6 addresses at interfaces.

Very useful suggestions on how to assign IP addresses When doing experiments, or even in the real life of a network manager, it is most useful to assign IP addresses with some logic that allows associating easily an IP address with the router to which it belongs. This will save us lots of time in configuring and debugging our experiments! In this book we will use the following rules:

- The last octet of an IP address is always the router number. For example, 222.222.20.3 says that this IPv4 address belongs to router R3, and 2001:a:a:10::1 says that this IPv6 address belongs to router R1.

- Except when using private addresses or domain-external addresses, our IPv4 addresses will be of the form 222.222.X.Y/24 where, as explained above, Y equals the router number and X identifies the actual subnet the address

No.	IOS command	Role
1	R1#configure terminal	enter global configuration mode
2	R1(config)#interface f0/0	enter interface configuration mode of f0/0
3	R1(config-if)#ip address 222.222.10.1 255.255.255.0	configure IPv4 address and subnet mask at f0/0 interface
4	R1(config-if)#no shutdown	switch on f0/0 interface
5	R1(config-if)#interface s0/0	enter interface configuration mode of s0/0
6	R1(config-if)#ip address 222.222.20.1 255.255.255.0	configure IPv4 address and subnet mask at s0/0 interface
7	R1(config-if)#no shutdown	switch on s0/0 interface
8	R1(config-if)#end	exit global configuration mode
9	R1#write	save configuration in startup-config file

(a)

No.	IOS command	Role
1	R1#configure terminal	enter global configuration mode
2	R1(config)#interface f0/0	enter interface configuration mode of f0/0
3	R1(config-if)#ipv6 address 2001:a:a:10::1/64	configure IPv6 address and prefix length at f0/0 interface
4	R1(config-if)#no shutdown	switch on f0/0 interface
5	R1(config-if)#interface s0/0	enter interface configuration mode of s0/0
6	R1(config-if)#ipv6 address 2001:a:a:10::1/64	configure IPv6 address and prefix length at s0/0 interface
7	R1(config-if)#no shutdown	switch on s0/0 interface
8	R1(config-if)#end	exit global configuration mode
9	R1#write	save configuration in startup-config file

(b)

FIGURE 8.4: Configuration of interfaces and IP addresses in case of (a) IPv4 and (b) IPv6.

belongs to. We will use multiples of 10 for X; thus, our IPv4 subnets will take the form 222.222.10.0/24, 222.222.20.0/24, and so on.

- Our IPv6 addresses will be of the form 2001:a:a:X::Y/64, where X and Y have the same meaning as in the IPv4 addresses. Moreover, when assigning multiple prefixes to the same link, we will change the second and third hextets (:a:a:) into the next alphabet letter. For example, two prefixes assigned to the same transit shared link can be 2001:a:a:X::/64 and 2001:b:b:X::/64.

- The RIDs that need to be configured in OSPFv2 and OSPFv3 will be of the form X.X.X.X., where X is the router number.

Interface costs One important parameter of an interface is the interface cost, which allows setting up specific routing solutions. In OSPFv2, the interface cost can be configured using the command **ip ospf cost**, in OSPFv3 using the command **ipv6 ospf cost**, and in IS-IS using the command **isis metric**. The default interface costs are always 10 in IS-IS. In OSPF, they are a function of the interface link speed, and are 10 for Fast Ethernet interfaces and 64 for serial interfaces.

```
!R1
configure terminal
interface f0/0
ip address 222.222.10.1 255.255.255.0
no shutdown
interface s0/0
ip address 222.222.20.1 255.255.255.0
no shutdown
end
write
```

FIGURE 8.5: Entering IOS commands in a batch.

8.3.3   Additional IOS features

The show command The **show** command allows visualizing the various types of information that Cisco IOS can provide, and is entered in the EXEC level. Try for yourself the commands **show interfaces**, **show interfaces f0/0**, **show ip interface**, **show ip interface f0/0**, **show ip interface brief**, **show ip interface brief f0/0**, and **show ip route**, to learn the kind of information you can get from them.

Seeing the contents of the running-config and startup-config files You can look at the contents of the running-config and startup-config files using the **show running-config** and **show startup-config** commands. One good exercise is to enter the commands of Figure 8.4 and to try to identify them in the running-config and startup-config files using the corresponding **show** commands.

Editing the startup-config file Sometimes it is easier to make changes directly in the startup-config files. You can edit these files inside GNS3. Just right-click on a router and select `Edit config`; a window from where you can edit the contents of the startup-config file will open.

Entering the commands in a batch You will find this feature very useful! Instead of entering the commands one-by-one at the console, you can write a group of commands in a word processor and copy-and-paste them to the console, all at the same time. The order of the commands must be same as if the commands were entered one-by-one. For example, to configure router R1 as in Figure 8.4.a, write the commands in a word processor (as in Figure 8.5), and copy-and-paste all commands at the same time to the console. The first line is a comment. This feature will save us lots of time!

Overwriting configurations When writing a configuration over another, the older configuration may not be completely replaced; some commands of the older configuration may remain and will need negation to be deleted. For example, when entering a new OSPF configuration, the networks advertised

```
R1(config-if)#ip ospf ?
  <1-65535>           Process ID
  authentication      Enable authentication
  authentication-key  Authentication password (key)
  cost                Interface cost
  database-filter     Filter OSPF LSA during synchronization and flooding
  dead-interval       Interval after which a neighbor is declared dead
  demand-circuit      OSPF demand circuit
  flood-reduction     OSPF Flood Reduction
  hello-interval      Time between HELLO packets
  lls                 Link-local Signaling (LLS) support
  message-digest-key  Message digest authentication password (key)
  mtu-ignore          Ignores the MTU in DBD packets
  network             Network type
  priority            Router priority
  resync-timeout      Interval after which adjacency is reset if oob-resync is
                      not started
  retransmit-interval Time between retransmitting lost link state
                      advertisements
  transmit-delay      Link state transmit delay
```

FIGURE 8.6: Using the help system to learn about the OSPFv2 options available in the interface mode.

by OSPF in the old configuration will remain in the new configuration, and this may cause lots of problems if you don't pay attention. So, be careful!

Negating a command Negating a command is often done by prepending the **no** keyword to the command. For example, removing an IPv4 address is done through the **no ip address** command and switching on an interface through the **no shutdown** command.

Using abbreviations for the IOS commands The Cisco IOS commands can be abbreviated whenever the abbreviation is unambiguous. For example, **configure terminal** can be abbreviated to **conf t**, **interface** to **int**, **no shutdown** to **no shut**, **show interfaces** to **sh int**, and **write** to **wr**.

The help system Cisco IOS has a built-in *context-sensitive help* system. To get help just enter the question mark, and you will get the list of commands available *where you are in IOS*. When a command has several parts, and you don't remember the next part, you can enter "what you know", followed by a space and a question mark, to obtain the various options for the next part. For example, if you are in the interface mode and want to know what are the options available to configure OSPFv2, just type **ip ospf ?** and you will get the result of Figure 8.6. This tells you, for example, that in order to configure the OSPF cost of an interface, you will have to add the keyword **cost** to the command **ip ospf**.

The debug command IOS includes a **debug** command that helps tracing the evolution of events related to the different protocols implemented in IOS.

```
event manager applet DebugOSPF
event syslog pattern "SYS-5-RESTART"
action 1.0 cli command "enable"
action 2.0 cli command "debug ip ospf adj"
action 3.0 syslog msg "Debug OSPF adjacencies"
```

FIGURE 8.7: Debug commands to be inserted in the startup-config file.

For example, you can use the command **debug ip ospf** to trace the events related to OSPF and the command **debug isis** to trace the events related to IS-IS. Both commands have several filtering options. For example, to see only the events related to the Hello protocol of OSPF you must enter the command **debug ip ospf hello**.

Debugging from start Sometimes it is important to start debugging when a router is switched on. You will need this, for example, to study cold start scenarios. In this case, the debug commands cannot be given from the console and must be inserted in the startup-config file. To do this you will have to write, immediately before the keyword **end** of the startup-config file, the set of commands of Figure 8.7, adapted to specific needs. The example is for OSPF. The names **DebugOSPF** (first line) and **Debug OSPF adjacencies** (last line) can be changed. The only other thing that you can change is the **action 2.0** line, which configures the actual debug command. In this example we have included the **debug ip ospf adj** command, which traces events related to OSPF adjacencies.

8.4 Basic routing configuration

In this section, we explain the basic routing configurations of all technologies addressed in this book. The configurations will be illustrated using again the router of Figure 8.3, with one Fast Ethernet interface and one serial interface. We will assume that the interfaces and IP addresses have already been configured, according to what was explained in Section 8.3.2.

Configuring a routing protocol in Cisco IOS always involves creating a routing process. This is performed using the **router** command in the global configuration mode. Depending on the specific technology, the routing configuration may also involve configurations in the interface mode. Note also that a router can run several routing protocols at the same time.

Besides creating the routing process, the basic routing configuration involves configuring the router identifier (the RID), and defining the prefixes that must be advertised by the router. A router can only be configured to advertise prefixes assigned to itself or its interfaces.

In the configuration examples, we will include all commands required to configure routing from scratch, i.e. from entering the global configuration mode until the configuration is saved in the startup-config file. The basic routing configurations of all technologies are shown in Figure 8.8.

8.4.1   OSPFv2

Figure 8.8.a shows the basic routing configuration of OSPFv2. The command **router ospf 1** creates the OSPFv2 routing process and places IOS in the router configuration mode. That the current configuration mode is the router mode can be recognized by the keyword **config-router** in the prompts. In this mode, you can configure the RID and the advertised prefixes. The command **router-id 1.1.1.1** configures the RID as being 1.1.1.1; note that the RID configuration is not mandatory in OSPFv2. The **network** command defines the prefixes advertised by the router and the area they belong to. Since, in this example, there are two interfaces with IPv4 addresses 222.222.10.1/24 and 222.222.20.1/24, respectively (see Section 8.3.2), you need to configure the prefixes 222.222.10.0/24 and 222.222.10.0/24. Note that the subnet mask is entered as the complement of the usual mask. For example, in the present configuration the mask is 255.255.255.0 and you must enter 0.0.0.255.

8.4.2   OSPFv3

Figure 8.8.b shows the basic routing configuration of OSPFv3. The logic is a bit different than that of OSPFv2. The first configuration, i.e. the command **ipv6 unicast-routing**, enables IPv6 routing. This is not needed in IPv4 routing, but is mandatory in IPv6 routing; you will also see this command in the IPv6 IS-IS configuration. Then, the **ipv6 router ospf 1** command creates the OSPFv3 process and places IOS in the router configuration mode. The only configuration needed in this mode is the RID, which is performed exactly as in OSPFv2, i.e. with the command **router-id**. Then, OSPFv3 must be configured at each interface, using the command **ipv6 ospf 1 area 0**. This tells IOS that OSPFv3 must be enabled at the interface and that the interface belongs to area 0. The router will automatically advertise the prefix associated with the interface address (see Section 8.3.2); unlike OSPFv2, there is no need to declare these prefixes in the router configuration mode.

8.4.3   IPv4 IS-IS

Figure 8.8.c shows the basic routing configuration of IPv4 IS-IS. After entering the global configuration mode, the command **router isis** creates the IPv4 IS-IS process and places IOS in the router configuration mode. The command **net 49.0001.0000.0000.0001.00** configures the NET of the router. Recall from Section 5.2.1 that the NET includes both the area identifier and the router identifier (called SID in IS-IS). The area identifier occupies the first

No.	IOS command	Role
1	R1#configure terminal	enter global configuration mode
2	R1(config)#router ospf 1	create OSPFv2 process and enter router configuration mode
3	R1(config-router)# router-id 1.1.1.1	configure router ID
4	R1(config-router)# network 222.222.10.0 0.0.0.255 area 0	define prefix to be advertised by OSPFv2 process and area ID of interface
5	R1(config-router)# network 222.222.20.0 0.0.0.255 area 0	define prefix to be advertised by OSPFv2 process and area ID of interface
6	R1(config-if)#end	exit global configuration mode
7	R1#write	save configuration in startup-config file

(a)

No.	IOS command	Role
1	R1#configure terminal	enter global configuration mode
2	R1(config)#ipv6 unicast-routing	enable IPv6 unicast routing
3	R1(config)#ipv6 router ospf 1	create OSPFv3 process and enter router configuration mode
4	R1(config-rtr)# router-id 1.1.1.1	configure router ID
5	R1(config-rtr)# int f0/0	enter interface configuration mode of f0/0
6	R1(config-if)# ipv6 ospf 1 area 0	enable OSPFv3 process at interface f0/0 and define area ID of interface
7	R1(config-if)# int s0/0	enter interface configuration mode of s0/0
8	R1(config-if)# ipv6 ospf 1 area 0	enable OSPFv3 process at interface s0/0 and define area ID of interface
9	R1(config-if)#end	exit global configuration mode
10	R1#write	save configuration in startup-config file

(b)

No.	IOS command	Role
1	R1#configure terminal	enter global configuration mode
2	R1(config)#router isis	create IPv4 IS-IS process and enter router configuration mode
3	R1(config-router)# net 49.0001.0000.0000.0001.00	configure NET
4	R1(config-router)# is-type level-1	configure router as L1
5	R1(config-router)# int f0/0	enter interface configuration mode of f0/0
6	R1(config-if)# router isis	enable IPv4 IS-IS process at interface f0/0
7	R1(config-if)# int s0/0	enter interface configuration mode of s0/0
8	R1(config-if)# router isis	enable IPv4 IS-IS process at interface s0/0
9	R1(config-if)#end	exit global configuration mode
10	R1#write	save configuration in startup-config file

(c)

No.	IOS command	Role
1	R1#configure terminal	enter global configuration mode
2	R1(config)#ipv6 unicast-routing	enable IPv6 routing
3	R1(config)#ipv6 router isis 1	create IPv6 IS-IS process and enter router configuration mode
4	R1(config-router)# net 49.0001.0000.0000.0001.00	configure NET
5	R1(config-router)# is-type level-1	configure router as L1
6	R1(config-router)# int f0/0	enter interface configuration mode of f0/0
7	R1(config-if)# ipv6 router isis 1	enable IPv6 IS-IS process at interface f0/0
8	R1(config-if)# int s0/0	enter interface configuration mode of s0/0
9	R1(config-if)# ipv6 router isis 1	enable IPv6 IS-IS process at interface s0/0
10	R1(config-if)#end	exit global configuration mode
11	R1#write	save configuration in startup-config file

(d)

FIGURE 8.8: Basic routing configurations of (a) OSPFv2, (b) OSPFv3, (c) IPv4 IS-IS, and (d) IPv6 IS-IS.

three octets, the SID the next six octets, and the SEL the last octet. The SEL part is always zero. The first octet of the area identifier (0×49) is the AFI. The remaining parts indicate that the router belongs to area 1 and has an SID of 1. The next command, **is-type level-1**, configures the router to be an L1 router. Finally, similarly to OSPFv3, IS-IS must be enabled at each interface, and this is the role of the command **router isis**, entered in the interface configuration modes of f0/0 and s0/0.

As explained in Section 4.3.2, there are three types of IS-IS routers: L1 routers, L2 routers, and L1/L2 routers. In L1 routers, interfaces are all of L1-only type; in L2 routers, interfaces are all of L2-only type; and in L1/L2 routers, interfaces can be of any type, i.e. L1-only, L2-only, or L1/L2. The default configuration of IOS is to consider the routers as L1/L2 routers and the interfaces of L1/L2 type. Any departure from this default setting must be explicitly configured. As mentioned above, to turn a router into an L1 router, the command **is-type level-1** must be entered in the router mode. Likewise, to turn a router into an L2 router enter the command **is-type level-2-only** in the same mode. Moreover, to turn an interface into an L1-only or and L2-only interface enter the commands **isis circuit-type level-1** or **isis circuit-type level-2-only** in the interface mode.

8.4.4   IPv6 IS-IS

Figure 8.8.d shows the basic routing configuration of IPv6 IS-IS. As in the case of OSPFv3, the first command (**ipv6 unicast-routing**) enables IPv6 routing. Aside from this, the configuration is very similar to IPv4 IS-IS. The command **ipv6 router isis 1** creates the IPv6 IS-IS process and tells IOS to enter the router configuration mode. The configurations in this mode are the same as those of IPv4 IS-IS. Then, the interfaces must also be configured to enable IPv6 IS-IS, and this is done through the command **ipv6 router isis 1**.

8.5   List of IOS commands

We list here the main Cisco IOS commands that will be used in this part of the book. Figure 8.9 shows the list of generic Cisco IOS commands; Figures 8.10, 8.11, and 8.12 show the configuration commands of OSPFv2, OSPFv3 and IS-IS (IPv4 and IPv6). Finally, Figures 8.13 and 8.14 show the configuration commands of BGP and RIP protocols, which will be used in some experiments.

IOS command (generic)	Mode	Role
ipv6 unicast-routing	global	enables ipv6 forwarding at router
interface *type number*	global	enters interface configuration mode
ip address *ip-address mask*	interface	configure IPv4 address at interface
ipv6 address *ipv6-prefix/prefix-length*	interface	configure IPv6 address at interface
no shutdown	interface	switches on interface
exit		exits from current configuration mode to next highest configuration mode
end		returns to privileged EXEC mode
configure terminal	privileged EXEC	enters global configuration mode
show running-config	privileged EXEC	displays contents of running-config file
show startup-config	privileged EXEC	displays contents of startup-config file
write	privileged EXEC	saves configuration in startup-config file
show ip route	privileged EXEC	displays IPv4 forwarding table
show ipv6 route	privileged EXEC	displays IPv6 forwarding table
show ip interface brief	privileged EXEC	displays summary information on interfaces
show interfaces *type number*	privileged EXEC	displays summary information on specific interface
traceroute *ip-address*	privileged EXEC	executes traceroute

FIGURE 8.9: List of generic commands.

IOS command (OSPFv2)	Mode	Role
router ospf *process-id*	global	creates OSPFv2 process and enters router configuration mode
network *ip-address wildcard-mask* **area** *area-id*	router	defines interface on which OSPFv2 runs and its area ID
router-id *ip-address*	router	defines RID
redistribute BGP *autonomous-system-number*	router	redistributes BGP prefixes into OSPFv2
redistribute BGP *autonomous-system-number* **metric-type** *type-value* **metric** *metric-value*	router	redistributes BGP prefixes into OSPFv2 with specific metric type and metric value
redistribute rip	router	redistributes RIP prefixes into OSPFv2
default-information originate always	router	configures ASBR to generate a default route
ip ospf cost *cost*	interface	sets interface cost; default value depends on interface type
ip ospf hello-interval *seconds*	interface	sets HelloInterval; default value is 10 seconds
ip ospf dead-interval *seconds*	interface	sets RouterDeadInterval; default value is 40 seconds
ip ospf priority *number-value*	interface	sets interface priority; default value is 1
show ip ospf database	privileged EXEC	displays summary information about LSDB
show ip ospf database router	privileged EXEC	displays contents of Router-LSAs
show ip ospf database network	privileged EXEC	displays contents of Network-LSAs
show ip ospf database external	privileged EXEC	displays contents of AS-External-LSAs
show ip ospf neighbor	privileged EXEC	displays information about neighbors
show ip ospf interface brief	privileged EXEC	displays information about interfaces
clear ip ospf process *process-id*	privileged EXEC	clears OSPF process
debug ip ospf adj	privileged EXEC	displays information on adjacency events
debug ip ospf hello	privileged EXEC	displays information on HELLO events

FIGURE 8.10: List of OSPFv2 commands.

IOS command (OSPFv3)	Mode	Role
ipv6 router ospf *process-id*	global	creates OSPFv3 process and enters router configuration mode
router-id *ip-address*	router	defines RID
ipv6 ospf *process-id* **area** *area-id*	interface	defines interface on which OSPFv3 runs and its area ID
ipv6 enable	interface	enables IPv6 on interface with no assigned IPv6 address
ipv6 ospf cost *interface-cost*	interface	sets interface cost; default value depends on interface type
show ipv6 ospf database	privileged EXEC	displays summary information about LSDB
show ipv6 ospf database router	privileged EXEC	displays contents of Router-LSAs
show ipv6 ospf database network	privileged EXEC	displays contents of Network-LSAs
show ipv6 ospf database prefix	privileged EXEC	displays contents of Intra-Area-Prefix-LSAs
show ipv6 ospf database link	privileged EXEC	displays contents of Link-LSAs
show ip ospf database external	privileged EXEC	displays contents of AS-External-LSAs

FIGURE 8.11: List of OSPFv3 commands.

IOS command (IS-IS)	Mode	Role
router isis	global	creates IPv4 IS-IS process and enters router configuration mode
ipv6 router isis *area-name*	global	creates IPv6 IS-IS process and enters router configuration mode
ip router isis	interface	activates IPv4 IS-IS at interface
isis metric *metric*	interface	sets interface cost; default value is 10
isis priority *number-value*	interface	sets interface priority; default value is 64
isis circuit-type level-1	interface	defines interface as L1-only
isis circuit-type level-2-only	interface	defines interface as L2-only
is-type level-2-only	router	defines router as L1 router
is-type level-1	router	defines router as L2 router
net *network-entity-title*	router	assigns NET address to router
metric-style wide	router	configures router to process only new-style LSPs
redistribute isis ip level-2 into level-1 distribute-list *list-number*	router	redistributes L2 prefixes into L1 LSDB
redistribute bgp *autonomous-system-number*	router	redistributes BGP prefixes into IS-IS
show isis database	privileged EXEC	displays summary information about LSDB
show isis database detail	privileged EXEC	displays detailed information about LSDB
show isis neighbors	privileged EXEC	displays information on IS-IS neighbors

FIGURE 8.12: List of IS-IS commands.

IOS command (BGP)	Mode	Role
router bgp *autonomous-system-number*	global	creates BGP process and enters router configuration mode
neighbor *ip-address* **remote-as** *as-number*	router	defines BGP neighbor
redistribute isis level-2	router	redistributes IS-IS prefixes into BGP
redistribute connected	router	redistributes directly attached prefixes
network *ip-address*	router	defines prefix to be advertised by BGP
redistribute ospf *process-id*	router	redistributes OSPFv2 prefixes into BGP

FIGURE 8.13: List of BGP commands.

IOS command (RIP)	Mode	Role
router rip	global	creates RIP process and enters router configuration mode
network *ip-address*	router	defines prefix to be advertised by RIP
redistribute ospf *process-id* **metric** *metric-value*	router	redistributes OSPFv2 prefix into RIP with specific metric

FIGURE 8.14: List of RIP commands.

9

Experiments on Forwarding Tables and LSDB

In LSR protocols, routers maintain an LSDB containing the topological and addressing information required to build their forwarding tables. The LSDB structure of OSPF and IS-IS was introduced in Chapter 5. In this chapter, we first present the forwarding tables and analyze the path selection process in Section 9.1. Then, Section 9.2 illustrates the LSDB structure of OSPF and IS-IS under different scenarios.

9.1 Routes and forwarding tables

Routing protocols build forwarding tables at routers, which indicate how an incoming packet must be forwarded to the next layer-3 network element (router or host). In this section, we analyze the path selection process and the structure of the forwarding tables obtained with IPv4 and IPv6 routing protocols. We will consider the forwarding tables obtained with IPv4 IS-IS, OSPFv2, and OSPFv3. We skip IPv6 IS-IS since its forwarding tables are similar to OSPFv3.

Experimental setup This issue will be illustrated with the help of the network of Figure 9.1. The network has three routers, two serial links, one transit shared link, and one host. We selected this topology to ensure the existence of alternative paths from each router to each subnet. The host is included to make our traceroute experiments more clear. The routers need to be configured with the basic routing configuration (see Section 8.4.3 for the IPv4 IS-IS configuration, Section 8.4.1 for the OSPFv2 configuration, and Section 8.4.2 for the OSPFv3 configuration). The host uses the VPCS element of GNS3. Its configuration is simple: just type the command **ip 222.222.30.100/24 222.222.30.2** in the VPCS console. In the figure "DG" means Default Gateway.

The IPv4 forwarding tables can be visualized with the command **show ip route** and the IPv6 ones with the command **show ipv6 route**. The way forwarding tables are displayed by Cisco IOS varies according to the address family and routing protocol, without much logic among them. It is mostly a

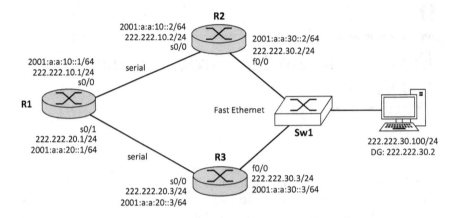

FIGURE 9.1: Forwarding tables experiments - Network topology.

matter of vendor taste! However, forwarding tables have one entry per destination subnet, including at least the subnet identifier (i.e. the prefix to which the entry applies), the IP address of the next-hop interface, and the name of the forwarding interface.

9.1.1   IPv4 forwarding tables obtained with IPv4 IS-IS

In Figure 9.2, we show the forwarding table of router R1 obtained with IPv4 IS-IS. The table has three entries, each corresponding to a destination subnet. The entry relative to 2222.222.30.0/24 has two lines, since there are two equal-cost shortest paths from router R1 to this subnet.

How was the routing information obtained? A forwarding table entry starts with a keyword indicating how the entry was obtained: "C" says the path is to a directly connected link and "i L1" says the path was obtained from the L1 LSDB of IS-IS. These codes are explained in the lines preceding the forwarding table.

Directly connected routes The entries relative to directly connected links include, after the keyword "C", the prefix that identifies the subnet, the text "is directly connected", and the name of the interface leading to the subnet. For example, the first line says that, if a packet arrives at R1 and is destined to 222.222.20.0/24, it should be forwarded through interface Serial0/1 (also known inside the router as s0/1).

IS-IS paths The entries obtained through routing protocols start with the keyword that identifies them (in our case "i L1") and contain four pieces of information. The first one is the prefix to which the entry applies. The second one is enclosed in square brackets and includes first the administrative distance of the protocol and, following the slash, the cost of the shortest path leading to the prefix. The third part includes the next-hop address, preceded by the word

```
R1#show ip route
Codes: C - connected, S - static, R - RIP, M - mobile, B - BGP
       D - EIGRP, EX - EIGRP external, O - OSPF, IA - OSPF inter area
       N1 - OSPF NSSA external type 1, N2 - OSPF NSSA external type 2
       E1 - OSPF external type 1, E2 - OSPF external type 2
       i - IS-IS, su - IS-IS summary, L1 - IS-IS level-1, L2 - IS-IS level-2
       ia - IS-IS inter area, * - candidate default, U - per-user static route
       o - ODR, P - periodic downloaded static route

Gateway of last resort is not set

C    222.222.20.0/24 is directly connected, Serial0/1
C    222.222.10.0/24 is directly connected, Serial0/0
i L1 222.222.30.0/24 [115/20] via 222.222.20.3, Serial0/1
                     [115/20] via 222.222.10.2, Serial0/0
```

FIGURE 9.2: IPv4 forwarding table obtained with IPv4 IS-IS at router R1.

"via". Finally, the fourth part is the name of the forwarding interface. When there is more than one shortest path leading to a destination, the entry has more than one line, and each line describes one alternative path. The initial keyword and the prefix information are only included in the first line. Thus, from our routing table, we learn that the administrative distance of IS-IS is 115, that there are two shortest paths leading to 222.222.30.0/24, both with cost 20, that one of the paths is through router R2 (the one defined by the forwarding interface Serial0/0 and the next-hop address 222.222.10.2), and that the other path is through router R3 (the one defined by the forwarding interface Serial0/1 and the next-hop address 222.222.20.3).

Next-hop addresses Recall that the next-hop address is the IPv4 address of the next interface that receives packets sent by a router. For example, in the third entry, the next-hop address of the first line is 222.222.30.3, since a packet destined to 222.222.30.0/24 and forwarded according to this line should be transmitted through interface Serial0/1 and received by the interface of the next router with IPv4 address 222.222.20.3.

Path cost calculations To understand the path cost value, recall that the path cost is obtained by adding the costs of the transmitting interfaces that belong to the path; recall also that the default interface costs of IS-IS are 10, irrespective of interface type. A packet sent from router R1 to 222.222.30/24 is transmitted by two interfaces, irrespective of the path it takes. If its goes via R2, it is transmitted through the interface s0/0 of R1 (which costs 10) and through interface f0/0 of R2 (which also costs 10). Thus, the cost of this path is $10 + 10 = 20$. The same reasoning applies to the other path. Note that the costs of the receiving interfaces are not included in the path cost calculation. Thus, in this example, the cost of the s0/0 interface of R2 and of

```
R1#traceroute 222.222.30.100 probe 6

Type escape sequence to abort.
Tracing the route to 222.222.30.100

 1 222.222.20.3 0 msec
   222.222.10.2 0 msec
   222.222.20.3 0 msec
   222.222.10.2 0 msec
   222.222.20.3 0 msec
   222.222.10.2 4 msec
 2 222.222.30.100 4 msec 8 msec 0 msec 12 msec 16 msec 8 msec
```

FIGURE 9.3: IPv4 forwarding table obtained with IPv4 IS-IS - Result of traceroute from R1 to host.

the s0/0 interface of R3, which receive packets from R1, are not included in the calculations of the path costs from R1 to 222.222.30.0/24.

Splitting traffic among equal-cost paths When there are multiple shortest paths leading to the same destination, the traffic is equally split among these paths. In our case, since there are two paths leading to 222.222.30.0/34, 50% of the traffic goes through the first path (via R2), and the other 50% goes through the second one (via R3). This can be checked using the **traceroute** command. Figure 9.3 shows the result of **traceroute 222.222.30.100 probe 6** performed at R1; the last part (**probe 6**) instructs Cisco IOS to repeat the traceroute six times before proceeding to the next router in the path (i.e. it sends six ICMP Echo Requests with a given TTL value). The figure shows that the traceroute alternates between going through R3 (the first one) and going through R2 (the second one).

9.1.2   IPv4 forwarding tables obtained with OSPFv2

In Figure 9.4, we show the forwarding table of router R1 obtained with OSPFv2. There are only three differences regarding the table obtained with IPv4 IS-IS. First, the paths learned through OSPF are identified by the keyword "O". Second, the path cost of the OSPF entries is now 74 (and not 20, as in IPv4 IS-IS). This is because, in OSPF, the default interface cost is 64 for serial links and 10 for Fast Ethernet links.

Finally, the entries relative OSPF include a time, placed between the next-hop address and the name of the forwarding interface. This is the time elapsed since the entry was last refreshed which, in the case of Figure 9.4, is 6 minutes and 13 seconds for both OSPF entries. Can you tell how much longer you have to wait until these entries are refreshed (i.e. the time is reset to zero)?

While the routing protocol is converging... The forwarding tables do not converge immediately to their final state. Routing protocols take some

```
R1#show ip route
Codes: C - connected, S - static, R - RIP, M - mobile, B - BGP
       D - EIGRP, EX - EIGRP external, O - OSPF, IA - OSPF inter area
       N1 - OSPF NSSA external type 1, N2 - OSPF NSSA external type 2
       E1 - OSPF external type 1, E2 - OSPF external type 2
       i - IS-IS, su - IS-IS summary, L1 - IS-IS level-1, L2 - IS-IS level-2
       ia - IS-IS inter area, * - candidate default, U - per-user static route
       o - ODR, P - periodic downloaded static route

Gateway of last resort is not set

C    222.222.20.0/24 is directly connected, Serial0/1
C    222.222.10.0/24 is directly connected, Serial0/0
O    222.222.30.0/24 [110/74] via 222.222.20.3, 00:06:13, Serial0/1
                      [110/74] via 222.222.10.2, 00:06:13, Serial0/0
```

FIGURE 9.4: IPv4 forwarding table obtained with OSPFv2 at router R1.

time to do their job! In the meantime, while a protocol is converging, the forwarding table may have missing, or even wrong, information. To observe this, restart the routers all at the same time (a cold start situation) and keep displaying the table of router R2 by successively entering the **show ip route** command. The results of this experiment are shown in Figure 9.5. We know what the final result should be regarding the path from R2 to 222.222.20.0/24: it should be via router R3, since the path cost via this router is 74, and the one via router R1 is 128 (because it includes two serial link interfaces, each with cost 64). However, the initial forwarding table (see Figure 9.5.a) indicates that the path is via R1 (the next-hop address is 222.222.10.1 and the forwarding interface is s0/0) and the path cost is 128. Thus, the path is still not the shortest one. If you keep entering **show ip route**, you will notice that it takes approximately 40 seconds until this OSPF entry is replaced by the correct one. The corresponding forwarding table is shown in Figure 9.5.b; it indicates that the path from R2 to 222.222.20.0/24 is through router R3 (the next hop address is 222.222.30.3 and the forwarding interface is f0/0) and has a cost of 74. Actually, this happens because, in OSPF, the convergence over point-point links is much faster than over shared links. In shared links, routers have to elect the DR and BDR before establishing adjacencies among themselves, and this takes approximately 40 seconds.

9.1.3 IPv6 forwarding tables

In Figure 9.6, we show the forwarding table of router R1 obtained with OSPFv3. This table seems more complex than its IPv4 counterpart, but it is not. The entries span more than one line, and this is what catches the eye; they occupy two lines if there is a single shortest path to the corresponding destination. All entries follow the same structure and include the following

(a)
O	222.222.20.0/24 [110/128] via 222.222.10.1, 00:00:04, Serial0/0
C	222.222.10.0/24 is directly connected, Serial0/0
C	222.222.30.0/24 is directly connected, FastEthernet0/0

(b)
O	222.222.20.0/24 [110/74] via 222.222.30.3, 00:00:02, FastEthernet0/0
C	222.222.10.0/24 is directly connected, Serial0/0
C	222.222.30.0/24 is directly connected, FastEthernet0/0

FIGURE 9.5: IPv4 forwarding tables obtained with OSPFv2 at router R2 - (a) Initial table, (b) table after network convergence.

parts: (i) keyword identifying how the routing information was obtained, (ii) prefix definition, (iii) administrative distance and path cost enclosed in square brackets, (iv) next-hop address preceded by "via" keyword, and (v) name of forwarding interface. The table includes local information, identified by the keyword "L", e.g. identifying the global IPv6 addresses assigned to the router interfaces, but this is less interesting information.

Directly connected routes Since the network topology is the same as in the IPv4 experiments, the paths are also the same. R1 is directly connected to the links with prefixes 2001:a:a:10::/64 and 2001:a:a:20::/64, and these paths are identified by the keyword "C". The first prefix is reached through

```
R1#show ipv6 route
IPv6 Routing Table - 7 entries
Codes: C - Connected, L - Local, S - Static, R - RIP, B - BGP
       U - Per-user Static route
       I1 - ISIS L1, I2 - ISIS L2, IA - ISIS interarea, IS - ISIS summary
       O - OSPF intra, OI - OSPF inter, OE1 - OSPF ext 1, OE2 - OSPF ext 2
       ON1 - OSPF NSSA ext 1, ON2 - OSPF NSSA ext 2
C  2001:A:A:10::/64 [0/0]
   via ::, Serial0/0
L  2001:A:A:10::1/128 [0/0]
   via ::, Serial0/0
C  2001:A:A:20::/64 [0/0]
   via ::, Serial0/1
L  2001:A:A:20::1/128 [0/0]
   via ::, Serial0/1
O  2001:A:A:30::/64 [110/74]
   via FE80::C002:34FF:FE0C:0, Serial0/0
   via FE80::C003:1EFF:FE6C:0, Serial0/1
L  FE80::/10 [0/0]
   via ::, Null0
L  FF00::/8 [0/0]
   via ::, Null0
```

FIGURE 9.6: IPv6 forwarding table obtained with OSPFv3 at router R1.

interface Serial0/0 (s0/0 for short), and the second one through interface Serial0/1 (s0/1 for short). Unlike the IPv4 tables, the directly connected routes now include the administrative distance and path cost information, showing that this type of path has an administrative distance of zero. The text "via::" means no next-hop address.

OSPFv3 paths The keyword "O" indicates that the corresponding entry was learned through OSPFv3. As in the IPv4 case, there are two equal-cost shortest paths from R1 to the transit shared link, which is now assigned prefix `2001:a:a:30::/64`.

One important feature of OSPFv3 is that the next-hop addresses are IPv6 link-local addresses. Actually, R1 learned these addresses from the Link-LSAs originated by R2 and R3 on the serial links. The first path to `2001:a:a:30::/64` is via router R2, having `fe80::c002:34ff:fe0c:0` as the next-hop address and Serial0/0 as the forwarding interface; the second path is via router R3, having `fe80::c002:34ff:fe6c:0` as the next-hop address and Serial0/1 as the forwarding interface.

9.1.4 Changing the paths

One of the most important tasks faced by a network manager is the definition of the paths of the various traffic flows crossing its network. This can be done through proper setting of the interface costs of OSPF or IS-IS. In previous sections, there were two equal-cost shortest paths from R1 to 222.222.30.0/24. Suppose we want to ensure that the traffic is completely routed through R2. To do so, the path cost through router R3 must be made higher than the one through R2. This can be implemented by increasing the cost of interface s0/1 of R1 or of interface f0/0 of R3; it may also be implemented by decreasing the cost of interface s0/0 of R1 or of interface f0/0 of R2. Suppose we increase the cost of interface s0/1 of R1 to 40. This can be done through the command **isis metric 40**, entered in the s0/1 interface mode. The resulting forwarding table is shown in Figure 9.7.a. There is now a single line relative to 222.222.30.0/24, indicating that the next hop is 222.222.10.2 (router R2) and the forwarding interface is s0/0. The traceroute from R1 to 222.222.30.100 (Figure 9.7.b) confirms that traffic is all routed through R2, since the first response always comes from 222.222.10.2.

9.1.5 Prevalence of directly connected routes

In Section 1.5 we discussed the issue of directly connected routes. These paths are preferred against the paths obtained through any routing protocol, due to their lower administrative distance. The example of the previous section confirms this behavior. Recall that we have increased the cost of interface s0/1 of R1 to 40. In this case, the path cost from R1 to 222.222.20.0/24 becomes shorter through R2 (i.e via interface s0/0). In this path, packets are transmitted through three interfaces (s0/0 of R1, f0/0 of R2, and s0/0 of R3)

```
     C   222.222.20.0/24 is directly connected, Serial0/1
(a)  C   222.222.10.0/24 is directly connected, Serial0/0
     i L1 222.222.30.0/24 [115/20] via 222.222.10.2, Serial0/0
```

```
     R1#traceroute 222.222.30.100 probe 6

     Type escape sequence to abort.
(b)  Tracing the route to 222.222.30.100

       1 222.222.10.2 0 msec 4 msec 0 msec 0 msec 0 msec 0 msec
       2 222.222.30.100 4 msec 12 msec 8 msec 12 msec 4 msec 16 msec
```

FIGURE 9.7: Changing the paths - (a) Forwarding table of R1 and (b) traceroute from R1 to host when path is through R2.

and, therefore, the path cost is 30. However, the forwarding table of Figure 9.7 indicates that the path to 222.222.20.0/24 is the direct one, through interface s0/1, which has cost 40.

9.2   LSDB structure of single-area networks

In this section, we analyze the LSDB structure of single-area networks. This issue was discussed in Section 2.2 and Chapter 5. We start by presenting, in Section 9.2.1, the LSDB structure of single-area networks not connected to other routing domains, i.e. without domain-external addressing information. Then, in Section 9.2.2, we show the LSDB structure of networks connected to other ASes and, in Section 9.2.3, we discuss how the LSDB accommodates default routes. Section 9.2.4 illustrates the OSPF E-type metrics. Finally, Section 9.2.5 shows the LSDB structure of OSPF networks connected to a RIP domain of the same AS.

9.2.1   LSDB structure without domain-external addressing information

This section discusses the LSDB structure of OSPFv2, OSPFv3, IPv4 IS-IS, and IPv6 IS-IS, when domain-external addressing information is not present. All experiments, except the last two, will be based on the network topology of Figure 9.8. This topology was selected since it has a small number of routers and contains all types of links: point-to-point links, transit shared links, and stub shared links.

Experimental procedure The experimental procedure is simple. For each technology, start by configuring the routers with the basic routing configura-

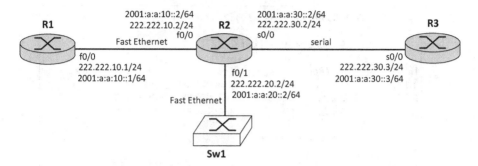

FIGURE 9.8: Experiments related to the LSDB structure of single-area networks, without domain-external addressing information - Network topology.

tion. Then, switch on the routers at the same time and wait until the network converges. The results of these experiments are discussed in the next sections.

9.2.1.1 OSPFv2

A summary of the OSPFv2 LSDB information is shown in Figure 9.9. It can be visualized with the command **show ip ospf database**. The structure of the OSPFv2 LSDB was discussed in Section 5.5.

Summary of OSPFv2 LSDB The LSDB summary displayed by Cisco IOS includes, for each LSA, information on the `Link State ID` (`Link ID` column), `Advertising Router` (`ADV Router` column), `LS Age` (`Age` column), `LS Sequence Number` (`Seq#` column), and `LS Checksum` fields (`Checksum` column); these fields all belong to the LSA header (see Figure 5.9). In the case of Router-LSAs, it also includes the `Number of Links` field (`Link count` column).

The display is organized in two parts, the first related to the Router-LSAs (`Router Link States`) and the second one with the Network-LSAs (`Net Link States`).

There are four different LSAs: three Router-LSAs, each describing one router and its interfaces, and one Network-LSA, describing the transit shared link, i.e. the one with prefix 222.222.10.0/24. Notice that the stub shared link is not represented by a Network-LSA.

The age of these LSAs indicate that the routers were switched on a few minutes ago (judging from the Router-LSA of R3 a bit more than 2 minutes ago). Notice that the Router-LSAs of R1 and R2 have an SN higher than the initial value (which is 0×80000001). This is because new Router-LSA instances are created whenever a new neighborhood relationship is created, and this happens several times when a router is switched on.

The Router-LSAs have equal values in the `Link State ID` (`Link ID`) and `Advertising Router` (`ADV Router`) fields. This is a feature of Router-LSAs; since a router can only originate one Router-LSA, the `Link State ID` is not

OSPF Router with ID (1.1.1.1) (Process ID 1)					
Router Link States (Area 0)					
Link ID	ADV Router	Age	Seq#	Checksum	Link count
1.1.1.1	1.1.1.1	83	0x80000003	0x00523B	1
2.2.2.2	2.2.2.2	84	0x80000004	0x0052E0	4
3.3.3.3	3.3.3.3	128	0x80000001	0x00C9FF	2
Net Link States (Area 0)					
Link ID	ADV Router	Age	Seq#	Checksum	
222.222.10.2	2.2.2.2	84	0x80000001	0x0093C8	

FIGURE 9.9: LSDB without domain-external addressing information - Summary of OSPFv2 LSDB.

needed to uniquely identify the LSA in the LSDB, and ends up having the same value of `Advertising Router`.

Notice that, in the Router-LSAs of R2 and R3, the `Number of Links` (`Link Count`) is one plus the number of interfaces these routers have. This is due to the point-to-point links, which require two link descriptions per interface. We will address this feature next.

Router-LSAs The three Router-LSAs are shown in Figure 9.10. They can be displayed through the command **show ip ospf database router**.

Router-LSA of R2 The Router-LSA of R2 includes all types of interfaces: to point-to-point links, to transit shared links, and to stub shared links. The LSA has four link descriptions. The first two refer to the point-to-point link interface. The first description (`Link connected to: another Router`) is a type 1 description, which provides topological information, and the second (`Link connected to: a Stub Network`) is a type 3 description, which provides addressing information. Cisco IOS names the link descriptions differently from the specification, which may be confusing. From the first description, we learn that this point-to-point link connects router R2 to router R3 (since the `Link ID` is 3.3.3.3) using an interface with IPv4 address 222.222.30.2 (`Link Data`), which has a cost of 64 (`Metric`). From the second description, we learn that the point-to-point link is assigned the prefix 222.222.30.0/24 (`Link ID` and `Link Data`). The cost information included in this type 3 description is irrelevant, since it was already provided in the type 1 description. We know that the first two descriptions are associated with each other (i.e. that the prefix 222.222.30.0/24 of the second description is assigned to the point-to-point link of the first description) since the router interface address of the first description (222.222.30.2) is contained in the prefix range defined by 222.222.30.0/24. Note that the same is not true for the prefix range of the third link description (222.222.20.0/24).

LS age: 139
Options: (No TOS-capability, DC)
LS Type: Router Links
Link State ID: 2.2.2.2
Advertising Router: 2.2.2.2
LS Seq Number: 80000004
Checksum: 0x52E0
Length: 72
Number of Links: 4

Link connected to: another Router (point-to-point)
(Link ID) Neighboring Router ID: 3.3.3.3
(Link Data) Router Interface address: 222.222.30.2
Number of TOS metrics: 0
TOS 0 Metrics: 64

Link connected to: a Stub Network
(Link ID) Network/subnet number: 222.222.30.0
(Link Data) Network Mask: 255.255.255.0
Number of TOS metrics: 0
TOS 0 Metrics: 64

Link connected to: a Stub Network
(Link ID) Network/subnet number: 222.222.20.0
(Link Data) Network Mask: 255.255.255.0
Number of TOS metrics: 0
TOS 0 Metrics: 10

Link connected to: a Transit Network
(Link ID) Designated Router address: 222.222.10.2
(Link Data) Router Interface address: 222.222.10.2
Number of TOS metrics: 0
TOS 0 Metrics: 10

LS age: 116
Options: (No TOS-capability, DC)
LS Type: Router Links
Link State ID: 1.1.1.1
Advertising Router: 1.1.1.1
LS Seq Number: 80000003
Checksum: 0x523B
Length: 36
Number of Links: 1

Link connected to: a Transit Network
(Link ID) Designated Router address: 222.222.10.2
(Link Data) Router Interface address: 222.222.10.1
Number of TOS metrics: 0
TOS 0 Metrics: 10

LS age: 154
Options: (No TOS-capability, DC)
LS Type: Router Links
Link State ID: 3.3.3.3
Advertising Router: 3.3.3.3
LS Seq Number: 80000001
Checksum: 0xC9FF
Length: 48
Number of Links: 2

Link connected to: another Router (point-to-point)
(Link ID) Neighboring Router ID: 2.2.2.2
(Link Data) Router Interface address: 222.222.30.3
Number of TOS metrics: 0
TOS 0 Metrics: 64

Link connected to: a Stub Network
(Link ID) Network/subnet number: 222.222.30.0
(Link Data) Network Mask: 255.255.255.0
Number of TOS metrics: 0
TOS 0 Metrics: 64

FIGURE 9.10: LSDB without domain-external addressing information -
OSPFv2 Router-LSAs.

The third link description (Link connected to: a Stub Network) is
again a type 3 link description, indicating that the prefix assigned to the
stub shared link is 222.222.20/24 (Link ID and Link Data), and that the
cost of the interface attaching the router with the link is 10 (Metric).

Finally, the fourth link description (Link connected to: a Transit
Network) is a type 2 description that characterizes the interface with the transit shared link. From this description, we learn that router R2 is connected
to a link identified by 222.222.10.2 (Link ID), through its interface with IPv4
address 222.222.10.2 (Link Data), that has a cost of 10 (Metric). The fact
that the link identifier and the interface address are the same indicates that
the router itself is the DR of the transit shared link. Recall that in OSPFv2 a

transit shared link is identified by the address of the DR interface attaching
to the link.

Router-LSA of R1 The Router-LSA of R1 shows that R1 has only one
interface, which is attached to a transit shared link. The LSA has a single link description of type 2 (`Link connected to: a Transit Network`),
telling that router R1 is connected to a link identified by 222.222.10.2 (`Link
ID`), through its interface with IPv4 address 222.222.10.1 (`Link Data`) and
cost 10 (`Metric`). The link identifier is the same as in the fourth link description of the Router-LSA of R2, since the two interfaces referred by these
descriptions are attached to the same transit shared link.

Router-LSA of R3 Router R3 is just attached to a point-to-point link.
Thus, its Router-LSA has two link descriptions, one conveying topological information and the other addressing information. The first link description is
of type 1 (`Link connected to: another Router`), and tells that the neighbor of R3 on the link is R2 (since `Link ID` is 2.2.2.2), the IPv4 address of
the interface attaching to the link is 222.222.30.3, (`Link Data`), and the interface cost is 64 (`Metric`). The second description is of type 3 (`Link connected
to: a Stub Network`), and tells that the point-to-point link is assigned prefix
222.222.30.0/24 (`Link ID` and `Link Data`). Notice that the prefix information
is exactly the same as in the Router-LSA of R2 (second link description). This
shows that a prefix assigned to a point-to-point link is advertised by its two
end routers.

Network-LSA Figure 9.11 shows the Network-LSA that describes the shared
link. It can be obtained through the command **show ip ospf database network**. This LSA provides both topological and addressing information. From
the `Link State ID` field we learn that the link identifier is 222.222.10.2, and
from the `Attached Router` fields we learn that routers R1 (1.1.1.1) and R2
(2.2.2.2) are attached to the link; this is topological information. The prefix
assigned to the link, i.e. the addressing information, is obtained by applying
the `Network Mask` field (/24) to the IP address contained in the `Link State
ID` field, which results in 222.222.10.0/24.

9.2.1.2 OSPFv3

Summary of OSPFv3 LSDB Figure 9.12 shows a summary of the OSPFv3
LSDB information. It can be visualized with the command **show ipv6 ospf
database**. The structure of the OSPFv3 LSDB was discussed in Sections 5.7
and 5.9.

In relation to OSPFv2 (see Figure 9.9) there are now many more LSAs.
The display has four parts, one for each type of LSA: `Router Link States`
for Router-LSAs, `Net Link States` for Network-LSAs, `Link (Type-8) Link
States` for Link-LSAs and `Intra Area Prefix Link States` for Intra-Area-
Prefix-LSAs. The information displayed by Cisco IOS varies according to the
type of LSA, but always includes the `Advertising Router` (`ADV Router` col-

```
Routing Bit Set on this LSA
LS age: 251
Options: (No TOS-capability, DC)
LS Type: Network Links
Link State ID: 222.222.10.2 (address of Designated Router)
Advertising Router: 2.2.2.2
LS Seq Number: 80000001
Checksum: 0x93C8
Length: 32
Network Mask: /24
    Attached Router: 2.2.2.2
    Attached Router: 1.1.1.1
```

FIGURE 9.11: LSDB without domain-external addressing information - OSPFv2 Network-LSA.

```
                Router Link States (Area 0)

ADV Router   Age    Seq#           Fragment ID   Link count   Bits
1.1.1.1      60     0x80000005     0             1            None
2.2.2.2      65     0x80000005     0             2            None
3.3.3.3      101    0x80000002     0             1            None


                Net Link States (Area 0)

ADV Router   Age    Seq#           Link ID       Rtr count
2.2.2.2      65     0x80000001     4             2


                Link (Type-8) Link States (Area 0)

ADV Router   Age    Seq#           Link ID       Interface
2.2.2.2      104    0x80000002     5             Fa0/1
2.2.2.2      104    0x80000002     6             Se0/0
3.3.3.3      103    0x80000002     6             Se0/0
1.1.1.1      104    0x80000002     4             Fa0/0
2.2.2.2      104    0x80000002     4             Fa0/0


                Intra Area Prefix Link States (Area 0)

ADV Router   Age    Seq#           Link ID       Ref-lstype   Ref-LSID
2.2.2.2      65     0x80000002     0             0x2001       0
2.2.2.2      69     0x80000001     4096          0x2002       4
3.3.3.3      107    0x80000001     0             0x2001       0
```

FIGURE 9.12: LSDB without domain-external addressing information - Summary of OSPFv3 LSDB.

umn), the LS Age (Age column), and the LS Sequence Number (Seq# column) fields.

The Router-LSAs and the Network-LSA are similar to OSPFv2, except that now they only provide topological information. From the Link count column (Number of Links field) one concludes that, in all routers and con-

trarily to OSPFv2, the number of link descriptions coincides with the number
of interfaces. The Network-LSA summary includes the `Link State ID` (`Link
ID` column), since it is a piece of the link identifier.

The addressing information is provided by the Intra-Area-Prefix-LSAs.
The Intra-Area-Prefix-LSA summary includes also the `Link State ID` (`Link
ID` column), the `Referenced LSA Type` (`Ref-lstype` column), and the
`Referenced Link State ID` (`Ref-LSID` column) fields. From the information
on these LSAs, we learn that the LSDB has three Intra-Area-Prefix-LSAs, two
of them referring to a Router-LSA (the ones where `Referenced LSA Type` is
0×2001) and one referring to a Network-LSA (with `Referenced LSA Type`
0×2002). We also learn that Router R2 originates two LSAs of this type, discriminated by the `Link State ID` field, which is 0 in one LSA and 4096 in
the other.

The Link-LSAs provide the link information and are also an add-on relative to OSPFv2. The Link-LSA summary information includes also the `Link
State ID` (`Link ID` column) field, which contains the number of the interface
to which the LSA is associated, and the interface designation in the `Interface`
column. We learn that there are five Link-LSAs, which coincide with the total
number of interfaces seen by router R2. R2 originates three LSAs of this type,
and R1 and R3 originated only one each.

Router-LSAs Figure 9.13 shows the Router-LSAs of routers R1 and R2. This
information can be visualized using the command **show ipv6 ospf database
router**. These LSAs are simpler than the corresponding ones in OSPFv2 (see
Figure 9.10) because they only convey topological information.

In the Router-LSA of R2 there are only two link descriptions. The first one
(`Link connected to: another Router`) is a type 1 description that characterizes the interface with the point-to-point link. From this description, we
learn that (i) the point-to-point link connects this router with router R3 (since
`Neighbor Router ID` is 3.3.3.3), (ii) the neighbor attaches to the link through
its interface number 6 (since `Neighbor Interface ID` is 6), (iii) this router
attaches to the link through its interface number 6 (since `Local Interface
ID` is 6), and (iv) the cost of this interface is 64 (since `Metric` is 64). Note that
the `Interface ID` is a local number and needs only be unique inside each
router.

The second link description (`Link connected to: a Transit Network`)
is a type 2 description that characterizes the interface with the transit shared
link. The link identifier is placed in `Neighbor Router ID` and `Neighbor
Interface ID` fields. The former carries the RID of the link DR, and the
latter the number of the interface attaching the DR to the link. From this
description, we learn that (i) the link identifier is 2.2.2.2 (`Neighbor Router
ID`) concatenated with 4 (`Neighbor Interface ID`), (ii) this router attaches
to the link through its interface number 4 (`Local Interface ID`), and (iii)
the cost of this interface is 10 (`Metric`). Thus, we conclude that this router is
precisely the link DR, since `Neighbor Router ID` is the same as `Advertising
Router`, and `Neighbor Interface ID` is the same as `Local Interface ID`.

```
LS age: 120
Options: (V6-Bit E-Bit R-bit DC-Bit)
LS Type: Router Links
Link State ID: 0
Advertising Router: 2.2.2.2
LS Seq Number: 80000005
Checksum: 0x85EA
Length: 56
Number of Links: 2

  Link connected to: another Router (point-to-point)
   Link Metric: 64
   Local Interface ID: 6
   Neighbor Interface ID: 6
   Neighbor Router ID: 3.3.3.3

  Link connected to: a Transit Network
   Link Metric: 10
   Local Interface ID: 4
   Neighbor (DR) Interface ID: 4
   Neighbor (DR) Router ID: 2.2.2.2
```

```
LS age: 115
Options: (V6-Bit E-Bit R-bit DC-Bit)
LS Type: Router Links
Link State ID: 0
Advertising Router: 1.1.1.1
LS Seq Number: 80000005
Checksum: 0x805D
Length: 40
Number of Links: 1

  Link connected to: a Transit Network
   Link Metric: 10
   Local Interface ID: 4
   Neighbor (DR) Interface ID: 4
   Neighbor (DR) Router ID: 2.2.2.2
```

```
LS age: 159
Options: (V6-Bit E-Bit R-bit DC-Bit)
LS Type: Router Links
Link State ID: 0
Advertising Router: 3.3.3.3
LS Seq Number: 80000002
Checksum: 0xA1FD
Length: 40
Number of Links: 1

  Link connected to: another Router (point-to-point)
   Link Metric: 64
   Local Interface ID: 6
   Neighbor Interface ID: 6
   Neighbor Router ID: 2.2.2.2
```

FIGURE 9.13: LSDB without domain-external addressing information - OSPFv3 Router-LSAs.

Router R1 has only one interface, which attaches the transit shared link. Thus, its Router-LSA has only one link description, of type 2 (`Link connected to: a Transit Network`), similar to the second link description of the Router-LSA of R2. The `Neighbor Router ID` and `Neighbor Interface ID` fields carry again the link identifier (2.2.2.2 concatenated with 4). We also learn that the router attaches to the link through its interface number 4 (`Local Interface ID`), and the cost of the interface is 10 (`Metric`).

Router R3 just interfaces the point-to-point link. Thus, its Router-LSA has

```
LS age: 287
Options: (V6-Bit E-Bit R-bit DC-Bit)
LS Type: Network Links
Link State ID: 4 (Interface ID of Designated Router)
Advertising Router: 2.2.2.2
LS Seq Number: 80000001
Checksum: 0x3ED
Length: 32
     Attached Router: 2.2.2.2
     Attached Router: 1.1.1.1
```

FIGURE 9.14: LSDB without domain-external addressing information -
OSPFv3 Network-LSA.

only one link description, of type 1 (`Link connected to: another Router`),
similar to the first link description of the Router-LSA of R2. It tells that the
neighbor on the other side of the link is R2 (`Neighbor Router ID` is 2.2.2.2)
and attaches to the link through its interface number 6 (`Neighbor Interface
ID` is 6), and that R3 attaches to the link through its interface number 6 (`Local
Interface ID` is 6), with a cost of 64 (`Metric` is 64). It is just a coincidence
that the local and neighbor interfaces have the same number.

Note that there are no IPv6 addresses in these LSAs. As mentioned above,
OSPFv3 Router-LSAs only convey topological information, and this information resorts only to topological identifiers, which, in this case, are IPv4
addresses. The `Link State ID` field gets a value of zero since, as in the case
of OSPFv2, it has no role in Router-LSAs.

Network-LSA In Figure 9.14, we show the Network-LSA that describes
the transit shared link. This information can be visualized using the command **show ipv6 ospf database network**. This LSA is very similar to its
OSPFv2 counterpart (see Figure 9.11). The LSA is originated by router R2
(`Advertising Router`), since R2 is the link DR. The link identifier is defined
by the `Advertising Router` and `Link State ID` fields. Thus, we learn that
the link identifier is 2.2.2.2 concatenated with 4. The association between this
LSA and the interfaces attached to the link it represents is made through the
link identifier. As noted above, the link identifier is included in the Router-
LSA of R2, namely in the type 2 link description that characterizes the shared
link interface. From the `Attached Router` fields, we learn that the routers attached to the link are R1 (1.1.1.1) and R2 (2.2.2.2).

As in the case of Router-LSAs, there are no IPv6 addresses in this LSA
since it only conveys topological information. This is also why, in relation to
the OSPFv2 Network-LSA, the network mask has been suppressed.

Intra-Area-Prefix-LSA Figure 9.15 shows the Intra-Area-Prefix-LSAs. This
information can be visualized using the command **show ipv6 ospf database
prefix**. Router R2 originates two LSAs of this type, and router R3 originates

```
Routing Bit Set on this LSA
LS age: 230
LS Type: Intra-Area-Prefix-LSA
Link State ID: 0
Advertising Router: 2.2.2.2
LS Seq Number: 80000002
Checksum: 0x98CA
Length: 56
Referenced LSA Type: 2001
Referenced Link State ID: 0
Referenced Advertising Router: 2.2.2.2
Number of Prefixes: 2
Prefix Address: 2001:A:A:20::
Prefix Length: 64, Options: None, Metric: 10
Prefix Address: 2001:A:A:30::
Prefix Length: 64, Options: None, Metric: 64


LS age: 271
LS Type: Intra-Area-Prefix-LSA
Link State ID: 0
Advertising Router: 3.3.3.3
LS Seq Number: 80000001
Checksum: 0xCB3D
Length: 44
Referenced LSA Type: 2001
Referenced Link State ID: 0
Referenced Advertising Router: 3.3.3.3
Number of Prefixes: 1
Prefix Address: 2001:A:A:30::
Prefix Length: 64, Options: None, Metric: 64
```

```
Routing Bit Set on this LSA
LS age: 230
LS Type: Intra-Area-Prefix-LSA
Link State ID: 4096
Advertising Router: 2.2.2.2
LS Seq Number: 80000001
Checksum: 0x6DEE
Length: 44
Referenced LSA Type: 2002
Referenced Link State ID: 4
Referenced Advertising Router: 2.2.2.2
Number of Prefixes: 1
Prefix Address: 2001:A:A:10::
Prefix Length: 64, Options: None, Metric: 0
```

FIGURE 9.15: LSDB without domain-external addressing information - OSPFv3 Intra-Area-Prefix-LSAs.

one. Router R2 needs to originate two Intra-Area-Prefix-LSAs because it advertises prefixes that are related to different topological LSAs.

The first Intra-Area-Prefix-LSA originated by R2 points to the Router-LSA of R2. We know that because the contents of **Referenced LSA Type**, **Referenced Link State ID**, and **Referenced Advertising Router** equal the **LS Type**, **Link State ID** and **Advertising Router** fields of the Router-LSA. This Intra-Area-Prefix-LSA advertises two prefixes, as indicated in the **Number of Prefixes** field: the prefix assigned to the point-to-point link (2001:a:a:30::/64) and the one assigned to the stub shared link (2001:a:a:20::/64); this is done using the **Address Prefix** and **Prefix Length** fields. The **Metric** field includes the cost of the interface attaching to the link to which the prefix was assigned, i.e. 64 for the point-to-point link and 10 for the shared link. The former is unnecessary since it is also advertised in the Router-LSA of R2.

The prefix assigned to the point-to-point link (2001:a:a:30::/64) is also advertised through the Intra-Area-Prefix-LSA originated by the other link end router, i.e. router R3. Recall that, as in OSPFv2, a prefix assigned to a point-to-point link is advertised by its two end routers. This LSA points to the Router-LSA originated by R3. Again, this can be checked through the contents of the pointer fields (`Referenced LSA Type`, `Referenced Link State ID` and `Referenced Advertising Router`). The `Metric` value is again 64, since the interface from R3 to the point-to-point link has again cost 64.

Finally, the prefix assigned to the transit shared link (`2001:a:a:10::/64`) is advertised through an Intra-Area-Prefix-LSA originated by the link DR, which is router R2. This LSA points to the Network-LSA that describes the link, and this can again be checked through the contents of the pointer fields. Now, `Referenced LSA Type` equals the `LS Type` of a Network-LSA, (0×2002) and `Referenced Link State ID` and `Referenced Advertising Router` contain the shared link identifier (2.2.2.2 concatenated with 4). The `Metric` field equals zero and is of no use here.

Link-LSAs of R2 Figure 9.16 shows the Link-LSAs of router R2. This information can be visualized using the command **show ipv6 ospf database link**. The Link-LSAs provide link information and may differ across routers, since they have only link scope (i.e. they are not flooded outside the link). However, since in our example network, R2 is connected to all links, its LSDB includes all Link-LSAs. There is one Link-LSA per interface, giving a total of five LSAs. The interface is identified through the `Advertising Router` and `Link State ID` fields; the latter contains the interface number. Each LSA contains the link local address (`Link Local Address`) and the global prefixes assigned to the interface, defined in the `Prefix Address` and `Prefix Length` fields, as in the case of Intra-Area-Prefix-LSAs. Notice, for example, that router R2 originates three `Link LSAs`, one associated with interface number 5 (which advertises `2001:a:a:20/64`), another with interface 6 (which advertises `2001:a:a:30/64`), and a third one with interface 4 (which advertises `2001:a:a:10/64`).

9.2.1.3   OSPFv3 specific features

In this experiment, we illustrate three important features of OSPFv3 related to the separation between topological and addressing information: (i) the possibility of assigning more than one subnet to a link, (ii) the possibility of assigning routable prefixes only to some links, and (iii) the possibility of declaring a prefix at just one of the routers attached to the link.

Experimental procedure In these experiments, we use the network topology of Figure 9.8. However, as shown in Figure 9.17, we (i) remove prefix `2001:a:a:10::/64` from interface f0/0 of R2 (so that only R1 has this prefix configured), (ii) remove prefix `2001:a:a:30::/64` from interfaces s0/0 of both R2 and R3 (so that the serial link has no routable address assigned to it), and (iii) add prefix `2001:b:b:20::/64` to interface f0/1 of R2 (so that the stub shared link has two prefixes assigned to it). Note that the interfaces

```
LS age: 374
Options: (V6-Bit E-Bit R-bit DC-Bit)
LS Type: Link-LSA (Interface: FastEthernet0/1)
Link State ID: 5 (Interface ID)
Advertising Router: 2.2.2.2
LS Seq Number: 80000002
Checksum: 0xC30E
Length: 56
Router Priority: 1
Link Local Address: FE80::C002:39FF:FE18:1
Number of Prefixes: 1
Prefix Address: 2001:A:A:20::
Prefix Length: 64, Options: None


LS age: 374
Options: (V6-Bit E-Bit R-bit DC-Bit)
LS Type: Link-LSA (Interface: Serial0/0)
Link State ID: 6 (Interface ID)
Advertising Router: 2.2.2.2
LS Seq Number: 80000002
Checksum: 0x6BB
Length: 56
Router Priority: 1
Link Local Address: FE80::C002:39FF:FE18:0
Number of Prefixes: 1
Prefix Address: 2001:A:A:30::
Prefix Length: 64, Options: None


LS age: 374
Options: (V6-Bit E-Bit R-bit DC-Bit)
LS Type: Link-LSA (Interface: Serial0/0)
Link State ID: 6 (Interface ID)
Advertising Router: 3.3.3.3
LS Seq Number: 80000002
Checksum: 0xBD34
Length: 56
Router Priority: 1
Link Local Address: FE80::C003:3FFF:FEDC:0
Number of Prefixes: 1
Prefix Address: 2001:A:A:30::
Prefix Length: 64, Options: None
```

```
LS age: 377
Options: (V6-Bit E-Bit R-bit DC-Bit)
LS Type: Link-LSA (Interface: FastEthernet0/0)
Link State ID: 4 (Interface ID)
Advertising Router: 1.1.1.1
LS Seq Number: 80000002
Checksum: 0xF99A
Length: 56
Router Priority: 1
Link Local Address: FE80::C001:39FF:FE6C:0
Number of Prefixes: 1
Prefix Address: 2001:A:A:10::
Prefix Length: 64, Options: None


LS age: 378
Options: (V6-Bit E-Bit R-bit DC-Bit)
LS Type: Link-LSA (Interface: FastEthernet0/0)
Link State ID: 4 (Interface ID)
Advertising Router: 2.2.2.2
LS Seq Number: 80000002
Checksum: 0x558E
Length: 56
Router Priority: 1
Link Local Address: FE80::C002:39FF:FE18:0
Number of Prefixes: 1
Prefix Address: 2001:A:A:10::
Prefix Length: 64, Options: None
```

FIGURE 9.16: LSDB without domain-external addressing information - OSPFv3 Link-LSAs of router R2.

with no prefix assigned should now include the **ipv6 enable** command, so that the interface keeps generating a link-local address. Figure 9.18 shows the configuration of the three routers for this experiment.

Forwarding table of R3 We first show the IPv6 forwarding table of R3 (Figure 9.19). All three prefixes are present in the routing table, including the two prefixes superposed on the stub shared link, even though no prefix is

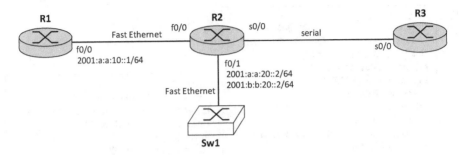

FIGURE 9.17: OSPFv3 specific features experiment - Network topology.

assigned to the serial link. The next-hop address of all entries is the link-local address of interface s0/0 of R2, i.e. `fe80::c002:39ff:fe18:0`. This shows that, in OSPFv3, the paths can be completely determined from the topological information provided by the Router-LSAs and Network-LSAs, and the link-local addresses provided by the Link-LSAs; recall that these addresses can be generated automatically without the network manager intervention. No

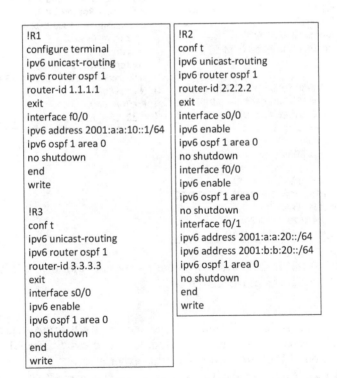

```
!R1                                    !R2
configure terminal                     conf t
ipv6 unicast-routing                   ipv6 unicast-routing
ipv6 router ospf 1                     ipv6 router ospf 1
router-id 1.1.1.1                      router-id 2.2.2.2
exit                                   exit
interface f0/0                         interface s0/0
ipv6 address 2001:a:a:10::1/64         ipv6 enable
ipv6 ospf 1 area 0                     ipv6 ospf 1 area 0
no shutdown                            no shutdown
end                                    interface f0/0
write                                  ipv6 enable
                                       ipv6 ospf 1 area 0
!R3                                    no shutdown
conf t                                 interface f0/1
ipv6 unicast-routing                   ipv6 address 2001:a:a:20::/64
ipv6 router ospf 1                     ipv6 address 2001:b:b:20::/64
router-id 3.3.3.3                      ipv6 ospf 1 area 0
exit                                   no shutdown
interface s0/0                         end
ipv6 enable                            write
ipv6 ospf 1 area 0
no shutdown
end
write
```

FIGURE 9.18: Router configurations - OSPFv3 specific features experiment.

```
O   2001:A:A:10::/64 [110/74]
        via FE80::C002:39FF:FE18:0, Serial0/0
O   2001:A:A:20::/64 [110/74]
        via FE80::C002:39FF:FE18:0, Serial0/0
O   2001:B:B:20::/64 [110/74]
        via FE80::C002:39FF:FE18:0, Serial0/0
```

FIGURE 9.19: OSPFv3 specific features experiment - Forwarding table of router R3.

routable prefixes are needed for this purpose. We only need to know what the destinations in our network are, i.e. what routable prefixes have been assigned to the network and to which network elements. This is precisely the information provided by the Intra-Area-Prefix-LSAs. So, routable prefixes have been decoupled from the network topology, which is great!

Intra-Area-Prefix-LSAs The Intra-Area-Prefix-LSAs are shown in Figure 9.20. There are only two LSAs and both are originated by R2. The first LSA is associated with the Router-LSA of R2 and advertises the two prefixes assigned to the stub shared link. The second LSA is associated with the Network-LSA that describes the transit shared link. It is originated by R2 since R2 is the link DR. However, note that the prefix was not configured at R2 but at R1; it was communicated to R2 through the Link-LSA originated by R1.

Link-LSAs of transit shared link The Link-LSAs of the transit shared link are shown in Figure 9.21. The Link-LSA originated by R1 includes the routable prefix 2001:a:a:10::/64 configured at this router. It is through this LSA that R2 learns about the prefix. The Link-LSA originated by R2 has no

```
Routing Bit Set on this LSA
LS age: 157
LS Type: Intra-Area-Prefix-LSA
Link State ID: 0
Advertising Router: 2.2.2.2
LS Seq Number: 80000002
Checksum: 0x2285
Length: 56
Referenced LSA Type: 2001
Referenced Link State ID: 0
Referenced Advertising Router: 2.2.2.2
Number of Prefixes: 2
Prefix Address: 2001:A:A:20::
Prefix Length: 64, Options: None, Metric: 10
Prefix Address: 2001:B:B:20::
Prefix Length: 64, Options: None, Metric: 10
```

```
Routing Bit Set on this LSA
LS age: 128
LS Type: Intra-Area-Prefix-LSA
Link State ID: 4096
Advertising Router: 2.2.2.2
LS Seq Number: 80000001
Checksum: 0x6DEE
Length: 44
Referenced LSA Type: 2002
Referenced Link State ID: 4
Referenced Advertising Router: 2.2.2.2
Number of Prefixes: 1
Prefix Address: 2001:A:A:10::
Prefix Length: 64, Options: None, Metric: 0
```

FIGURE 9.20: OSPFv3 specific features experiment - Intra-Area-Prefix-LSA.

LS age: 232	LS age: 237
Options: (V6-Bit E-Bit R-bit DC-Bit)	Options: (V6-Bit E-Bit R-bit DC-Bit)
LS Type: Link-LSA (Interface: FastEthernet0/0)	LS Type: Link-LSA (Interface: FastEthernet0/0)
Link State ID: 4 (Interface ID)	Link State ID: 4 (Interface ID)
Advertising Router: 1.1.1.1	Advertising Router: 2.2.2.2
LS Seq Number: 80000002	LS Seq Number: 80000001
Checksum: 0xF99A	Checksum: 0x393E
Length: 56	Length: 44
Router Priority: 1	Router Priority: 1
Link Local Address: FE80::C001:39FF:FE6C:0	Link Local Address: FE80::C002:39FF:FE18:0
Number of Prefixes: 1	Number of Prefixes: 0
Prefix Address: 2001:A:A:10::	
Prefix Length: 64, Options: None	

FIGURE 9.21: OSPFv3 specific features experiment - Link-LSAs of transit shared link.

routable prefix, since none has been configured at the interface. So, routable prefixes have been decoupled from router interfaces, which is really great!

9.2.1.4   IS-IS for IPv4

Summary of IPv4 IS-IS LSDB A summary of the IPv4 IS-IS LSDB information is shown in Figure 9.22. It can be visualized with the command **show isis database**. The structure of the IS-IS LSDB was discussed in Section 5.6.

The information displayed by Cisco IOS includes the LSP ID (LSPID column), the **Sequence Number** (LSP Seq Num column), the **Checksum** (LSP Checksum column), the **Remaining Lifetime** (LSP Holdtime column) fields, and the contents of the bits ATT, P, and OL (ATT/P/OL column).

First, note that Cisco IOS associates the SID of the router with the router name, due to the presence of the Dynamic Hostname TLV (see Section 5.6.4). For example, the LSP ID of the Nonpseudonode-LSP originated by R1 is displayed as R1.00-00, and not as 000000000001.00-00.

The LSDB includes four LSPs. There are three Nonpseudonode-LSPs (R1.00-00, R2.00-00, R3.00-00), each describing one router, and one Pseudonode-LSP (R2.01-00), describing the transit shared link. The Pseudonode-LSPs are characterized by having non-zero Pseudonode ID, which

LSPID	LSP Seq Num	LSP Checksum	LSP Holdtime	ATT/P/OL
R1.00-00	0x00000002	0x2E4F	1041	0/0/0
R2.00-00	* 0x00000003	0x47B1	1044	0/0/0
R2.01-00	* 0x00000001	0x7CD6	1044	0/0/0
R3.00-00	0x00000002	0xF55A	1040	0/0/0

FIGURE 9.22: LSDB without domain-external addressing information - Summary of IPv4 IS-IS LSDB.

LSPID	LSP Seq Num	LSP Checksum	LSP Holdtime	ATT/P/OL
R1.00-00	**0x00000002**	**0x2E4F**	**1003**	**0/0/0**
Area Address: 49.0001				
NLPID: 0xCC				
Hostname: R1				
IP Address: 222.222.10.1				
Metric: 10 IP 222.222.10.0 255.255.255.0				
Metric: 10 IS R2.01				
R2.00-00 * 0x00000003		**0x47B1**	**1007**	**0/0/0**
Area Address: 49.0001				
NLPID: 0xCC				
Hostname: R2				
IP Address: 222.222.20.2				
Metric: 10 IP 222.222.10.0 255.255.255.0				
Metric: 10 IP 222.222.30.0 255.255.255.0				
Metric: 10 IP 222.222.20.0 255.255.255.0				
Metric: 10 IS R2.01				
Metric: 10 IS R3.00				
R2.01-00 * 0x00000001		**0x7CD6**	**1004**	**0/0/0**
Metric: 0 IS R2.00				
Metric: 0 IS R1.00				
R3.00-00	**0x00000002**	**0xF55A**	**999**	**0/0/0**
Area Address: 49.0001				
NLPID: 0xCC				
Hostname: R3				
IP Address: 222.222.30.3				
Metric: 10 IP 222.222.30.0 255.255.255.0				
Metric: 10 IS R2.00				

FIGURE 9.23: LSDB without domain-external addressing information - Detailed information of IPv4 IS-IS LSDB.

is the penultimate octet of the LSP ID. From this information, we also learn that the DIS of the transit shared link is router R2, since the SID of the Pseudonode-LSP (R2.01-00) is precisely R2. Note that the stub shared link is not represented by a Pseudonode-LSP, similarly to OSPF where it is not represented by a Network-LSA.

Detailed IPv4 IS-IS LSDB Figure 9.23 shows detailed information of the IPv4 IS-IS LSDB. It can be visualized with the command **show isis database detail**. The information is not easy to read since some lines correspond to a complete TLV and others to part of a TLV. The first line related to each LSP (shown with boldface) includes the same contents as the summary information of Figure 9.22 (i.e. LSPID, LSP Seq Num, LSP Checksum, LSP Holdtime, and ATT/P/OL).

Nonpseudonode-LSP of R2 Let us concentrate on the Nonpseudonode-LSP of R2 (i.e. the one with LSP ID R2.00-00). The second line refers to the Area Addresses TLV; it displays the Area ID configured in the router, which is 0×490001. The third line refers to the Protocols Supported TLV and indicates that the network layer protocol associated with the LSP is IPv4, since the NLPID (Network Layer Protocol Identifier) equals 0×cc. The fourth line refers to the Dynamic Hostname TLV, which associates the SID of the originating router with its name; the name placed in this TLV is the name given by Cisco IOS to the router, which is R2. The fifth line refers to the IP Interface Address TLV and includes the IPv4 address assigned to one of the router interfaces.

The next three lines correspond to the three IP Internal Reachability Information TLVs that provide the addressing information. The keyword used to identify these TLVs is "IP". Each TLV advertises a prefix assigned to a router interface. The prefixes are identified in the IP Address and Subnet Mask fields. The TLVs also include the interface costs in the Default Metric field (displayed as Metric), which are all 10 in this case.

The topological information is provided by the IS Neighbors TLV, displayed in the ninth and tenth lines. The keyword used to identify these TLVs is "IS". The neighbors are identified using their Node IDs (SID concatenated with Pseudonode ID). The entries indicate that the router is connected to R2.01, i.e. to the transit shared link, and to R3.00, i.e. to router R3 via a point-to-point link. The cost of the interfaces with these links is also included (in the Default Metric field).

Pseudonode-LSP The Pseudonode-LSP (R2.01-00) only includes the IS Neighbors TLV. This TLV indicates that the transit shared link is connected to routers R2 (with Node ID R2.00) and R1 (with Node ID R1.00). The Default Metric field has value zero in both entries, but this information is useless.

Note that the Pseudonode-LSP has no addressing information. In fact, this type of LSP is not allowed to carry IP Internal Reachability Information TLVs. Because of that, the prefix assigned to the transit shared link (222.222.10.0/24) must be advertised both by R1 (in the R1.00-00 LSP) and R2 (in the R2.00-00 LSP). This feature contrasts with OSPF, where prefixes assigned to transit shared links are only advertised by the link DR, using a Network-LSA (OSPFv2) or an Intra-Area-Prefix-LSA (OSPFv3).

Nonpseudonode-LSPs of R1 and R3 The Nonpseudonode-LSPs of R1 and R3 are simpler than the one of R2. The IS Neighbors TLV of R1 indicates that the router is connected to the transit shared link (R2.01-00) and, as mentioned above, the IP Internal Reachability Information TLV advertises the prefix assigned to this link. The IS Neighbors TLV of R3 indicates that the router is connected to R2 through a point-to-point link (R2.00-00) and the IP Internal Reachability Information TLV advertises the prefix assigned to this link (222.222.30.0/24). Note that this prefix is advertised by the two link end routers (R2 and R3). This feature is similar to OSPF.

LSPID	LSP Seq Num	LSP Checksum	LSP Holdtime	ATT/P/OL
R1.00-00	0x00000003	0xC55C	1053	0/0/0
Area Address: 49.0001				
NLPID: 0x8E				
Hostname: R1				
IPv6 Address: 2001:A:A:10::1				
Metric: 10 IPv6 2001:A:A:10::/64				
Metric: 10 IS R2.01				
R2.00-00	0x00000004	0x4DBB	1059	0/0/0
Area Address: 49.0001				
NLPID: 0x8E				
Hostname: R2				
IPv6 Address: 2001:A:A:20::2				
Metric: 10 IPv6 2001:A:A:10::/64				
Metric: 10 IPv6 2001:A:A:30::/64				
Metric: 10 IPv6 2001:A:A:20::/64				
Metric: 10 IS R2.01				
Metric: 10 IS R3.00				
R2.01-00	0x00000001	0x7CD6	1054	0/0/0
Metric: 0 IS R2.00				
Metric: 0 IS R1.00				
R3.00-00	* 0x00000003	0xDCFF	1055	0/0/0
Area Address: 49.0001				
NLPID: 0x8E				
Hostname: R3				
IPv6 Address: 2001:A:A:30::3				
Metric: 10 IPv6 2001:A:A:30::/64				
Metric: 10 IS R2.00				

FIGURE 9.24: LSDB without domain-external addressing information - Detailed information of IPv6 IS-IS LSDB.

9.2.1.5 IS-IS for IPv6

Detailed IPv6 IS-IS LSDB The detailed information of the IPv6 IS-IS LSDB is shown in Figure 9.24. It can be visualized with the command **show isis database detail**, as in the case of IPv4 IS-IS. The structure of the IPv6 IS-IS LSDB was discussed in Section 5.6.

This LSDB is almost the same as the IP4 IS-IS LSDB (see Figure 9.23). The code of the Protocols Supported TLV (NLPID) changed to $0 \times 8e$, since the network layer protocol is now IPv6. The LSPs now include an IPv6 Interface Address TLV, which has the same role of the IP Interface address TLV: it advertises an IPv6 address assigned to the router (which cannot be an IPv6 link-local address). The topological information, which is provided by the IS Neighbors TLVs, is exactly the same, since the network topology has not changed. The most significant change is in the addressing information. The prefixes are now described by IPv6 Reachability TLVs, using the `Prefix` and `Prefix Length` fields, and are identified in the Cisco IOS display by the keyword "IPv6". These TLVs also include the cost information in the `Metric`

LSPID	LSP Seq Num	LSP Checksum	LSP Holdtime	ATT/P/OL
R1.00-00	0x00000002	0xFD80	940	0/0/0
R2.00-00	* 0x00000002	0xB942	941	0/0/0
R3.00-00	0x00000002	0xF55A	940	0/0/0

FIGURE 9.25: Point-to-point adjacencies on shared links - Summary of IPv4 IS-IS LSDB when the Fast Ethernet connecting routers R1 and R2 is configured as a point-to-point link.

field. Note that the criterion for advertising prefixes is the same as in IPv4 IS-IS. In particular, a prefix assigned to a link (point-to-point or shared) is advertised by all routers attached to the link through their Nonpseudonode-LSPs. For example, router R1 advertises 2001:a:a:10::/64 despite not being the link DIS.

9.2.1.6 Point-to-point adjacencies on shared links

IS-IS includes a very interesting feature, which is the possibility of suppressing pseudonodes (Pseudonode-LSPs) from the network representation when the shared links they represent have only two attached routers. When this is done, the shared link behaves like a point-to-point link, i.e. no DIS is elected for the link and CSNPs are not periodically transmitted on the link.

In fact, this feature highlights that link types (point-to-point versus shared) are just abstractions for the purpose of the network representation and need not be tied to specific technologies. In our case, a Fast Ethernet link, which is a shared link technology in the sense that it allows connecting multiple devices (hosts or routers), will be represented in the LSDB as a point-to-point link (since it is being used to connect just two routers).

Experimental procedure For this experiment, we will again resort to the network topology of Figure 9.8. The routers must be configured with the basic IPv4 IS-IS routing configuration, modified at interfaces f0/0 of routers R1 and R2. In particular, the command **isis network point-to-point** must be added to these interfaces.

Summary of IPv4 IS-IS LSDB Figure 9.25 shows the summary information of the IPv4 LSDB. It can be seen that the LSDB includes only Nonpseudonode-LSPs, and the Pseudonode-LSP R2.01-00 that before represented the transit shared link (see Figure 9.22) is no longer present.

9.2.1.7 Use of local tags to identify links

In this section, we analyze in detail how parallel point-to-point links and transit shared links with the same DR are described in the LSDB. We will concentrate on OSPFv3 and IPv4 IS-IS since these technologies resort to locally generated tags to uniquely identify links. In OSPFv2, this issue is less complex since links are always identified using IPv4 addresses. IPv6 IS-IS has

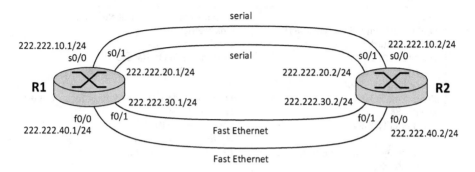

FIGURE 9.26: Experiments related to the use of local tags to identify links - Network topology.

a similar structure to IPv4 IS-IS in this respect; thus it is not worth repeating the experiment.

The network topology for this experiment is shown in Figure 9.26. It includes just two routers, but four links between them: two point-to-point links and two transit shared links (Fast Ethernets). The routers must be configured with the basic router configuration. Moreover, since we want to concentrate on topological aspects, there is no need to configure routable addresses in the OSPFv3 network. Finally, to ease the analysis of the IS-IS experiment, it is important that the interface costs of the point-to-point links are different. The default interface cost is 10, and we will configure a cost of 50 at the s0/0 interfaces. To do that, insert the command **isis metric 50** in the s0/0 interface mode of both routers.

In these experiments, we want to make sure that R2 becomes the DR in both transit shared links. This can be easily achieved if the two routers are started at the same time, since R2 has a higher RID.

OSPFv3 The OSPFv3 LSDB has two Router-LSAs, describing the two routers and its interfaces, and two Network-LSAs, describing the two transit shared links. It also has six Link-LSAs, advertising the IPv6 link-local addresses assigned to the six interfaces. It has no Intra-Area-Prefix-LSAs, since no routable IPv6 addresses have been configured at the routers. We can check that using the **show ipv6 database** command. In Figure 9.27, we show the complete Router-LSA of R1. It includes four links descriptions, one for each interface. The first two descriptions (`Link connected to: a Transit Network`) are type 2 descriptions characterizing the transit shared links. The link identifier is placed in the `Neighbor Router ID` and `Neighbor Interface ID` fields. Since the DR in both transit shared links is R2, `Neighbor Router ID` is 2.2.2.2 in both descriptions. The links identifiers are differentiated by the number of the interface attaching R2 to each link, contained in the `Neighbor Interface ID` field: it is 5 for the first interface and 4 for the second one.

The last two link descriptions (`Link connected to: another Router`)

LS age: 32
Options: (V6-Bit E-Bit R-bit DC-Bit)
LS Type: Router Links
Link State ID: 0
Advertising Router: 1.1.1.1
LS Seq Number: 80000006
Checksum: 0x6E91
Length: 88
Number of Links: 4
 	
  Link connected to: a Transit Network
  Link Metric: 10
  Local Interface ID: 5
  Neighbor (DR) Interface ID: 5
  Neighbor (DR) Router ID: 2.2.2.2
	Link connected to: a Transit Network
	  Link Metric: 10
	  Local Interface ID: 4
	  Neighbor (DR) Interface ID: 4
	  Neighbor (DR) Router ID: 2.2.2.2
	
	Link connected to: another Router (point-to-point)
	  Link Metric: 64
	  Local Interface ID: 7
	  Neighbor Interface ID: 7
	  Neighbor Router ID: 2.2.2.2
	
	Link connected to: another Router (point-to-point)
	  Link Metric: 64
	  Local Interface ID: 6
	  Neighbor Interface ID: 6
	  Neighbor Router ID: 2.2.2.2

FIGURE 9.27: Use of local tags to identify links - OSPFv3 Router-LSA of R1.

are type 1 descriptions characterizing the two point-to-point links. The links are identified by the RID of the neighbor (placed in the Neighbor Router ID) and the number of the interface attaching the router with the link (placed in the Local Interface ID field). Since the neighbor is the same for both links (2.2.2.2) the identifiers are differentiated by the local interface number, which is 7 in the first link, and 6 in the second one.

IPv4 IS-IS The detailed IPv4 IS-IS LSDB is shown in Figure 9.28. There are two Nonpseudonode-LSPs representing the two routers (R1.00-00 and R2.00-00) and two Pseudonode-LSPs representing the two transit shared links (R2.01-00 and R2.02-00). Since the DIS is the same in both shared links, the link identifiers are differentiated by the Pseudonode ID value, which is 1 in the first link and 2 in the second one. In the Nonpseudonode-LSPs, the topological information is given by the IS Neighbors TLV. Notice that, in both LSPs, there are only three TLVs of this type (despite having four links). Two of these TLVs indicate the neighborhood with the transit shared links (R2.01-00 and R2.02-00). Regarding the point-to-point links, only one is represented, the one with interface cost 10. This shows that, in IS-IS, parallel point-to-point links are represented–topologically–through a single IS Neighbors TLV that advertises the link with the least interface cost. In fact, this is all that is needed to compute shortest paths. Nevertheless, the interface costs are included in the reachability TLVs that describe the prefixes assigned to the links (identified by the "IP" keyword). Notice that the reachability TLVs describing the prefix 222.222.10.0/24 includes a Metric of 50.

LSPID	LSP Seq Num	LSP Checksum	LSP Holdtime	ATT/P/OL
R1.00-00	0x00000007	0xFEB9	1048	0/0/0
Area Address: 49.0001				
NLPID: 0xCC				
Hostname: R1				
IP Address: 222.222.20.1				
Metric: 10	IP 222.222.30.0 255.255.255.0			
Metric: 50	IP 222.222.10.0 255.255.255.0			
Metric: 10	IP 222.222.40.0 255.255.255.0			
Metric: 10	IP 222.222.20.0 255.255.255.0			
Metric: 10	IS R2.00			
Metric: 10	IS R2.02			
Metric: 10	IS R2.01			
R2.00-00	* 0x00000005	0x9226	1177	0/0/0
Area Address: 49.0001				
NLPID: 0xCC				
Hostname: R2				
IP Address: 222.222.20.2				
Metric: 10	IP 222.222.30.0 255.255.255.0			
Metric: 50	IP 222.222.10.0 255.255.255.0			
Metric: 10	IP 222.222.40.0 255.255.255.0			
Metric: 10	IP 222.222.20.0 255.255.255.0			
Metric: 10	IS R2.02			
Metric: 10	IS R2.01			
Metric: 10	IS R1.00			
R2.01-00	* 0x00000001	0x7CD6	863	0/0/0
Metric: 0	IS R2.00			
Metric: 0	IS R1.00			
R2.02-00	* 0x00000001	0x75DC	839	0/0/0
Metric: 0	IS R2.00			
Metric: 0	IS R1.00			

FIGURE 9.28: Use of local tags to identify links - Detailed IPv4 IS-IS LSDB.

9.2.2 LSDB structure when connected to another AS

We analyze now the structure of the LSDB when the routing domain is connected directly to another AS. The network topology for the experiments carried out in this section is shown in Figure 9.29. The topology is similar to the one of Figure 9.8, but the routers are now placed in different ASes: R1 and R2 in AS 200, and R3 in AS 100. The link between the ASBRs, i.e. between R2 and R3, uses private addresses from the range 10.0.0.0/24, and the routers are connected through BGP. AS 200 includes prefix 222.222.10.0/24 and AS 100 includes prefix 150.150.0.0/16. Each prefix must be exported to the neighboring AS using BGP. AS 200 uses an intra-domain routing protocol, which is either OSPFv2 or IPv4 IS-IS. We will not include the IPv6 routing protocols in these experiments, since the differences are not significant.

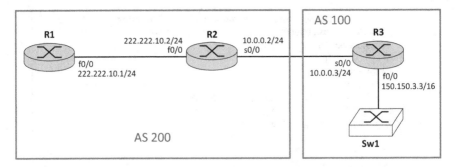

FIGURE 9.29: LSDB structure experiments when routing domain is connected directly to another AS - Network topology.

9.2.2.1   OSPFv2

Router configurations The novelty in relation to the previous LSDB experiments are the configurations required to export addressing information to other ASes and to disseminate internally the addressing information imported from other ASes. In Figure 9.30 we show the complete configurations of R2 and R3. We highlight in bold the commands that are new in relation to the basic OSPFv2 configuration. R1 just needs the basic configuration.

First, note that R2 is configured with OSPFv2, but only at its interface f0/0. The s0/0 interface is used to connect to another AS, and this commu-

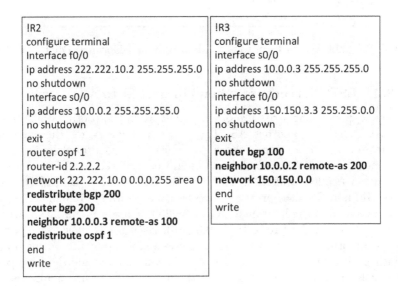

FIGURE 9.30: LSDB structure when connected to another AS - OSPFv2 configurations of R2 and R3.

nication is through BGP, and not through OSPF (which is an intra-domain routing protocol).

To configure BGP, we need to create a routing process, just as in the case of OSPF. This is done through the commands **router bgp 200** at R2 and **router bgp 100** at R3. The number in these commands is the AS number of the AS where the router is located. The second command in the BGP configuration configures the connection with the BGP neighbor (located on the other AS). At router R2, the command is **neighbor 10.0.0.3 remote-as 200**, and at router R3 it is **neighbor 10.0.0.2 remote-as 100**. This command indicates the IPv4 address of the BGP neighbor's interface and the BGP neighbor's AS number. The next configurations are different at R2 and R3. At R3, the command **network 150.150.0.0** tells the router to export this prefix to its BGP neighbor. At R2, the command **redistribute ospf 1** tells the router to export to its neighbor the prefixes it learned internally from OSPFv2. Finally, the OSPFv2 configuration of R2 needs a command instructing the router to disseminate through OSPFv2 what it learned from BGP; this is the role of the command **redistribute bgp 200**. It is this command that prompts R2 to originate the AS-External-LSA that advertises 150.150.0.0/16 inside AS 200. The routers can be started once these configurations are done.

LSDB and forwarding table Figure 9.31.a shows the summary of the OSPFv2 LSDB (which can be obtained with the command **show ip ospf database**). The summary now has three parts. The first two parts list the Router-LSAs and Network-LSAs. There are two Router-LSAs, describing R1 and R2, and one Network-LSA, describing the transit-shared link of 222.222.10.0/24. Note that contrarily to the experiment of Section 9.2.1.1 (see Figure 9.9), there is no Router-LSA representing R3, since R3 is now outside the OSPF routing domain. The last part of the LSDB summary (Type-5 AS External Link States) refers to the AS-External-LSAs and indicates that there is one LSA of this type in the LSDB, originated by R2 (ADV Router column) and advertising a prefix with subnet address 150.150.0.0 (Link ID column).

The AS-External-LSA is shown in Figure 9.31.b (it can be obtained with the command **show ip ospf database external**). In addition to the summary information, it can be seen that the subnet mask is /16 (Network Mask field), the metric type is E2 (Metric Type field, which corresponds to the E flag), and the external cost (Metric field) is 1.

Finally, Figure 9.31.c shows the forwarding table of R1. The second entry refers to the domain-external prefix 150.150.0.0/16, which became known at R1 through the AS-External-LSA injected by R2. The keyword "O E2" indicates that this entry was learned from OSPF and is an E2 external route.

9.2.2.2   IPv4 IS-IS

Router configurations The configuration of this experiment has a few changes in relation to the basic IS-IS routing configuration. We show it in Fig-

(a)

Router Link States (Area 0)

Link ID	ADV Router	Age	Seq#	Checksum	Link count
1.1.1.1	1.1.1.1	100	0x80000003	0x00523B	1
2.2.2.2	2.2.2.2	101	0x80000003	0x001A68	1

Net Link States (Area 0)

Link ID	ADV Router	Age	Seq#	Checksum
222.222.10.2	2.2.2.2	101	0x80000001	0x0093C8

Type-5 AS External Link States

Link ID	ADV Router	Age	Seq#	Checksum	Tag
150.150.0.0	2.2.2.2	82	0x80000001	0x0047D2	100

(b)

```
Routing Bit Set on this LSA
LS age: 176
Options: (No TOS-capability, DC)
LS Type: AS External Link
Link State ID: 150.150.0.0 (External Network Number )
Advertising Router: 2.2.2.2
LS Seq Number: 80000001
Checksum: 0x47D2
Length: 36
Network Mask: /16
    Metric Type: 2 (Larger than any link state path)
    TOS: 0
    Metric: 1
    Forward Address: 0.0.0.0
    External Route Tag: 100
```

(c)

```
C   222.222.10.0/24 is directly connected, FastEthernet0/0
O E2 150.150.0.0/16 [110/1] via 222.222.10.2, 00:00:23, FastEthernet0/0
```

FIGURE 9.31: LSDB structure when connected to another AS - (a) Summary of OSPFv2 LSDB, (b) OSPFv2 AS-External-LSA, and (c) forwarding table of R1.

ure 9.32 for all three routers. First, R1 and R2 are configured as L2 routers. Second, the command **metric-style wide** is inserted in the router isis mode. This command prompts the use of the extended TLVs, i.e. the Extended IS Reachability TLV (type 22) and the Extended IP Reachability TLV (type 135). This is required when connecting to external ASes, at least in Cisco implementations. The designation **metric-style wide** comes from the fact that the Metric length in these TLVs is much larger than the one of its predecessors (the IS Neighbors TLV, the IP Internal Reachability Information TLV, and the IP External Reachability Information TLV) (see Figure 5.12).

The BGP configuration is similar to the one in Section 9.2.2.1 (see Figure 9.30). At R3, it is exactly the same. At R2, and in its bgp router mode,

```
!R1                                      !R2
configure terminal                       configure terminal
interface f0/0                           interface f0/0
ip address 222.222.10.1 255.255.255.0    ip address 222.222.10.2 255.255.255.0
ip router isis                           ip router isis
no shutdown                              no shutdown
exit                                     interface s0/0
router isis                              ip address 10.0.0.2 255.255.255.0
net 49.0001.0000.0000.0001.00            no shutdown
metric-style wide                        exit
is-type level-2                          router isis
end                                      net 49.0001.0000.0000.0002.00
write                                    metric-style wide
                                         redistribute bgp 200
!R3                                      is-type level-2
configure terminal                       router bgp 200
interface s0/0                           neighbor 10.0.0.3 remote-as 100
ip address 10.0.0.3 255.255.255.0        redistribute isis level-2
no shutdown                              redistribute connected
interface f0/0                           end
ip address 150.150.3.3 255.255.0.0       write
no shutdown
exit
router bgp 100
neighbor 10.0.0.2 remote-as 200
network 150.150.0.0
end
write
```

FIGURE 9.32: LSDB structure when connected to another AS - IPv4 IS-IS router configurations.

the command **redistribute isis level-2** must be inserted, instead of **redistribute ospf 1**. Moreover, the command **redistribute connected** must also be inserted. The first command instructs the router to export all prefixes learned through level-2 IS-IS, and the second instructs to export the prefixes assigned to links the router is directly attached to (in fact, only the latter command is required, since AS 200 has only one prefix, and this prefix is assigned to the transit shared link, directly attached to R2).

LSDB and forwarding table The IPv4 IS-IS LSDB is not very different from the LSDB of Section 9.2.1.4. The LSDB (see Figure 9.33.a) is now an L2 LSDB, but its structure is the same as the L1 LSDB. The LSDB has no longer a Nonpseudonode-LSP of router R3, since R3 does not belong to AS 200. Regarding the TLVs, there are two differences. One is the presence of the Extended IS Reachability TLV (type 22) instead of the IS Neighbors TLV (type 2), which is identified through the keyword "IS Extended". As explained above, this change was due to the **metric-style wide** command. The second

LSPID	LSP Seq Num	LSP Checksum	LSP Holdtime	ATT/P/OL
R1.00-00	*** 0x00000002**	**0x7AD5**	**923**	**0/0/0**
Area Address: 49.0001				
NLPID: 0xCC				
Hostname: R1				
IP Address: 222.222.10.1				
Metric: 10 IS-Extended R2.01				
Metric: 10 IP 222.222.10.0/24				
R2.00-00	**0x00000003**	**0x43C4**	**984**	**0/0/0**
Area Address: 49.0001				
NLPID: 0xCC				
Hostname: R2				
IP Address: 222.222.10.2				
Metric: 10 IS-Extended R2.01				
Metric: 0 IP 150.150.0.0/16				
Metric: 10 IP 222.222.10.0/24				
R2.01-00	**0x00000001**	**0x7E4B**	**919**	**0/0/0**
Metric: 0 IS-Extended R2.00				
Metric: 0 IS-Extended R1.00				

(a)

(b)

```
C  222.222.10.0/24 is directly connected, FastEthernet0/0
i L2 150.150.0.0/16 [115/10] via 222.222.10.2, FastEthernet0/0
```

FIGURE 9.33: LSDB structure when connected to another AS - (a) Detailed IPv4 IS-IS LSDB, and (b) forwarding table of R1.

is the presence of the Extended IP Reachability TLV (type 135) that advertises the domain-external prefix 150.150.0.0/16; this TLV is highlighted in bold in the figure. Because of this TLV, the forwarding table of R1 (see Figure 9.33.b) now includes an entry to 150.150.0.0/16. The keyword "i L2" indicates that the entry was built from the IS-IS L2 LSDB.

9.2.3   LSDB structure with default routes

As discussed in Section 2.2.4, injecting domain-external prefixes in an AS should be avoided and is not needed when the AS has a single outgoing ASBR. Instead, default routes should be used. The default route is a forwarding table entry with zero subnet address and mask, i.e. 0.0.0.0/0, which is used as a last resort in forwarding packets when their destination addresses do not match any other entries. Both OSPF and IS-IS include the possibility of disseminating default routes from the domain border routers, i.e. in the same way domain-external prefixes are injected in an AS. We will illustrate this feature using again the network topology of Section 9.2.2 (see Figure 9.29). Only small modifications are needed in the router configurations in relation to the ones in Section 9.2.2. One just needs to replace the redistribution of BGP into the intra-domain routing protocol (OSPF or IS-IS) at the domain border router by the injection of a default route.

(a)
```
Routing Bit Set on this LSA
LS age: 139
Options: (No TOS-capability, DC)
LS Type: AS External Link
Link State ID: 0.0.0.0 (External Network Number )
Advertising Router: 2.2.2.2
LS Seq Number: 80000001
Checksum: 0xFEAB
Length: 36
Network Mask: /0
    Metric Type: 2 (Larger than any link state path)
    TOS: 0
    Metric: 1
    Forward Address: 0.0.0.0
    External Route Tag: 1
```

(b)
```
C   222.222.10.0/24 is directly connected, FastEthernet0/0
O*E2 0.0.0.0/0 [110/1] via 222.222.10.2, 00:03:10, FastEthernet0/0
```

FIGURE 9.34: LSDB structure with default routes - (a) OSPFv2 AS-External-LSA and (b) forwarding table of R1 obtained with OSPFv2.

9.2.3.1   OSPFv2

Router configurations In relation to the configuration of Section 9.2.2.1, just replace, at R2, the **redistribute bgp 200** command of the router ospf mode (see Figure 9.30), by the command **default-information originate always**.

AS-External-LSA and forwarding table of R1 The default route is disseminated within the AS in the same way as domain-external prefixes, i.e. using an AS-External-LSA. This LSA is shown in Figure 9.34.a. The `Link State ID` and `Network Mask` fields define the prefix to be installed, and they are now 0.0.0.0 and 0, respectively. The forwarding table of R1 (Figure 9.34.b) confirms that the default route was indeed installed.

9.2.3.2   IPv4 IS-IS

Router configurations In relation to the configuration of Section 9.2.2.2, just replace, at R2, the **redistribute bgp 200** command of the router isis mode (see Figure 9.32), by the command **default-information originate**.

Forwarding table of R1 The consequence of this change is the replacement, at the forwarding table of R1, of the entry relative to 150.150.0.0/16 by the default route (see Figure 9.35). This information was disseminated within AS 200 through an Extended IP Reachability TLV, the same that disseminates domain-external prefixes.

C   222.222.10.0/24 is directly connected, FastEthernet0/0
i*L2 0.0.0.0/0 [115/10] via 222.222.10.2, FastEthernet0/0

FIGURE 9.35: LSDB structure with default routes - Forwarding table of R1 obtained with IPv4 IS-IS.

9.2.4   OSPF E-type metrics

In this section, we perform several experiments that illustrate the use of the E-type metric of OSPF. This issue was discussed in Section 5.8.3. The experiments will be based on the network topology of Figure 9.36. In this network, there are again two ASes (100 and 200), but AS 200 now has two ASBRs (R3 and R4) and two internal routers (R1 and R2).

Router configurations With a few exceptions, the routers must be configured according to the instructions of Section 9.2.2.1 (see Figure 9.30). The

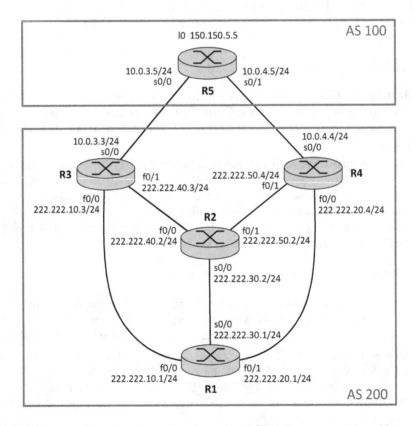

FIGURE 9.36: Experiments related to the OSPF E-type metric - Network topology.

```
!R1
configure terminal
interface f0/0
ip address 222.222.10.1 255.255.255.0
no shutdown
interface f0/1
ip address 222.222.20.1 255.255.255.0
no shutdown
interface s0/0
ip address 222.222.30.1 255.255.255.0
ip ospf cost 10
no shutdown
exit
router ospf 1
router-id 1.1.1.1
network 222.222.10.0 0.0.0.255 area 0
network 222.222.20.0 0.0.0.255 area 0
network 222.222.30.0 0.0.0.255 area 0
end
Write

!R5
configure terminal
interface s0/0
ip address 10.0.3.5 255.255.255.0
no shutdown
interface s0/1
ip address 10.0.4.5 255.255.255.0
no shutdown
interface l0
ip address 150.150.5.5 255.255.0.0
no shutdown
exit
router bgp 100
neighbor 10.0.4.4 remote-as 200
neighbor 10.0.3.3 remote-as 200
network 150.150.0.0
end
write
```

```
!R3
configure terminal
interface f0/0
ip address 222.222.10.3 255.255.255.0
no shutdown
interface f0/1
ip address 222.222.40.3 255.255.255.0
no shutdown
interface s0/0
ip address 10.0.3.3 255.255.255.0
no shutdown
exit
router ospf 1
router-id 3.3.3.3
network 222.222.10.0 0.0.0.255 area 0
network 222.222.40.0 0.0.0.255 area 0
redistribute bgp 200
router bgp 200
neighbor 10.0.3.5 remote-as 100
redistribute ospf 1
end
write
```

FIGURE 9.37: OSPF E-type metrics - Configurations of routers R1, R3, and R5.

complete configurations of routers R1, R3, and R5 are shown in Figure 9.37. We highlight in bold the new commands in relation to the basic configurations.

To simplify the topology, we now configure a loopback address at R5; this avoids including a Fast Ethernet interface connected to an Ethernet switch (as we did in AS 100 of Section 9.2.2.1). Configuring a loopback interface is not different from configuring other types of interfaces, only the interface name

changes; **l0** is the abbreviation of **loopback 0**. The loopback address assigned
to the interface is 150.150.5.5.

We want also to force the cost of interface s0/0 of router R1 to be 10 (its
default value is 64); for that, the command **ip ospf cost 10** must be entered
at this interface. Notice that in the BGP configuration of R5 there are now
two neighbors being declared: one is R3 (with IPv4 address 10.0.3.3), and the
other is R4 (with IPv4 address 10.0.4.4). We will now discuss several cases.

Case 1 The first case corresponds to the configurations just described. In
Figure 9.38, we show the LSDB summary, the forwarding table of R1, and
the result of a traceroute from R1 to R5 (150.150.5.5). The LSDB summary
(Figure 9.38.a) shows that there are four Router-LSAs, since AS 200 has four
routers, and four Network-LSAs, since there are four transit shared links inside
this AS; the remaining link is a point-to-point link between R1 and R2 and,
therefore, is described through the Router-LSAs of R1 and R2. In addition to
these LSAs, there are two AS-External-LSAs, both advertising the domain-
external prefix 150.150.0.0/16, one originated by R3 and the other by R4.

With the configurations described above, the ASBRs have a metric type of
E2 and an external cost of 1. We can check this by looking to the full contents
of the AS-External-LSAs using the command **show ip ospf database external**; the metric type can be found in the `Metric Type` line and the external
cost in the `Metric` line. In this case, there is no preferred outgoing ASBR
for AS 200, and R1 ends up installing two entries leading to 150.150.0.0/16
in its forwarding table (Figure 9.38.b): one via R3 (with next-hop address
222.222.10.3) and another via R4 (with next-hop address 222.222.20.4). Note
that there are three different paths from R1 to R3 or R4. For example, the
paths from R1 to R3 are R1→R3 (the direct route), R1→R2→R3 (with cost
20), and R1→R4→R2→R3 (with cost 30); the least cost path is R1→R3.
However, the routing table entries relative to 150.150.0.0/16 both have a path
cost of one. This is because, since the metric type is E2, routers ignore the
cost of the internal sub-path (from an internal router to an ASBR) and the
path cost is the lowest among the external costs advertised by the ASBRs (in
this case, both ASBRs advertise an external cost of 1).

We can confirm that R1 uses indeed two paths to reach R1 through the
traceroute command (see Figure 9.38.c.) Notice that the paths alternate
between going through R3 or R4.

Case 2 Now let us change the external cost of R4 to 2, while keeping the metric
type as E2. The goal is to ensure that R3 becomes the preferred outgoing
ASBR for the domain. We can do that by entering the command **redistribute
bgp 200 metric-type 2 metric 2**, in the ospf router mode of R4. This
replaces the previous command that was simply **redistribute bgp 200**. The
result of this experiment is displayed in Figure 9.39. Router R1 still receives
two AS-External-LSAs (we can confirm that using **show ip ospf database
external**), but the one originated by R4 now has an external cost of 2. Thus,
R3 becomes the best ASBR and the forwarding table of R1 (Figure 9.39.a)
now includes only one path to 150.150.0.0/16, the one through R3 (with next-

Router Link States (Area 0)

Link ID	ADV Router	Age	Seq#	Checksum	Link count
1.1.1.1	1.1.1.1	284	0x80000003	0x008057	4
2.2.2.2	2.2.2.2	283	0x80000004	0x006AEC	4
3.3.3.3	3.3.3.3	285	0x80000004	0x003953	2
4.4.4.4	4.4.4.4	290	0x80000003	0x00E177	2

Net Link States (Area 0)

(a)

Link ID	ADV Router	Age	Seq#	Checksum
222.222.10.3	3.3.3.3	290	0x80000001	0x008DC5
222.222.20.4	4.4.4.4	288	0x80000001	0x001927
222.222.40.3	3.3.3.3	284	0x80000001	0x0074BC
222.222.50.4	4.4.4.4	288	0x80000001	0x00FF1E

Type-5 AS External Link States

Link ID	ADV Router	Age	Seq#	Checksum	Tag
150.150.0.0	3.3.3.3	270	0x80000001	0x0029EC	100
150.150.0.0	4.4.4.4	270	0x80000001	0x000B07	100

(b)
```
O    222.222.50.0/24 [110/20] via 222.222.30.2, 00:06:01, Serial0/0
                     [110/20] via 222.222.20.4, 00:06:01, FastEthernet0/1
C    222.222.20.0/24 is directly connected, FastEthernet0/1
C    222.222.10.0/24 is directly connected, FastEthernet0/0
O    222.222.40.0/24 [110/20] via 222.222.30.2, 00:06:01, Serial0/0
                     [110/20] via 222.222.10.3, 00:06:01, FastEthernet0/0
C    222.222.30.0/24 is directly connected, Serial0/0
O E2 150.150.0.0/16 [110/1] via 222.222.20.4, 00:05:57, FastEthernet0/1
                    [110/1] via 222.222.10.3, 00:05:57, FastEthernet0/0
```

(c)
```
R1#traceroute 150.150.5.5

Type escape sequence to abort.
Tracing the route to 150.150.5.5

1 222.222.20.4 36 msec
  222.222.10.3 32 msec
  222.222.20.4 84 msec
2 10.0.3.5 28 msec
  10.0.4.5 76 msec
  10.0.3.5 48 msec
```

FIGURE 9.38: OSPF E-type metrics (case 1) - (a) Summary of OSPFv2 LSDB, (b) forwarding table of R1, and (c) traceroute from R1 to 150.150.5.5.

hop address 222.222.10.3). This is confirmed through the traceroute of Figure 9.39.b.

Case 3 In the two previous experiments, the internal sub-path towards R3 was always the direct route R1→R3, via the f0/0 interface of R1. To see that

(a)
```
O   222.222.50.0/24 [110/20] via 222.222.30.2, 00:02:50, Serial0/0
                    [110/20] via 222.222.20.4, 00:02:50, FastEthernet0/1
C   222.222.20.0/24 is directly connected, FastEthernet0/1
C   222.222.10.0/24 is directly connected, FastEthernet0/0
O   222.222.40.0/24 [110/20] via 222.222.30.2, 00:02:50, Serial0/0
                    [110/20] via 222.222.10.3, 00:02:50, FastEthernet0/0
C   222.222.30.0/24 is directly connected, Serial0/0
O E2 150.150.0.0/16 [110/1] via 222.222.10.3, 00:00:16, FastEthernet0/0
```

(b)
```
R1#traceroute 150.150.5.5

Type escape sequence to abort.
Tracing the route to 150.150.5.5

1 222.222.10.3 76 msec 48 msec 60 msec
2 10.0.3.5 128 msec 92 msec 120 msec
```

FIGURE 9.39: OSPF E-type metrics (case 2) - (a) Forwarding table of R1 and (b) traceroute from R1 to 150.150.5.5.

the preferred outgoing ASBR is kept when the internal sub-path changes, we will now increase the OSPF cost of interface f0/0 of R1 to 40. We can do that by entering the command **ip ospf cost 40** in the interface f0/0 mode. The forwarding table of R1 (Figure 9.40.a) shows that there is again just one entry to 150.150.0.0/16, but the next-hop router is now R2 (next-hop address is 222.222.30.2). The traceroute (Figure 9.40.b) shows that the path to R5 is now R1→R2→R3→R5. Thus, R3 is still the outgoing ASBR since its external cost is lower than the one of R4. Moreover, the cost of the internal sub-path R1→R2→R3 is 20. However, the path cost indicated in the 150.150.0.0/16 is still the external cost (still 1), since the metric type is E2.

Case 4 We will now force the shortest path from R1 to R3 to cross R4. One possibility to implement this routing solution is to increase the cost of interface s0/0 of router R1 to 30. When this is done, the shortest path from R1 to R3 becomes R1→R4→R2→R3, with a cost of 30. Figure 9.41.a shows that the next-hop to 150.150.0.0/16 changed to R4 (the next-hop address is now 222.222.20.4). However, according to the traceroute (Figure 9.41.b) packets leave AS 200 through R4 (and not through R3). Although the intention of R1 is that the packets leave the AS through R3 (because the AS-External-LSA originated by R3 has a lower external cost), the fact is that R4 deviates the packets directly to R5. Indeed, under the next-hop routing paradigm of Internet routing, routers have no control over the full paths; they control only the path to the next hop. This can be checked through the forwarding table of R4 (Figure 9.41.c). According to the table, packets that arrive at this router and are destined to 150.150.5.5 are forwarded through its last entry, which was learned through BGP and has R5 as the next-hop (the next-hop address is 10.0.4.5). Notice that router R4 also contains in its LSDB the AS-External-

(a)
```
O   222.222.50.0/24 [110/20] via 222.222.30.2, 00:00:10, Serial0/0
                    [110/20] via 222.222.20.4, 00:00:10, FastEthernet0/1
C   222.222.20.0/24 is directly connected, FastEthernet0/1
C   222.222.10.0/24 is directly connected, FastEthernet0/0
O   222.222.40.0/24 [110/20] via 222.222.30.2, 00:00:10, Serial0/0
C   222.222.30.0/24 is directly connected, Serial0/0
O E2 150.150.0.0/16 [110/1] via 222.222.30.2, 00:00:10, Serial0/0
```

(b)
```
R1#traceroute 150.150.5.5

Type escape sequence to abort.
Tracing the route to 150.150.5.5

 1 222.222.30.2 0 msec 0 msec 12 msec
 2 222.222.40.3 36 msec 60 msec 60 msec
 3 10.0.3.5 52 msec 108 msec 56 msec
```

FIGURE 9.40: OSPF E-type metrics (case 3) - (a) Forwarding table of R1 and (b) traceroute from R1 to 150.150.5.5.

(a)
```
O   222.222.50.0/24 [110/20] via 222.222.20.4, 00:02:55, FastEthernet0/1
C   222.222.20.0/24 is directly connected, FastEthernet0/1
C   222.222.10.0/24 is directly connected, FastEthernet0/0
O   222.222.40.0/24 [110/40] via 222.222.20.4, 00:02:55, FastEthernet0/1
C   222.222.30.0/24 is directly connected, Serial0/0
O E2 150.150.0.0/16 [110/1] via 222.222.20.4, 00:02:55, FastEthernet0/1
```

(b)
```
R1#traceroute 150.150.5.5

Type escape sequence to abort.
Tracing the route to 150.150.5.5

 1 222.222.20.4 72 msec 32 msec 80 msec
 2 10.0.4.5 72 msec 36 msec 88 msec
```

(c)
```
C   222.222.50.0/24 is directly connected, FastEthernet0/1
C   222.222.20.0/24 is directly connected, FastEthernet0/0
O   222.222.10.0/24 [110/30] via 222.222.50.2, 00:22:08, FastEthernet0/1
O   222.222.40.0/24 [110/20] via 222.222.50.2, 00:22:08, FastEthernet0/1
    10.0.0.0/24 is subnetted, 1 subnets
C      10.0.4.0 is directly connected, Serial0/0
O   222.222.30.0/24 [110/20] via 222.222.50.2, 00:22:08, FastEthernet0/1
B   150.150.0.0/16 [20/0] via 10.0.4.5, 01:22:56
```

FIGURE 9.41: OSPF E-type metrics (case 4) - (a) Forwarding table of R1, (b) traceroute from R1 to 150.150.5.5, and (c) forwarding table of R4.

LSA originated by R3 (we can check this through the command **show ip ospf database external**). Thus, R4 knows that there is a path to 150.150.0.0/16 through R3. However, when deciding which path to use, R4 prefers the path

(a)
```
O   222.222.50.0/24 [110/20] via 222.222.30.2, 00:00:27, Serial0/0
                    [110/20] via 222.222.20.4, 00:00:27, FastEthernet0/1
C   222.222.20.0/24 is directly connected, FastEthernet0/1
C   222.222.10.0/24 is directly connected, FastEthernet0/0
O   222.222.40.0/24 [110/20] via 222.222.30.2, 00:00:27, Serial0/0
                    [110/20] via 222.222.10.3, 00:00:27, FastEthernet0/0
C   222.222.30.0/24 is directly connected, Serial0/0
O E1 150.150.0.0/16 [110/11] via 222.222.20.4, 00:00:22, FastEthernet0/1
                    [110/11] via 222.222.10.3, 00:00:22, FastEthernet0/0
```

(b)
```
R1#traceroute 150.150.5.5

Type escape sequence to abort.
Tracing the route to 150.150.5.5

1 222.222.20.4 68 msec
  222.222.10.3 68 msec
  222.222.20.4 68 msec
2 10.0.3.5 36 msec
  10.0.4.5 96 msec
  10.0.3.5 72 msec
```

FIGURE 9.42: OSPF E-type metrics (case 5) - (a) Forwarding table of R1 and (b) traceroute from R1 to 150.150.5.5.

it directly learned from its BGP peer, because external BGP has a lower administrative distance than OSPF: OSPF has 110 and external BGP has 20. These administrative distances are also displayed in the forwarding table.

Case 5 We will now change the metric type to E1 and return all interface costs to 10. To change the metric type, enter the command **redistribute bgp 200 metric-type 1 metric 1** in the ospf router mode of both R3 and R4. The net result is the same as in case 1. In the forwarding table of R1 (Figure 9.42.a) there are again two entries to 150.150.0.0/16, and the traceroute (Figure 9.42.b) confirms that the paths alternate between going through R3 and R4. However, the entries indicate that the metric is now of type E1 (keyword "O E1"). Moreover, the cost is now 11, which corresponds to the sum of the shortest path cost from R1 to the ASBR with the external cost configured at the ASBR.

Case 6 To change the outgoing ASBR, we can either change the cost of the internal sub-path or the external cost. We start with the second option. We will increase the external cost configured at R4 to 10 (using the command **redistribute bgp 200 metric-type 1 metric 10**). The forwarding table of R1 (Figure 9.43.a) and the traceroute (Figure 9.43.b) show that the path is now through R3, and has an end-to-end cost of 11.

Case 7 Let us now change again the outgoing ASBR, but through the manipulation of the internal sub-path. We will increase the cost of interface f0/0 of R1 to 20. In this case, the cost of the end-to-end path to R5 becomes 21

(a)
```
O   222.222.50.0/24 [110/20] via 222.222.30.2, 00:13:32, Serial0/0
                    [110/20] via 222.222.20.4, 00:13:32, FastEthernet0/1
C   222.222.20.0/24 is directly connected, FastEthernet0/1
C   222.222.10.0/24 is directly connected, FastEthernet0/0
O   222.222.40.0/24 [110/20] via 222.222.30.2, 00:13:32, Serial0/0
                    [110/20] via 222.222.10.3, 00:13:32, FastEthernet0/0
C   222.222.30.0/24 is directly connected, Serial0/0
O E1 150.150.0.0/16 [110/11] via 222.222.10.3, 00:04:24, FastEthernet0/0
```

(b)
```
R1#traceroute 150.150.5.5

Type escape sequence to abort.
Tracing the route to 150.150.5.5

  1 222.222.10.3 64 msec 64 msec 68 msec
  2 10.0.3.5 60 msec 80 msec 52 msec
```

FIGURE 9.43: OSPF E-type metrics (case 6) - (a) Forwarding table of R1 and (b) traceroute from R1 to 150.150.5.5.

(a)
```
O   222.222.50.0/24 [110/20] via 222.222.30.2, 00:00:01, Serial0/0
                    [110/20] via 222.222.20.4, 00:00:01, FastEthernet0/1
C   222.222.20.0/24 is directly connected, FastEthernet0/1
C   222.222.10.0/24 is directly connected, FastEthernet0/0
O   222.222.40.0/24 [110/20] via 222.222.30.2, 00:00:01, Serial0/0
C   222.222.30.0/24 is directly connected, Serial0/0
O E1 150.150.0.0/16 [110/20] via 222.222.20.4, 00:00:01, FastEthernet0/1
```

(b)
```
R1#traceroute 150.150.5.5

Type escape sequence to abort.
Tracing the route to 150.150.5.5

  1 222.222.20.4 84 msec 72 msec 128 msec
  2 10.0.4.5 120 msec 64 msec 48 msec
```

FIGURE 9.44: OSPF E-type metrics (case 7) - (a) Forwarding table of R1 and (b) traceroute from R1 to 150.150.5.5.

through R3: the internal cost is 20 either directly via interface f0/0 or via interface s0/0, and the external cost is 1. However, the end-to-end cost through R4 is 20, and R4 becomes the preferred outgoing ASBR. This can be confirmed through the forwarding table of R1 (Figure 9.44.a) and the traceroute (Figure 9.44.b).

9.2.5 OSPF LSDB structure when connected to RIP domain

In this section, we discuss the LSDB structure when an OSPFv2 domain is directly connected to a RIP domain and both domains are in the same AS.

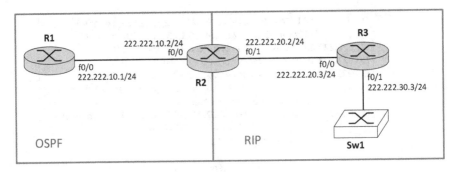

FIGURE 9.45: Experiments related to the OSPF LSDB structure when connected to RIP domain - Network topology.

Our main purpose is to show that the LSDB structure is the same as if the routing domain was connected to another AS (which we discussed in Section 9.2.2.1).

The network topology for this experiment is shown in Figure 9.45. Router R2 is on the border between the two routing domains (OSPF calls it an ASBR even in this case). Router R1 and subnet 222.222.10.0/24 belong to the OSPF domain (as well as one half of router R2); router R3 and subnets 222.222.20.0/24 and 222.222.30.0/24 belong to the RIP domain (as well as the other half of router R2).

Router configurations In Figure 9.46, we show the configurations of routers R2 and R3. Router R1 just needs to be configured with the basic OSPFv2 routing configuration. R3 must be configured with RIP. The configuration of RIP is very similar to the OSPFv2 one. We just need to create a RIP process, using the command **router rip** in the global configuration mode, and declare the subnets the router is directly attached to (and that are to be advertised through RIP). We do not need (nor is it possible) to give a number to the RIP process, as we would in OSPF.

In router R2, both an OSPFv2 and a RIP process must be created and configured. In the OSPFv2 process, we must declare the subnets directly attached to the router (that are to be advertised through OSPFv2), and in the RIP process we must do the same thing. The router must also be configured to redistribute prefixes learned by one protocol into the other. This is similar to the BGP configuration, except that it occurs in the same router. Thus, (i) the **redistribute rip** command included in the **router ospf 1** mode tells R2 to inject into the OSPF process the prefixes learned through RIP, and (ii) the **redistribute ospf 1 metric 1** included in the **router rip** process tells R2 to inject in the RIP process the prefixes learned through OSPF, giving them a metric of 1.

LSDB and forwarding table Figure 9.47 shows the summary of the OSPFv2 LSDB and the forwarding table of router R1. There are two Router-LSAs, describing R1 and R2 and their interfaces, and one Network-LSA,

```
!R2
configure terminal
interface f0/0
ip address 222.222.10.2 255.255.255.0
no shutdown
interface f0/1
ip address 222.222.20.2 255.255.255.0
no shutdown
exit
router ospf 1
router-id 2.2.2.2
network 222.222.10.0 0.0.0.255 area 0
redistribute rip
router rip
network 222.222.20.0
redistribute ospf 1 metric 1
end
write
```

```
!R3
configure terminal
interface f0/0
ip address 222.222.20.3 255.255.255.0
no shutdown
interface f0/1
ip address 222.222.30.3 255.255.255.0
no shutdown
exit
router rip
network 222.222.20.0
network 222.222.30.0
end
write
```

FIGURE 9.46: OSPF LSDB structure when connected to RIP domain - Configurations of routers R2 and R3.

(a)

Router Link States (Area 0)

Link ID	ADV Router	Age	Seq#	Checksum	Link count
1.1.1.1	1.1.1.1	55	0x80000002	0x00543A	1
2.2.2.2	2.2.2.2	56	0x80000002	0x001C67	1

Net Link States (Area 0)

Link ID	ADV Router	Age	Seq#	Checksum
222.222.10.2	2.2.2.2	56	0x80000001	0x0093C8

Type-5 AS External Link States

Link ID	ADV Router	Age	Seq#	Checksum	Tag
222.222.20.0	2.2.2.2	96	0x80000001	0x000BBB	0
222.222.30.0	2.2.2.2	61	0x80000001	0x009C20	0

(b)
```
O E2 222.222.20.0/24 [110/20] via 222.222.10.2, 00:00:20, FastEthernet0/0
C    222.222.10.0/24 is directly connected, FastEthernet0/0
O E2 222.222.30.0/24 [110/20] via 222.222.10.2, 00:00:20, FastEthernet0/0
```

FIGURE 9.47: OSPF LSDB structure when connected to RIP domain - (a) Summary of OSPFv2 LSDB, and (b) forwarding table of R1.

describing the transit shared link and its assigned prefix (222.222.10.0/24). Besides that, the two domain-external prefixes imported from the RIP domain (222.222.20.0/24 and 222.222.30.0/24) are described through two AS-

External-LSAs (just as if the prefixes were imported from another AS using BGP). In the forwarding table of R1, the entries relative to these prefixes start with the keyword "O E2", indicating that they are domain-external prefixes and were learned through OSPF with E2 type metric. Note that the E2 metric value is 20. This is the default value given to the E2 metric when the redistribution is from RIP. As shown in Section 9.2.4, the metric value would be one in case the redistribution is from BGP, but this value is configurable. We can then conclude that the LSDB structure is the same whether the external-addressing information is imported from another AS or from another routing domain of the same AS.

10

Experiments on Synchronization Mechanisms

The synchronization mechanisms establish the neighborhood relationships among routers and keep the LSDB updated and synchronized. This issue was discussed in Chapter 6. In this chapter, we present experiments that illustrate the operation of the various mechanisms. We start by addressing the Hello protocol in Section 10.1 and the designated router election in Section 10.2. Sections 10.3 and 10.4 cover the flooding and deletion procedures. Finally, in Section 10.5 we illustrate the initial LSDB synchronization process.

10.1   Hello protocol

In this section we perform several experiments that illustrate the operation of the Hello protocol. The network topology of Figure 10.1 is used in all experiments, except in the one of Section 10.1.4 related to IS-IS point-to-point links. It includes four routers attached to the same shared link using an Ethernet hub. The routers must be configured initially with the basic routing configuration. In the case of OSPFv3, there is no need to configure IPv6 addresses at the interfaces. In the case of IS-IS, the routers must be configured as L1/L2 routers, with no restrictions on the interface type (in which case, the interfaces will also be of the L1/L2 type). We do not show experiments with IPv6 IS-IS since the behavior is the same as IPv4 IS-IS.

Why routers must be connected through a hub? We highlight that the routers must be connected through a hub and not through a switch. This is important for the OSPF experiments, since some packets are unicasted during the designated router election process. Switches filter unicast traffic and, therefore, this type of traffic may not be observed by Wireshark probes located outside the traffic's path. For example, a probe installed at interface f0/0 of router R1 would not see unicast traffic exchanged between routers R2 and R3.

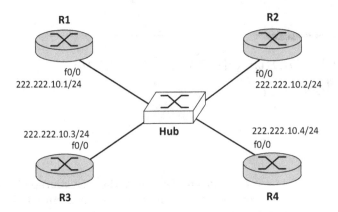

FIGURE 10.1: Hello protocol experiments - Network topology.

10.1.1 Periodic transmission of HELLO packets

We analyze first the periodic transmission of HELLO packets in OSPF and IS-IS under stable conditions, i.e. in a converged network.

Experimental procedure The experimental procedure is very simple in this case: (i) start all routers at the same time and wait until the network converges (at least 40 seconds), (ii) install a Wireshark probe at the shared link with an **ospf** or **isis** filter, depending on the case.

Wireshark capture OSPFv2 Figure 10.2 shows the OSPFv2 Wireshark capture. The HELLO packets are all transmitted using the OSPFv2 AllSPFRouters multicast address (224.0.0.5). Notice that each router transmits HELLO packets on the link with a periodicity of approximately 10 seconds.

A close look at an OSPFv2 HELLO Figure 10.3 shows a HELLO packet sent by R2. Notice first the layer-2 and layer-3 encapsulating headers, i.e. the Ethernet header and the IPv4 header. Since the network is stable, the packet lists all neighbors of R2 in the Active Neighbor fields, using their RIDs; the neighbors are R1 (RID=1.1.1.1), R3 (RID=3.3.3.3), and R4 (RID=4.4.4.4). Moreover, it indicates that router R4 is the DR and router R3 the BDR. However, it does not identify the DR and BDR using their RIDs, but using the IPv4 addresses assigned to their interfaces. The IPv4 address of the DR acts as the shared link identifier. The OSPFv2 HELLO also includes the Network Mask, Hello Interval, Router Dead Interval, and Router Priority values, besides the various options flags.

A close look at an OSPFv3 HELLO The periodic transmission of HELLO packets in OSPFv3 is similar to OSPFv2 (so, we do not show it). However, the contents of HELLO packets is slightly different. Figure 10.4 shows a HELLO packet sent by R2. It is interesting to compare with the OSPFv2 HELLO of

No.	Time	Source	Destination	Protocol	Info
5	3.882868	222.222.10.2	224.0.0.5	OSPF	Hello
6	3.976596	222.222.10.3	224.0.0.5	OSPF	Hello
7	4.054703	222.222.10.1	224.0.0.5	OSPF	Hello
8	4.413994	222.222.10.4	224.0.0.5	OSPF	Hello
13	13.896455	222.222.10.2	224.0.0.5	OSPF	Hello
14	13.946853	222.222.10.3	224.0.0.5	OSPF	Hello
15	14.047661	222.222.10.1	224.0.0.5	OSPF	Hello
16	14.441228	222.222.10.4	224.0.0.5	OSPF	Hello
21	23.895568	222.222.10.2	224.0.0.5	OSPF	Hello
22	23.98119	222.222.10.3	224.0.0.5	OSPF	Hello
23	24.031086	222.222.10.1	224.0.0.5	OSPF	Hello
24	24.458952	222.222.10.4	224.0.0.5	OSPF	Hello

FIGURE 10.2: Periodic transmission of OSPFv2 HELLO packets under stable conditions - Wireshark capture.

Figure 10.3. The layer-3 encapsulating header is now an IPv6 header; notice that the source address is the IPv6 link-local address assigned to the f0/0 interface of R2, and the destination address is the IPv6 AllSPFRouters address (ff02::5). The neighbors of R2 are identified as in OSPFv2, i.e. by including their RIDs in the Active Neighbor fields. The main difference is in the identification of the DR and BDR that, in OSPFv3, is also done using the RIDs. The Interface ID is the interface number of the sending interface, which in this case is 4. Recall that the Interface ID of the DR, together with its RID, is what defines the shared link identifier. The Hello Interval, Router Dead Interval, and Router Priority fields also exist in the OSPFv2 HELLO.

```
> Ethernet II, Src: c2:02:08:60:00:00 (c2:02:08:60:00:00), Dst: I
> Internet Protocol Version 4, Src: 222.222.10.2, Dst: 224.0.0.5
v Open Shortest Path First
  > OSPF Header
  v OSPF Hello Packet
      Network Mask: 255.255.255.0
      Hello Interval [sec]: 10
    > Options: 0x12 ((L) LLS Data block, (E) External Routing)
      Router Priority: 1
      Router Dead Interval [sec]: 40
      Designated Router: 222.222.10.4
      Backup Designated Router: 222.222.10.3
      Active Neighbor: 1.1.1.1
      Active Neighbor: 3.3.3.3
      Active Neighbor: 4.4.4.4
  > OSPF LLS Data Block
```

FIGURE 10.3: Periodic transmission of OSPFv2 HELLO packets under stable conditions - OSPFv2 HELLO packet transmitted by router R2.

```
Ethernet II, Src: c2:02:04:3c:00:00 (c2:02:04:3c:00:00), Dst: IPv6mcas
Internet Protocol Version 6, Src: fe80::c002:4ff:fe3c:0, Dst: ff02::5
Open Shortest Path First
> OSPF Header
v OSPF Hello Packet
      Interface ID: 4
      Router Priority: 1
   > Options: 0x000013 (R, E, V6)
      Hello Interval [sec]: 10
      Router Dead Interval [sec]: 40
      Designated Router: 4.4.4.4
      Backup Designated Router: 3.3.3.3
      Active Neighbor: 3.3.3.3
      Active Neighbor: 1.1.1.1
      Active Neighbor: 4.4.4.4
```

FIGURE 10.4: Periodic transmission of OSPFv3 HELLO packets under stable conditions - OSPFv3 HELLO packet transmitted by router R2.

However, the `Network Mask` was removed since OSPFv3 HELLOs carry no addressing information.

Wireshark capture IS-IS Figure 10.5 shows the IS-IS Wireshark capture. In relation to the original capture, we only kept the HELLO packets; recall that, in IS-IS shared links, CSNPs are also transmitted periodically. Since the router interfaces are of the L1/L2 type, each interface sends both L1 and L2 LAN IS-IS HELLOs. The source and destination indicated by Wireshark are the MAC addresses, since IS-IS control packets are directly encapsulated in layer-2 packets (in this case Ethernet packets). Notice that the L1 type packets are sent using the `AllL1ISs` multicast address (indicated as ISIS-all-level-1-IS's in Wireshark), and the L2 type packets are sent using the `AllL2ISs` address (indicated as ISIS-all-level-2-IS's in Wireshark). Regarding the periodicity of HELLO transmissions, note that R4 transmits HELLOs more frequently than other routers, since it is the link DIS. The periodicity is approximately 3 seconds for R4 and approximately 10 seconds (sometimes less) for the other routers.

A close look at an L1 LAN IS-IS HELLO Figure 10.6 shows an L1 LAN IS-IS HELLO packet sent by router R2. First notice the layer-2 header, which includes the `AllL1ISs` multicast address (`01:80:c2:00:00:14`) as the destination address. We do not show the IS-IS packet header, but that is where the packet type is indicated (L1 LAN IS-IS HELLOs have a packet type of 15). In the HELLO packet there are many things to notice. The `Circuit Type` bits confirm that this packet was sent from an L1/L2 interface (its value is 3). The `Source ID` field (indicated as SystemID Sender of PDU) confirms that the packet was sent by router R2. The `LAN ID` field (indicated as SystemID Designated DIS) confirms that router R4 is the DIS. We also learn that the `Holding Timer` is 30 seconds and interface `Priority` is 64. The remaining contents of the HELLO are TLVs: the Protocols Supported, the Area Ad-

No.	Time	Source	Destination	Protocol	Info
1	0	c2:02:2d:68:00:00	ISIS-all-level-1-IS's	ISIS HELLO	L1, SID=R2
2	0.853716	c2:04:44:4c:00:00	ISIS-all-level-1-IS's	ISIS HELLO	L1, SID=R4
3	0.897599	c2:01:2f:40:00:00	ISIS-all-level-2-IS's	ISIS HELLO	L2 HELLO, SID=R1
4	1.356372	c2:04:44:4c:00:00	ISIS-all-level-2-IS's	ISIS HELLO	L2 HELLO, SID=R4
5	2.520259	c2:03:0f:f8:00:00	ISIS-all-level-1-IS's	ISIS HELLO	L1, SID=R3
6	3.595476	c2:04:44:4c:00:00	ISIS-all-level-1-IS's	ISIS HELLO	L1, SID=R4
8	4.318541	c2:03:0f:f8:00:00	ISIS-all-level-2-IS's	ISIS HELLO	L2 HELLO, SID=R3
10	4.66262	c2:04:44:4c:00:00	ISIS-all-level-2-IS's	ISIS HELLO	L2 HELLO, SID=R4
14	4.908961	c2:02:2d:68:00:00	ISIS-all-level-2-IS's	ISIS HELLO	L2 HELLO, SID=R2
16	6.556554	c2:04:44:4c:00:00	ISIS-all-level-1-IS's	ISIS HELLO	L1, SID=R4
17	7.269646	c2:01:2f:40:00:00	ISIS-all-level-1-IS's	ISIS HELLO	L1, SID=R1
18	7.640654	c2:04:44:4c:00:00	ISIS-all-level-2-IS's	ISIS HELLO	L2 HELLO, SID=R4
22	8.921228	c2:02:2d:68:00:00	ISIS-all-level-1-IS's	ISIS HELLO	L1, SID=R2
24	9.823814	c2:04:44:4c:00:00	ISIS-all-level-1-IS's	ISIS HELLO	L1, SID=R4
25	10.53192	c2:04:44:4c:00:00	ISIS-all-level-2-IS's	ISIS HELLO	L2 HELLO, SID=R4
26	10.74635	c2:01:2f:40:00:00	ISIS-all-level-2-IS's	ISIS HELLO	L2 HELLO, SID=R1
27	12.24434	c2:03:0f:f8:00:00	ISIS-all-level-2-IS's	ISIS HELLO	L2 HELLO, SID=R3
28	12.50265	c2:03:0f:f8:00:00	ISIS-all-level-1-IS's	ISIS HELLO	L1, SID=R3
30	12.90158	c2:04:44:4c:00:00	ISIS-all-level-1-IS's	ISIS HELLO	L1, SID=R4
32	13.65856	c2:04:44:4c:00:00	ISIS-all-level-2-IS's	ISIS HELLO	L2 HELLO, SID=R4
33	14.50928	c2:02:2d:68:00:00	ISIS-all-level-2-IS's	ISIS HELLO	L2 HELLO, SID=R2
34	15.61133	c2:01:2f:40:00:00	ISIS-all-level-1-IS's	ISIS HELLO	L1, SID=R1

R1=0000.0000.0001, R2=0000.0000.0002, R3=0000.0000.0003, R4=0000.0000.0004

FIGURE 10.5: Periodic transmission of LAN IS-IS HELLO packets under stable conditions - Wireshark capture.

dresses, the IP Interface Address, the Restart Signaling (not covered in this book), and the IS Neighbors (of type 6). The packet also includes several Padding TLVs, which we do not include in the figure. Regarding the TLVs, only the contents of the IS Neighbors TLV is shown. This TLV identifies the neighbors of R2 using their MAC addresses. According to the Wireshark capture of Figure 10.5, we can check that `c2:04:44:4c:00:00` belongs to R4, `c2:03:0f:f8:00:00` belongs to R3, and `c2:01:2f:40:00:00` belongs to R1.

10.1.2 Detection of router failures

We will now simulate a router failure and observe the reaction. The behavior is similar in OSPF and IS-IS. Thus, we only illustrate the case of OSPFv2.

Experimental procedure The procedure is the following: (i) switch on all routers at the same time and wait until the network converges; (ii) install a Wireshark probe at the shared link with an **ospf** filter; (iii) switch off router R1.

```
∨  IEEE 802.3 Ethernet
    >  Destination: ISIS-all-level-1-IS's (01:80:c2:00:00:14)
    >  Source: c2:02:2d:68:00:00 (c2:02:2d:68:00:00)
       Length: 1500
>  Logical-Link Control
>  ISO 10589 ISIS InTRA Domain Routeing Information Exchange Protocol
∨  ISIS HELLO
       .... ..11 = Circuit type: Level 1 and 2 (0x3)
       0000 00.. = Reserved: 0x00
       SystemID {Sender of PDU}: 0000.0000.0002
       Holding timer: 30
       PDU length: 1497
       .100 0000 = Priority: 64
       0... .... = Reserved: 0
       SystemID {Designated IS}: 0000.0000.0004.01
    >  Protocols Supported (t=129, l=1)
    >  Area address(es) (t=1, l=4)
    >  IP Interface address(es) (t=132, l=4)
    >  Restart Signaling (t=211, l=3)
    ∨  IS Neighbor(s) (t=6, l=18)
          Type: 6
          Length: 18
          IS Neighbor: c2:04:44:4c:00:00 (c2:04:44:4c:00:00)
          IS Neighbor: c2:01:2f:40:00:00 (c2:01:2f:40:00:00)
          IS Neighbor: c2:03:0f:f8:00:00 (c2:03:0f:f8:00:00)
```

FIGURE 10.6: Periodic transmission of LAN IS-IS HELLO packets under stable conditions - L1 LAN IS-IS HELLO packet transmitted by router R2.

Wireshark capture Figure 10.7 shows the Wireshark capture. In relation to the original capture, we only kept the HELLO packets. The capture starts with the last HELLO sent by R1, before the failure (packet 31); it lists R2, R3, and R4 as neighbors. Then, it follows a period where the remaining routers keep advertising R1 as neighbor (from packet 32 to packet 56). The first router recognizing that R1 is no longer active is router R4 in packet 57, and then R2 and R3 also recognize it in packets 67 and 68.

It is interesting to look in detail at the timing of events (see Figure 10.8). The time to detect a failure is `RouterDeadInterval`, which has a default value of 40 seconds. When R2, R3, and R4 receive the last HELLO from R1, they reset their Hello liveness timers. Since they will no longer receive HELLOs from R1, the timers will keep advancing until they reach 40 seconds, at which point they declare R1 as dead. This happens at approximately second 30.06+40=70.06. The time at which the first HELLO packet without R1 listed is transmitted depends on the relative timing of the HELLO transmissions of the various routers. The capture shows that the R4 transmissions occur immediately after the R1 transmission, and the R2 and R3 transmissions occur immediately before. Thus, when the Hello liveness timers expire (and they all will at second 70.06), R4 has a scheduled HELLO transmission to happen in less than a second (packet 57 at second 70.28), but the next HELLO

No.	Time	Source	Destination	Protocol	Info
31	30.063489	222.222.10.1	224.0.0.5	OSPF	Hello, AN=R2,R3,R4
32	30.295869	222.222.10.4	224.0.0.5	OSPF	Hello, AN=R1,R2,R3
37	40.025866	222.222.10.3	224.0.0.5	OSPF	Hello, AN=R1,R2,R4
38	40.025866	222.222.10.2	224.0.0.5	OSPF	Hello, AN=R1,R3,R4
39	40.302128	222.222.10.4	224.0.0.5	OSPF	Hello, AN=R1,R2,R3
43	49.99972	222.222.10.2	224.0.0.5	OSPF	Hello, AN=R1,R3,R4
44	50.035624	222.222.10.3	224.0.0.5	OSPF	Hello, AN=R1,R2,R4
45	50.299918	222.222.10.4	224.0.0.5	OSPF	Hello, AN=R1,R2,R3
49	60.029483	222.222.10.2	224.0.0.5	OSPF	Hello, AN=R1,R3,R4
50	60.068379	222.222.10.3	224.0.0.5	OSPF	Hello, AN=R1,R2,R4
51	60.341651	222.222.10.4	224.0.0.5	OSPF	Hello, AN=R1,R2,R3
55	70.01032	222.222.10.2	224.0.0.5	OSPF	Hello, AN=R1,R3,R4
56	70.031264	222.222.10.3	224.0.0.5	OSPF	Hello, AN=R1,R2,R4
57	70.281595	222.222.10.4	224.0.0.5	OSPF	Hello, AN=R2,R3
67	80.002633	222.222.10.2	224.0.0.5	OSPF	Hello, AN=R3,R4
68	80.025572	222.222.10.3	224.0.0.5	OSPF	Hello, AN=R2,R4

AN = Active Neighbor, R1 = 1.1.1.1, R2 = 2.2.2.2, R3 = 3.3.3.3, R4 = 4.4.4.4

FIGURE 10.7: Detection of router failure in OSPFv2 - Wireshark capture.

transmissions of R2 and R3 happen only 10 seconds later (packets 67 and 68 at second 80).

10.1.3 Verifying bidirectional relationships in OSPF

In this experiment, we analyze how routers verify that they can communicate in both directions with neighbors. The process is the same for all technologies and link types, except for IS-IS point-to-point links. In this section, we illustrate the process for the case of OSPFv2 shared links. IS-IS point-to-point links will be discussed in Section 10.1.4.

Experimental procedure In this experiment, we use Wireshark and the debug feature of Cisco IOS. The procedure is the following: (i) start routers

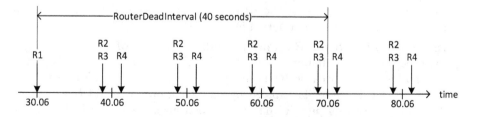

FIGURE 10.8: Detection of router failure in OSPFv2 - Relative timing of HELLO transmissions and liveness timer expiration.

(a)

No.	Time	Source	Destination	Protocol	Info
32	39.23812	222.222.10.1	224.0.0.5	OSPF	Hello, no neighbors
41	40.540638	222.222.10.4	224.0.0.5	OSPF	Hello, AN=R1,R2,R3
42	40.552606	222.222.10.1	222.222.10.4	OSPF	Hello, AN=R4
46	42.903831	222.222.10.2	224.0.0.5	OSPF	Hello, AN=R1,R3,R4
47	42.91181	222.222.10.1	222.222.10.2	OSPF	Hello, AN=R2,R4
49	45.129883	222.222.10.3	224.0.0.5	OSPF	Hello, AN=R1,R2,R4
77	49.138171	222.222.10.1	224.0.0.5	OSPF	Hello, AN=R2,R3,R4

AN = Active Neighbor, R1 = 1.1.1.1, R2 = 2.2.2.2, R3 = 3.3.3.3, R4 = 4.4.4.4

(b)

1	*Mar 1 00:03:43.515: OSPF: Rcv hello from 1.1.1.1 area 0 from FastEthernet0/0 222.222.10.1
2	*Mar 1 00:03:43.515: OSPF: Send immediate hello to nbr 1.1.1.1, src address 222.222.10.1, on FastEthernet0/0
3	*Mar 1 00:03:43.519: OSPF: Send hello to 222.222.10.1 area 0 on FastEthernet0/0 from 222.222.10.4
4	*Mar 1 00:03:43.519: OSPF: End of hello processing
5	*Mar 1 00:03:44.795: OSPF: Send hello to 224.0.0.5 area 0 on FastEthernet0/0 from 222.222.10.4
6	*Mar 1 00:03:44.827: OSPF: Rcv hello from 1.1.1.1 area 0 from FastEthernet0/0 222.222.10.1
7	*Mar 1 00:03:44.827: OSPF: 2 Way Communication to 1.1.1.1 on FastEthernet0/0, state 2WAY

FIGURE 10.9: Verifying bidirectional relationships in OSPFv2 - (a) Wireshark capture and (b) debug output messages of R4.

R2, R3, and R4 and wait until the network converges; (ii) insert the commands **debug ip ospf hello** and **debug ip ospf adj** in router R4; (iii) install a Wireshark probe at the shared link with an **ospf** filter; (iv) switch on router R1.

Wireshark capture and debug output Figure 10.9 shows the Wireshark capture and the debug output of router R4. In relation to the original Wireshark capture, we only kept the HELLO packets. In our discussion, we will use these two pieces of information in parallel.

Initial HELLO from R1 and periodic HELLO of R4 When R1 is switched on, it immediately sends a HELLO packet (packet 32). This packet lists no neighbors, because R1 hasn't yet received any HELLO packet. The reception of this packet is signaled by the debug output of R4 (message 1). According to this output, R4 decides to send an immediate HELLO to R1 (message 2), which in fact it doesn't. The immediate HELLOs are unicasted in OSPF, and the Wireshark capture shows that there is no unicasted HELLO from R4 to R1. We guess that this transmission was canceled since there was

Relationships	2-Way state
R1 → R2	when packet 46 received
R1 → R3	when packet 49 received
R1 → R4	when packet 41 received
R2 → R1	when packet 47 received
R3 → R1	when packet 77 received
R4 → R1	when packet 42 received

FIGURE 10.10: Verifying bidirectional relationships in OSPFv2 - When neighborhood relationships become bidirectional.

a periodic HELLO transmission scheduled for the same time (packet 41 and message 3). In packet 41, router R4 already lists R1 as neighbor.

Reaction to packet 41 When R1 receives packet 41, its neighborhood relationship with R4 moves to the 2-Way state (because R1 sees itself listed, for the first time, in a HELLO sent by R4). R1 also decides to send an immediate HELLO to R4 (packet 42); recall from the discussion in Section 6.2.7, that a router sends an immediate HELLO to a neighbor when it receives a HELLO from that neighbor and the state of their relationship is less than 2-Way; this is precisely the case of R1 when it received the HELLO from R4 (packet 41). Note that in packet 42, R1 only lists R4 as neighbor.

Reaction to packet 42 Message 6 of the debug output of R4 signals the reception of packet 42 sent by R1. At this point, the relationship between R4 and R1 (as seen by R4) reaches the 2-Way state, and this event is signaled by message 7 of the debug output.

Reaction to packet 46 Packet 46 is a periodic HELLO sent by R2 where R1 is already listed. Again, when R1 receives this packet, its relationship with R2 moves to the 2-Way state, and R1 also decides to send an immediate HELLO to R2 (packet 47). In this packet, R1 still doesn't list R3 as neighbor.

Reaction to packet 47 When R2 receives packet 47, the state of its relationship with R1 moves to the 2-Way state.

Packets 49 and 77 Packet 49 is the first periodic HELLO sent by R3, which already lists R1 as neighbor, and packet 77 is the first periodic HELLO sent by R1 where all neighbors are listed.

When each relationship becomes bidirectional Figure 10.10 summarizes the events whereby each relationship moves to the 2-Way state. Only the six relationships involving R1 are listed, since all other relationships were formed before R1 has been switched on.

No.	Time	Source	Destination	Protocol	Info
9	29.92082	N/A	N/A	ISIS HELLO	P2P HELLO, SID=R1, Adj: Down
15	34.48821	N/A	N/A	ISIS HELLO	P2P HELLO, SID=R2, Adj: Down
16	34.49219	N/A	N/A	ISIS HELLO	P2P HELLO, SID=R1, Adj: Initializing
17	34.49618	N/A	N/A	ISIS HELLO	P2P HELLO, SID=R2, Adj: Initializing
18	34.50021	N/A	N/A	ISIS HELLO	P2P HELLO, SID=R1, Adj: Up
19	34.5042	N/A	N/A	ISIS HELLO	P2P HELLO, SID=R2, Adj: Up

R1=0000.0000.0001, R2=0000.0000.0002, R3=0000.0000.0003, R4=0000.0000.0004,

Adj = Adjacency

FIGURE 10.11: Verifying bidirectional relationships in IS-IS point-to-point links - Wireshark capture.

10.1.4 Verifying bidirectional relationships in IS-IS point-to-point links

The verification of bidirectional relationships is done differently in IS-IS point-to-point links. As explained in Section 6.2.6, it uses the Point-to-Point Three-Way Adjacency TLV, which advertises the state of the interface that transmits the POINT-TO-POINT IS-IS HELLO packet (instead of its identifier).

Experimental set up and procedure To test this feature just set up a simple network of two routers connected by a point-to-point link and configure the routers with the basic IS-IS routing configuration (restricting the interfaces to being L1-only). In GNS3, you will only be able to install a Wireshark probe at a serial link if the routers are switched on. Thus, the experimental procedure must be the following: (i) switch on the two routers, (ii) shutdown one router interface and let the network converge, (iii) install a Wireshark probe at the link, and (iv) switch on the interface again.

Wireshark capture The Wireshark capture is shown in Figure 10.11. We only show the packets until the interface reached the final state (which is the Up state). In first two packets (packets 9 and 15), the interfaces declare the Down state. When interfaces receive these packets, they move to the Initializing state and send HELLO packets declaring this fact (packets 16 and 17). Interfaces are in this state when they have received a HELLO from a neighbor, but are not sure that the neighbor received its own HELLO. Finally, when the interfaces receive HELLOs declaring the Initializing state, they move to the Up state and send HELLOs declaring this fact (packets 18 and 19). From then on, the interfaces will keep sending HELLOs periodically, where they declare the Up state. The number of messages could be less, as in the example of Figure 6.9.b. However, in this experiment the two routers initiate the process at the same time, and that is why six messages are exchanged.

Figure 10.12 shows the contents of packet 17, sent by R2 when it was in the Initializing state. The packet includes information about the Circuit Type (L1-only in this case) and the SID of the sending router (indicated in SystemID Sender of PDU field). Notice also the presence of the Point-to-

```
Cisco HDLC
ISO 10589 ISIS InTRA Domain Routeing Information Exchange Protocol
ISIS HELLO
    .... ..01 = Circuit type: Level 1 only (0x1)
    0000 00.. = Reserved: 0x00
    SystemID {Sender of PDU}: 0000.0000.0002
    Holding timer: 30
    PDU length: 1499
    Local circuit ID: 0
 >  Restart Signaling (t=211, l=3)
 ⌄  Point-to-point Adjacency State (t=240, l=1)
        Type: 240
        Length: 1
        Adjacency State: Initializing (1)
 >  Protocols Supported (t=129, l=1)
 >  Area address(es) (t=1, l=4)
 >  IP Interface address(es) (t=132, l=4)
```

FIGURE 10.12: Verifying bidirectional relationships in IS-IS point-to-point links - POINT-TO-POINT IS-IS HELLO sent by router R2.

Point Three-Way Adjacency TLV (type 240) where the `Initializing` state is declared (coded with value 1).

10.1.5 OSPF adjacencies

We will now check that OSPF adjacencies cannot be established between two interfaces if they are not configured with the same `Area ID`, `HelloInterval`, or `RouterDeadInterval`. This issue was discussed in Section 6.2.3.2. The three rules mentioned above apply to both OSPFv2 and OSPFv3, and we will illustrate them using an OSPFv2 network. We leave the rules that are specific of each technology as additional experiments.

Experimental procedure We will perform a sequence of modifications to the basic routing configuration. Thus, switch on all routers at the same time and wait until the network converges. To start with, run the **show ip ospf neighbor** command at router R4 to check that R4 recognizes R1, R2, and R3 as adjacent neighbors (we will keep our eye on R4; R2 and R3 follow the same behavior). The result of this command is shown in Figure 10.13.a. We will now perform three experiments.

Change of HelloInterval At interface f0/0 of R1, enter **ip ospf hello-interval 20**. This sets the `HelloInterval` value to 20 seconds. Now, go to the console of R4 and wait until you see a warning message indicating that the adjacency with R1 finished (you will have to wait approximately 40 seconds). Then, run the **show ip ospf neighbor** command at R4 to check that R1 is no longer considered neighbor. This is shown in Figure 10.13.b.

Change of RouterDeadInterval Again at interface f0/0 of R1, enter **no ip ospf hello-interval**; this will return the `HelloInterval` back to the default

(a)

Neighbor ID	Pri	State	Dead Time	Address	Interface
1.1.1.1	1	FULL/DROTHER	00:00:32	222.222.10.1	FastEthernet0/0
2.2.2.2	1	FULL/DROTHER	00:00:33	222.222.10.2	FastEthernet0/0
3.3.3.3	1	FULL/BDR	00:00:32	222.222.10.3	FastEthernet0/0

(b)

R4#
*Mar 1 00:08:36.247: %OSPF-5-ADJCHG: Process 1, Nbr 1.1.1.1 on FastEthernet0/0 from FULL to DOWN, Neighbor Down: Dead timer expired

R4#show ip ospf neighbor

Neighbor ID	Pri	State	Dead Time	Address	Interface
2.2.2.2	1	FULL/DROTHER	00:00:32	222.222.10.2	FastEthernet0/0
3.3.3.3	1	FULL/BDR	00:00:39	222.222.10.3	FastEthernet0/0

FIGURE 10.13: OSPF adjacencies between R4 and other routers (a) before and (b) after changing the HelloInterval of R1.

value (10 seconds). Use the **show ip ospf neighbor** command to check that the adjacency between R1 and R4 was reestablished very quickly. Now, enter **ip ospf dead-interval 100** at the same interface. This sets the value of RouterDeadInterval to 100 seconds. The result is the same as the previous experiment: after a waiting period of approximately 40 seconds, R1 ceases to be considered neighbor of R4 (check this again with the **show ip ospf neighbor** command).

Change of Area ID Again at interface f0/0 of R1, enter **no ip ospf dead-interval** to reset RouterDeadInterval to its default value. We will now change the area of interface f0/0 of R1 to area 1. To do that, enter the command **network 222.222.10.0 0.0.0.255 area 1** in the router mode. You will see a similar behavior as the one of the two previous experiments.

Now change also the area of interface f0/0 of R2 to area 1. You will notice that R2 becomes adjacent R1 (because they have matching parameters), and R3 and R4 stay as adjacent neighbors. Thus, there has been a partition in neighborhood relationships. However, this configuration is not allowed in OSPF since a link must belong to just one area.

10.1.6   IS-IS adjacencies

The goal in this section is to explore the various types of adjacencies in IS-IS, as determined by the interface type (L1-only, L2-only, and L1/L2), and the declared addresses (in the Area Addresses TLV).

Experimental procedure We perform a sequence of four changes to the configuration of router R1 and observe the results at router R4, starting from the basic IS-IS routing configuration, with all routers and interfaces configured as L1/L2. We use Wireshark in one experiment. To start with, switch on all routers at the same time and wait until the network converges.

	System Id	Type	Interface	IP Address	State	Holdtime	Circuit Id
	R3	L1	Fa0/0	222.222.10.3	UP	23	R4.01
	R3	L2	Fa0/0	222.222.10.3	UP	24	R4.01
(a)	R2	L1	Fa0/0	222.222.10.2	UP	24	R4.01
	R2	L2	Fa0/0	222.222.10.2	UP	23	R4.01
	R1	L1	Fa0/0	222.222.10.1	UP	25	R4.01
	R1	L2	Fa0/0	222.222.10.1	UP	24	R4.01

	System Id	Type	Interface	IP Address	State	Holdtime	Circuit Id
	R3	L1	Fa0/0	222.222.10.3	UP	23	R4.01
	R3	L2	Fa0/0	222.222.10.3	UP	23	R4.01
(b)	R2	L1	Fa0/0	222.222.10.2	UP	25	R4.01
	R2	L2	Fa0/0	222.222.10.2	UP	29	R4.01
	R1	L1	Fa0/0	222.222.10.1	UP	28	R4.01

	System Id	Type	Interface	IP Address	State	Holdtime	Circuit Id
	R3	L1	Fa0/0	222.222.10.3	UP	21	R4.01
(c)	R3	L2	Fa0/0	222.222.10.3	UP	22	R4.01
	R2	L1	Fa0/0	222.222.10.2	UP	21	R4.01
	R2	L2	Fa0/0	222.222.10.2	UP	28	R4.01

	System Id	Type	Interface	IP Address	State	Holdtime	Circuit Id
	R1	L2	Fa0/0	222.222.10.1	UP	27	R4.01
	R2	L1	Fa0/0	222.222.10.2	UP	28	R4.01
(d)	R2	L2	Fa0/0	222.222.10.2	UP	26	R4.01
	R3	L1	Fa0/0	222.222.10.3	UP	22	R4.01
	R3	L2	Fa0/0	222.222.10.3	UP	29	R4.01

FIGURE 10.14: IS-IS adjacencies seen by R4 when (a) all interfaces are L1/L2, (b) R1 interface is L1-only, (c) R1 interface is L1-only and R1 is in a different area, and (d) R1 interface is L2-only and R1 is in a different area.

All interfaces of L1/L2 type Figure 10.14.a displays the result of the **show isis neighbors** at R4. Since all interfaces are of L1/L2 type, R4 establishes three L1 adjacencies and three L2 adjacencies with its neighbors.

R1 interface is L1-only Now configure the interface f0/0 of router R1 as L1-only. To do that, enter the command **isis circuit-type level-1** at this interface. Since an L1-only interface can only establish L1 adjacencies, the L2 adjacency that existed previously between R1 and R4 is destroyed; only the L1 adjacency remains. This is shown in 10.14.b.

R1 interface is L1-only and R1 has two configured areas The next experiment is to add area 2 to router R1. To do that, insert the command **net 49.0002.0000.0000.0001.00** in the router isis mode of R1. Now, R1 is configured with two areas and will advertise this fact in the L1 HELLO packets it transmits. This is shown in Figure 10.15. The Area Addresses TLV advertises two areas, area 1 and area 2. Note also that the `Circuit Type` of the sending interface is L1-only. In this case, there is no change in the adjacencies

```
IEEE 802.3 Ethernet
Logical-Link Control
ISO 10589 ISIS InTRA Domain Routeing Information Exchange Protocol
ISIS HELLO
    .... ..01 = Circuit type: Level 1 only (0x1)
    0000 00.. = Reserved: 0x00
    SystemID {Sender of PDU}: 0000.0000.0001
    Holding timer: 30
    PDU length: 1497
    .100 0000 = Priority: 64
    0... .... = Reserved: 0
    SystemID {Designated IS}: 0000.0000.0004.01
>   Protocols Supported (t=129, l=1)
v   Area address(es) (t=1, l=8)
        Type: 1
        Length: 8
        Area address (3): 49.0001
        Area address (3): 49.0002
>   IP Interface address(es) (t=132, l=4)
>   Restart Signaling (t=211, l=3)
>   IS Neighbor(s) (t=6, l=18)
```

FIGURE 10.15: IS-IS adjacencies - L1 HELLO sent by R1 advertising two configured areas.

between R1 and R4, since they still have one area in common (which is area 1). You can check that using the command **show isis neighbors**.

R1 interface is L1-only and R1 in a different area Now, let us remove area 1 from router R1. To do that, enter the command **no net 49.0001.0000.0000.0001.00** in the router isis mode of R1. The result is shown in Figure 10.14.c. There are no longer adjacencies between R1 and R4 since (i) R1 only transmits L1 HELLOs and therefore can only establish L1 adjacencies with neighbors, and (ii) it has no area in common with its neighbors.

R1 interface is L2-only and R1 in a different area Finally, we change the interface type of R1 to L2-only (without changing the configured area). To do that, enter the command **isis circuit-type level-2** at interface f0/0 of the router. Figure 10.14.d shows that R1 and R4 reestablished an adjacency of type L2. This is because L2 adjacencies are not restricted by the configured areas: they can be established between two interfaces that transmit L2 HELLOs (i.e. configured as L2-only or L1/L2) even if the corresponding routers have no area in common.

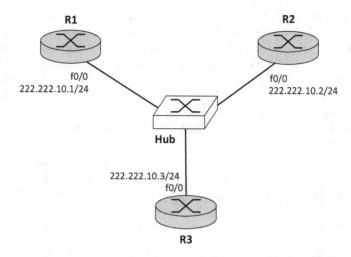

FIGURE 10.16: Experiments related to the election of the designated routers - Network topology.

10.2 Designated router election

This section presents several experiments that illustrate the designated router election process in OSPF and IS-IS. This issue was discussed in Section 6.3.

Experimental setup The experiments use the network topology of Figure 10.16, with three routers attached to a shared link. They will be performed with the IPv4 versions of both protocols, since the behavior of the IPv6 versions is similar. The routers must be initially configured with the basic routing configurations. Both Wireshark and the debug feature of Cisco IOS will be used in these experiments.

10.2.1 OSPF

We will address five different cases: (i) cold start, (ii) cold start with router priorities, (iii) router joining the link when DR and BDR already elected, (iv) only one router started at link, and (v) DR failure.

10.2.1.1 Cold start

The objective of this experiment is to analyze how a set of routers attached to a shared link elect the DR and BDR when started at approximately the same time.

Experimental procedure Before starting the routers install a Wireshark probe at one of the link interfaces (e.g. interface f0/0 of R1) and configure an **ospf** filter. Then start all routers at the same time. The experiment can be stopped when the correct DR and BDR get elected. This can be checked by looking at the contents of the HELLO packets; it can also be checked by running the **show ip ospf neighbor** command at each router. At the end of the experiment, the DR must be router R3 because it has the highest RID, and the BDR must be router R2 because it has the second highest RID. Note that in this experiment, all routers have the same priority. We can confirm the role of each interface by running the **show ip ospf neighbor** command at each router console.

Wireshark capture Figure 10.17 shows the Wireshark capture. In relation to the original capture, we kept only the HELLO packets and the first DB DESCRIPTION packet transmitted by each router. We kept the DB DE-SCRIPTION packets since they indicate the end of the election process at a router. We start by describing the main parts of the capture:

- **Packets 6, 11, 20** - First HELLO packets sent by each interface.

- **Packets 34 to 39** - HELLO packets sent when interfaces complete establishing their neighborhood relationships, i.e. when communication between each pair of routers becomes bidirectional; some immediate HELLOs are observed during this period.

- **Packets 43 to 45, 49 to 51, 55, 56, 60** - HELLO packets sent when all interfaces already recognize their two neighbors, but the DR and BDR were still not elected.

- **Packets 57 to 59, 61, 89, 90** - First DB DESCRIPTION packet transmitted by each router after the election of the DR and BDR.

- **Packets 88, 99, 101** - HELLO packets sent after the election of the DR and BDR.

First HELLOs The first HELLO packets sent by each interface (packets 6, 11, 20) advertise both a DR and a BDR of 0.0.0.0 and do not list active neighbors. The subsequent packets list at least one neighbor, and some of them are immediate HELLOs.

Why are immediate HELLOs less than expected? The transmission of immediate HELLO packets is not easy to trace during cold start. In principle, a router should send an immediate HELLO to a neighbor when it receives a periodic HELLO from that neighbor and the state of their relationship is less than 2-Way (see Section 6.2.7). For example, one would expect to see an immediate HELLO from R3 to R2 immediately after the reception of packet 11 (sent by R2). However, routers take some time to stabilize after being switched on and, during this period, routers (i) may delay the transmission of its packets, and (ii) may not receive packets transmitted by other routers. Moreover, we guess that implementations may cancel the transmission of a scheduled immediate HELLO if, in the meantime, the relationship with the

No.	Time	Source	Destination	Protocol	Info
6	1.418964	222.222.10.3	224.0.0.5	OSPF	Hello, DR: 0, BDR: 0
11	1.953022	222.222.10.2	224.0.0.5	OSPF	Hello, DR: 0, BDR: 0
20	3.128556	222.222.10.1	224.0.0.5	OSPF	Hello, DR: 0, BDR: 0
34	10.433464	222.222.10.3	224.0.0.5	OSPF	Hello, DR: 0, BDR: 0, AN: R1, R2
35	10.453523	222.222.10.2	222.222.10.3	OSPF	Hello, DR: 0, BDR: 0, AN: R1, R3
36	10.47358	222.222.10.1	222.222.10.3	OSPF	Hello, DR: 0, BDR: 0, AN: R3
37	11.287938	222.222.10.2	224.0.0.5	OSPF	Hello, DR: 0, BDR: 0, AN: R1, R3
38	11.29897	222.222.10.1	222.222.10.2	OSPF	Hello, DR: 0, BDR: 0, AN: R2, R3
39	13.036823	222.222.10.1	224.0.0.5	OSPF	Hello, DR: 0, BDR: 0, AN: R2, R3
43	20.08127	222.222.10.3	224.0.0.5	OSPF	Hello, DR: 0, BDR: 0, AN: R1, R2
44	21.015473	222.222.10.2	224.0.0.5	OSPF	Hello, DR: 0, BDR: 0, AN: R1, R3
45	22.481149	222.222.10.1	224.0.0.5	OSPF	Hello, DR: 0, BDR: 0, AN: R2, R3
49	29.673992	222.222.10.3	224.0.0.5	OSPF	Hello, DR: 0, BDR: 0, AN: R1, R2
50	30.277771	222.222.10.2	224.0.0.5	OSPF	Hello, DR: 0, BDR: 0, AN: R1, R3
51	32.462841	222.222.10.1	224.0.0.5	OSPF	Hello, DR: 0, BDR: 0, AN: R2, R3
55	39.259131	222.222.10.3	224.0.0.5	OSPF	Hello, DR: 0, BDR: 0, AN: R1, R2
56	39.304785	222.222.10.2	224.0.0.5	OSPF	Hello, DR: 0, BDR: 0, AN: R1, R3
57	41.409546	222.222.10.3	222.222.10.1	OSPF	DB Description, I=1, M=1, MS=1
58	41.409546	222.222.10.3	222.222.10.2	OSPF	DB Description, I=1, M=1, MS=1
59	41.949757	222.222.10.2	222.222.10.3	OSPF	DB Description, I=1, M=1, MS=1
60	42.446438	222.222.10.1	224.0.0.5	OSPF	Hello, DR: 0, BDR: 0, AN: R2, R3
61	43.114915	222.222.10.1	222.222.10.3	OSPF	DB Description, I=1, M=1, MS=1
88	48.343105	222.222.10.3	224.0.0.5	OSPF	Hello, DR: r3, BDR: r2, AN: R1, R2
89	48.351639	222.222.10.2	222.222.10.1	OSPF	DB Description, I=1, M=1, MS=1
90	48.351639	222.222.10.1	222.222.10.2	OSPF	DB Description, I=1, M=1, MS=1
99	49.225802	222.222.10.2	224.0.0.5	OSPF	Hello, DR: r3, BDR: r2, AN: R1, R3
101	52.395042	222.222.10.1	224.0.0.5	OSPF	Hello, DR: r3, BDR: r2, AN: R2, R3

0 = 0.0.0.0, AN = Active Neighbor, R1 = 1.1.1.1, R2 = 2.2.2.2, R3 = 3.3.3.3, r2 = 222.222.10.2, r3 = 222.222.10.3

FIGURE 10.17: Election of the DR and BDR in cold start scenario - Wireshark capture.

corresponding neighbor reached the 2-Way state, or when the time to send a periodic HELLO has come. These facts may explain why the number of observed immediate HELLOs is usually less than expected.

Explaining the immediate HELLOs In the Wireshark capture, immediate HELLO packets are only observed in the period around second 12 (packets 34 to 39). They are easily recognized by their unicast destination addresses. Packet 35 sent from R2 to R3, and packet 36 from R1 to R3, were probably a response to packet 34. In fact, when R2 and R1 receive packet 34, the state of their relationships with R3 goes to the 2-Way state (this is the perspective of R1 and R2 regarding their relationship with R3); they both react sending the immediate HELLOs to R3 because they want to make sure that the state of

Relationships	2-Way state
R1 → R2	when packet 37 received
R1 → R3	when packet 34 received
R2 → R1	when packet 38 received
R2 → R3	when packet 34 received
R3 → R1	when packet 36 received
R3 → R2	when packet 35 received

FIGURE 10.18: Election of the DR and BDR in cold start scenario - When neighborhood relationships become bidirectional.

the relationships seen by R3 (i.e. from R3 to R1 and from R3 to R2) evolves to the 2-Way state as fast as possible.

Packet 36 only lists R3 as neighbor. When this packet was scheduled for transmission at router R1, the router still did not recognize R2 as neighbor, which means that it missed packet 11. Note that a router needs to receive a HELLO from a neighbor in order to recognize it as neighbor; it does not suffice seeing the neighbor listed in a HELLO sent by some other router. For example, router R1 could not recognize R2 as neighbor based on the information provided by R3 in packet 34. Recognizing a neighbor means that the relationship with the neighbor changes to the Init state.

Packet 38, sent from R1 to R2, is another immediate HELLO. This packet was probably a response to packet 37 and confirms that R1 might have not received packet 11. Note that router R3 did not send an immediate HELLO to router R2 in response to packet 37. This is because their relationship was already in 2-Way state at R3 (following the reception of packet 35).

When relationships become bidirectional It is interesting to determine when each neighborhood relationship becomes bidirectional (i.e. reaches the 2-Way state). This happens when a router sees itself listed in a HELLO packet received from a neighbor. Note that the two neighbors engaged in a relationship do not reach this state at the same time. With three routers there is a total of six relationships. For example, regarding the relationship between R1 and R3, it reaches the 2-Way state at R1 when R1 receives packet 34 since this is the first HELLO sent by R3 where R1 is listed. However, the relationship only reaches the 2-Way state at R3 when R3 receives packet 36, since this is the first HELLO sent by R1 where R3 is listed. Figure 10.18 shows when the various neighborhood relationships reach the 2-Way state.

Periodic HELLOs are not affected by immediate HELLOs The periodic HELLO packets are transmitted approximately every 10 seconds, irrespective of the immediate ones. Observe, for example, that router R1 sends periodic HELLO packets at seconds 3.1 (packet 20), 13.0 (packet 39), 22.5 (packet 45), 32.5 (packet 51), and so on.

Waiting until the election starts By second 21 (packets 43 to 45), all interfaces have indicated in their HELLO packets that they recognize all their

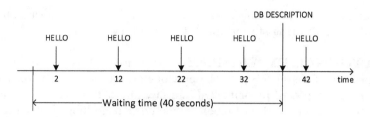

FIGURE 10.19: Election of the DR and BDR in cold start scenario - Desynchronization between the designated router election and the transmission of periodic HELLOs.

neighbors. However, the election process starts only when the `Waiting` state (of the interface state machine) is abandoned, which happens when the sojourn time in this state reaches 40 seconds. At this point in time, the interfaces elect the DR and BDR, move to the `ExStart` state with their two neighbors, and send them DB DESCRIPTION packets to initiate the LSDB synchronization process. Thus, the first DB DESCRIPTION packet transmitted by a router indicates that the router already elected the DR and BDR. In our case, since there are only three routers attached to the shared link, all interfaces must synchronize with each other: the DR and the BDR synchronize with each other, and each of them synchronizes with the non-DR router. Thus, all routers transmit DB DESCRIPTION packets.

The Wireshark capture shows that router R3 is the first to abandon the `Waiting` state and elect the DR and BDR, since it is the first to send DB DESCRIPTION packets (packets 57 and 58, at second 41.4). The remaining routers do it later.

Desynchronization between the designated router election and the transmission of periodic HELLOs The first HELLO packets indicating the correct DR and BDR are packets 88, 99, and 101. These packets are transmitted after the first DB DESCRIPTION packets, because the transmission of HELLO packets occurs periodically and is not synchronized with other events. When the `Waiting` period ends (in this case, through the `WaitTimer` event), the interfaces immediately elect the DR and BDR, move to the `ExStart` state with the neighbors that were selected to become adjacent, and send them DB DESCRIPTION packets. The announcement of the selected DR and BDR via periodic HELLO packets will have to wait until the next time scheduled for the transmission of these packets. This explains why packet 88 is sent several seconds after the DB DESCRIPTION packets sent by R3 (packets 57 and 58). Figure 10.19 illustrates the desynchronization between the designated router election and the transmission of periodic HELLOs.

Using the debug output The analysis of this experiment can be complemented with information provided by the debug output of the routers. With the **debug ip ospf adj** command one can follow the evolution of the interface and neighbor state machines. In particular, it can be seen that all routers abandon the `Waiting` state through the `WaitTimer` event.

ADDITIONAL EXPERIMENT - After network convergence, increase the priority of router R1 to 2 (the previous priority was 1). To do that enter the command **ip ospf priority 2** at interface f0/0 of this router. You can check with the `show ip ospf neighbor` command that nothing changed regarding who the DR and BDR are. In fact, changing the priority of one router triggers the `NeighborChange` event, and the election process is run again. However, the outcome of the election is the same as before: R3 keeps being the DR and R2 the BDR. That the election process is run again at each router can be checked using the **debug ip ospf adj** command.

10.2.1.2   Cold start with router priorities

This experiment repeats the one in the previous section but with preconfigured router priorities. We will give the highest priority to R1 and second highest priority to R2. Besides analyzing the outcome of the experiment, we will also be interested in tracing closely the evolution of the election process. We will use the debug output of Cisco IOS for that purpose. Wireshark captures will not be used this time.

Experimental procedure Before performing this experiment we need to change the basic routing configuration of R1 and R2. First, we must include in the startup-config file of R2 the commands required to debug the OSPF adjacencies from start (see Figure 8.7). Second, we must configure a router priority of 3 in interface f0/0 of R1 and a router priority of 2 in interface f0/0 of R2. This can also be done by editing the startup-config file. For example, to set the priority at R1, insert the command **ip ospf priority 3** in the interface mode of f0/0.

After performing these configurations we just need to repeat the procedure of the previous experiment: start the routers all at the same time and stop the experiment when the correct DR and BDR become elected.

Experiment outcome The outcome of the experiment is different from that in the previous section. The router with highest priority (router R1) becomes the DR and the router with second highest priority (router R2) becomes the BDR. This can be confirmed by running the command **show ip ospf neighbor** at R3 (see Figure 10.20). As in the previous experiment, the routers leave the `Waiting` state at approximately the same time and immediately run the election algorithm. Since each router had the chance to learn the RID and `Router Priority` of all its neighbors during the `Waiting` period, they all conclude that R1 must be the DR and R2 the BDR.

The election process at R2 and R3 It is interesting to notice that routers R2 and R3 take some time to arrive at the correct DR and BDR; they initially elect R1 as both DR and BDR, and only later elect R2 as BDR. We can confirm this by observing the debug output of R2 or R3. The behavior is the

```
R3#show ip ospf neighbor

Neighbor ID   Pri   State      Dead Time   Address        Interface
1.1.1.1        3    FULL/DR    00:00:39    222.222.10.1   FastEthernet0/0
2.2.2.2        2    FULL/BDR   00:00:39    222.222.10.2   FastEthernet0/0
```

FIGURE 10.20: Election of DR and BDR in cold start scenario with router priorities - OSPF neighbors of R3.

same in R2 and R3, but we only show, in Figure 10.21, the debug output of R2 (only the relevant messages are shown). Messages 1 and 2 indicate the time instants when the relationship with R1 and R3 reached the 2-Way state. Message 5 signals the end of the Waiting period (through the WaitTimer event). Messages 6 to 9 relate to the first election, immediately after R2 has abandoned the Waiting state. Router R2 elects R1 as BDR in step 2 of the election algorithm (message 7) and then R1 again as DR in step 3 (message 8). It then finishes the computation since its own role has not changed (message 9). Thus, for some time, R2 believes that R1 is both the DR and the BDR.

This situation is only changed when R2 receives a HELLO packet from R1 advertising R1 as DR and R2 as BDR. When the HELLO arrives, the NeighborChange event is triggered (message 35), and the election algorithm is run again (messages 35 to 41). In step 2, R2 is elected BDR (message 37), since R1 already advertised itself as DR and, therefore, must be removed from the computation. In step 3, R1 is confirmed as DR (message 38). The algorithm

1	*Feb 28 09:40:17.895: OSPF-1 ADJ Fa0/0: 2 Way Communication to 1.1.1.1, state 2WAY
2	*Feb 28 09:40:18.027: OSPF-1 ADJ Fa0/0: 2 Way Communication to 3.3.3.3, state 2WAY
5	*Feb 28 09:40:48.955: OSPF-1 ADJ Fa0/0: end of Wait on interface
6	*Feb 28 09:40:48.955: OSPF-1 ADJ Fa0/0: DR/BDR election
7	*Feb 28 09:40:48.955: OSPF-1 ADJ Fa0/0: Elect BDR 1.1.1.1
8	*Feb 28 09:40:48.959: OSPF-1 ADJ Fa0/0: Elect DR 1.1.1.1
9	*Feb 28 09:40:48.959: OSPF-1 ADJ Fa0/0: DR: 1.1.1.1 (Id) BDR: 1.1.1.1 (Id)
35	*Feb 28 09:40:56.967: OSPF-1 ADJ Fa0/0: Neighbor change event
36	*Feb 28 09:40:56.967: OSPF-1 ADJ Fa0/0: DR/BDR election
37	*Feb 28 09:40:56.967: OSPF-1 ADJ Fa0/0: Elect BDR 2.2.2.2
38	*Feb 28 09:40:56.967: OSPF-1 ADJ Fa0/0: Elect DR 1.1.1.1
39	*Feb 28 09:40:56.967: OSPF-1 ADJ Fa0/0: Elect BDR 2.2.2.2
40	*Feb 28 09:40:56.967: OSPF-1 ADJ Fa0/0: Elect DR 1.1.1.1
41	*Feb 28 09:40:56.967: OSPF-1 ADJ Fa0/0: DR: 1.1.1.1 (Id) BDR: 2.2.2.2 (Id)

FIGURE 10.21: Election of DR and BDR in cold start scenario with router priorities - Debug output messages of R2.

has to go through a second pass since the role of R2 changed (messages 39 and 40), but DR and BDR stay the same.

Note that this behavior is typical of the BDR and other non-DR routers in a cold start situation, and may sometimes be revealed by HELLO packets; you may see the BDR and non-DR routers advertising the same DR and BDR. However, this depends on the relative timing of the HELLO packet transmissions of the various routers.

ADDITIONAL EXPERIMENT - Try to observe HELLO packets where R2 or R3 advertise R1 as both the DR and BDR. You need to install a Wireshark probe at the shared link and repeat the previous experiment several times. Be persistent!

10.2.1.3   Router joining link when DR and BDR already elected

This experiment analyzes what happens when a router joins a shared link that already has a DR and a BDR. As in the previous experiment, we will use the debug output of Cisco IOS to trace the evolution of the election process.

Experimental procedure and outcome Before performing the experiment, include in the startup-config file of R3 the commands required to debug the OSPF adjacencies from the start (see Figure 8.7). Then start routers R1 and R2 at the same time and wait until the network converges. At this point, check that R2 was elected DR and R1 was elected BDR (using the **show ip ospf neighbor** command). Finally, start R3 and stop the experiment when the network converges again, i.e. when R3 elects the correct DR and BDR (based on the information provided by the debug output). You can run the **show ip ospf neighbor** command again to confirm that R2 is still the DR and R1 the BDR, despite the fact that R3 has a higher RID (and same `Router Priority`).

The election process at R3 The relevant messages of the debug output of R3 are shown in Figure 10.22 (some debug messages were deleted). The router first establishes bidirectional communication with R1 (message 1). When this happens, R3 abandons the `Waiting` state through the `BackupSeen` event, since R1 declared itself as BDR in its HELLO packet, and immediately runs the election algorithm. In the first execution (messages 3 to 6), the algorithm declares R1 as both DR and BDR. Note that the HELLO received from R1 already indicates R2 as the DR, but in step 3 of the algorithm a router is only considered in the first group if it declared itself as DR. Later, when R3 establishes bidirectional communication with R2 (message 23), the `NeighborChange` event is triggered (message 24), and the election algorithm is run again (messages 25 to 28). In this execution, R2 is finally elected as DR and R1 is confirmed as BDR. The missing debug messages (from 7 to 22) relate to the LSDB synchronization between R3 and R4, which starts as soon they become neighbors.

1	*Feb 28 13:09:27.123: OSPF-1 ADJ Fa0/0: 2 Way Communication to 1.1.1.1, state 2WAY
2	*Feb 28 13:09:27.127: OSPF-1 ADJ Fa0/0: Backup seen event before WAIT timer
3	*Feb 28 13:09:27.127: OSPF-1 ADJ Fa0/0: DR/BDR election
4	*Feb 28 13:09:27.131: OSPF-1 ADJ Fa0/0: Elect BDR 1.1.1.1
5	*Feb 28 13:09:27.131: OSPF-1 ADJ Fa0/0: Elect DR 1.1.1.1
6	*Feb 28 13:09:27.131: OSPF-1 ADJ Fa0/0: DR: 1.1.1.1 (Id) BDR: 1.1.1.1 (Id)
23	*Feb 28 13:09:30.891: OSPF-1 ADJ Fa0/0: 2 Way Communication to 2.2.2.2, state 2WAY
24	*Feb 28 13:09:30.895: OSPF-1 ADJ Fa0/0: Neighbor change event
25	*Feb 28 13:09:30.895: OSPF-1 ADJ Fa0/0: DR/BDR election
26	*Feb 28 13:09:30.895: OSPF-1 ADJ Fa0/0: Elect BDR 1.1.1.1
27	*Feb 28 13:09:30.899: OSPF-1 ADJ Fa0/0: Elect DR 2.2.2.2
28	*Feb 28 13:09:30.899: OSPF-1 ADJ Fa0/0: DR: 2.2.2.2 (Id) BDR: 1.1.1.1 (Id)

FIGURE 10.22: Election of DR and BDR when router joins the link with previously elected DR and BDR - Debug output messages of R3.

10.2.1.4 Only one router started at link

In this experiment, we illustrate what happens when only one router is started at a shared link. To analyze this experiment we will use both Wireshark and the debug feature of OSPF.

Experimental procedure The procedure is the following: (i) install a Wireshark probe at the shared link (e.g. at interface f0/0 of router R1), (ii) start router R1 and type the command **debug ip ospf adj** (do not stress; you will have 40 seconds to do that). The experiment can be stopped when the debug output indicates that the DR has been elected. We can confirm that R1 was indeed elected DR through the **show ip ospf interface brief** command.

Wireshark capture The Wireshark capture is shown in Figure 10.23. The HELLO packets up to packet 16 indicate a DR of 0.0.0.0, and the following ones indicate that the DR is router R1 (222.222.10.1). The first packet that advertises the elected DR is packet 18 sent at second 48.7. This packet is sent after the Waiting period of 40 seconds.

The debug output (Figure 10.24) shows that the interface abandons the Waiting state through the WaitTimer event (message 1). The election algorithm is then immediately run (messages 2 to 7). The algorithm performs two passes. In the first pass (messages 3 and 4), router R1 is elected both as DR and BDR. Thus, in view of condition 4, the algorithm must perform a second pass (messages 5 to 7). In step 2 of the second pass, the DR is excluded from the calculation and, since there is no router left, the BDR is set to 0.0.0.0. Finally, in step 3 of the second pass, R1 is confirmed as DR.

No.	Time	Source	Destination	Protocol	Info
4	1.566575	222.222.10.1	224.0.0.5	OSPF	Hello, DR: 0
10	10.71785	222.222.10.1	224.0.0.5	OSPF	Hello, DR: 0
12	19.75043	222.222.10.1	224.0.0.5	OSPF	Hello, DR: 0
14	29.26289	222.222.10.1	224.0.0.5	OSPF	Hello, DR: 0
16	39.0478	222.222.10.1	224.0.0.5	OSPF	Hello, DR: 0
18	48.66526	222.222.10.1	224.0.0.5	OSPF	Hello, DR: r1
20	58.46271	222.222.10.1	224.0.0.5	OSPF	Hello, DR: r1

0 = 0.0.0.0, r1 = 222.222.10.1

FIGURE 10.23: DR election when only one router at link - Wireshark capture.

10.2.1.5   DR failure

This experiment illustrates the OSPF behavior in case of a DR failure. We analyze the experiment with the help of Wireshark and the debug feature of Cisco IOS.

Experimental procedure and outcome The procedure is the following: (i) start all routers at the same time, (ii) enter the command **debug ip ospf adj** at routers R1 and R2, (iii) install a Wireshark probe at the shared link (e.g. at interface f0/0 of router R1), and (iv) switch off router R3 (to simulate its failure). The experiment can be stopped when the correct DR and BDR get elected (as indicated by the debug output). In this case it is expected that R2 becomes the new DR and R1 the new BDR. We can confirm this through the command **show ip ospf neighbors** issued at R1 and R2.

Wireshark capture and debug output of R1 Figure 10.25 shows the Wireshark capture and Figure 10.26 shows the debug output of router R1. No packets were removed from the Wireshark capture, except the initial and the final HELLO packets, transmitted while the network was in a stable state. We start by describing the main parts of the Wireshark capture:

• **Packets 8 to 10 - HELLO** packets sent before the DR failure.

1	*Feb 9 19:17:40.739: OSPF-1 ADJ Fa0/0: end of Wait on interface
2	*Feb 9 19:17:40.739: OSPF-1 ADJ Fa0/0: DR/BDR election
3	*Feb 9 19:17:40.739: OSPF-1 ADJ Fa0/0: Elect BDR 1.1.1.1
4	*Feb 9 19:17:40.739: OSPF-1 ADJ Fa0/0: Elect DR 1.1.1.1
5	*Feb 9 19:17:40.739: OSPF-1 ADJ Fa0/0: Elect BDR 0.0.0.0
6	*Feb 9 19:17:40.739: OSPF-1 ADJ Fa0/0: Elect DR 1.1.1.1
7	*Feb 9 19:17:40.739: OSPF-1 ADJ Fa0/0: DR: 1.1.1.1 (Id) BDR: none

FIGURE 10.24: DR election when only one router at link - Debug output messages of R1.

No.	Time	Source	Destination	Protocol	Info
8	9.173029	222.222.10.2	224.0.0.5	OSPF	Hello, DR: r3, BDR: r2, AN: R1, R3
9	10.08646	222.222.10.3	224.0.0.5	OSPF	Hello, DR: r3, BDR: r2, AN: R1, R2
10	10.20678	222.222.10.1	224.0.0.5	OSPF	Hello, DR: r3, BDR: r2, AN: R2, R3
14	19.13573	222.222.10.2	224.0.0.5	OSPF	Hello, DR: r3, BDR: r2, AN: R1, R3
15	19.74832	222.222.10.1	224.0.0.5	OSPF	Hello, DR: r3, BDR: r2, AN: R2, R3
19	29.0188	222.222.10.2	224.0.0.5	OSPF	Hello, DR: r3, BDR: r2, AN: R1, R3
20	29.39781	222.222.10.1	224.0.0.5	OSPF	Hello, DR: r3, BDR: r2, AN: R2, R3
23	38.45353	222.222.10.2	224.0.0.5	OSPF	Hello, DR: r3, BDR: r2, AN: R1, R3
24	38.99437	222.222.10.1	224.0.0.5	OSPF	Hello, DR: r3, BDR: r2, AN: R2, R3
26	48.01053	222.222.10.2	224.0.0.5	OSPF	Hello, DR: r3, BDR: r2, AN: R1, R3
27	48.24097	222.222.10.1	224.0.0.5	OSPF	Hello, DR: r3, BDR: r2, AN: R2, R3
29	50.58624	222.222.10.1	224.0.0.6	OSPF	LS Update, R1 Router-LSA SN=3
30	50.58624	222.222.10.2	224.0.0.5	OSPF	LS Update, R2 Router-LSA SN=3
31	50.62653	222.222.10.2	224.0.0.5	OSPF	LS Update, R1 Router-LSA SN=3, R2 Network-LSA SN=1
32	53.0869	222.222.10.2	224.0.0.5	OSPF	LS Acknowledge, R1 Router-LSA SN=3
33	53.09291	222.222.10.1	224.0.0.6	OSPF	LS Acknowledge, R2 Router-LSA SN=3, R2 Network-LSA SN=1
34	57.59704	222.222.10.1	224.0.0.5	OSPF	Hello, DR: r2, BDR: r2, AN: R2
36	58.00507	222.222.10.2	224.0.0.5	OSPF	Hello, DR: r2, BDR: r1, AN: R1
39	67.20786	222.222.10.1	224.0.0.5	OSPF	Hello, DR: r2, BDR: r1, AN: R2
41	67.87163	222.222.10.2	224.0.0.5	OSPF	Hello, DR: r2, BDR: r1, AN: R1

0 = 0.0.0.0, AN = Active Neighbor, R1 = 1.1.1.1, R2 = 2.2.2.2, R3 = 3.3.3.3, r1 = 222.222.10.1, r2 = 222.222.10.2, r3 = 222.222.10.3

FIGURE 10.25: Election of DR and BDR when DR fails - Wireshark capture.

- **Packets 14, 15, 19, 20, 23, 24, 26, 27** - HELLO packets sent after the DR failure but before other routers detected the failure.

- **Packets 29 to 33** - LS UPDATEs exchanged between R1 and R2 to update the LSDB and corresponding LS ACKNOWLEDGMENTs.

- **Packet 34** - HELLO packet where R1 advertises R2 as being both the DR and BDR.

- **Packets 36, 39, 41** - HELLO packets announcing the correct DR and BDR.

While R1 and R2 do not detect the failure... After the DR failure, R1 and R2 take `RouterDeadInterval` seconds to detect it (i.e. 40 seconds default). In the meantime, R1 and R2 keep announcing R3 as active neighbor and DR. This corresponds to packets 14 to 27 of the Wireshark capture.

Failure detection at R1 When the `RouterDeadInterval` timer expires at each router (R1 or R2), router R3 is declared dead, the state of the neighbor state machine with R3 goes to `Down`, and the `NeighborChange` event is

1	*Feb 9 23:28:52.943: OSPF-1 ADJ Fa0/0: 3.3.3.3 address 222.222.10.3 is dead
2	*Feb 9 23:28:52.943: OSPF-1 ADJ Fa0/0: 3.3.3.3 address 222.222.10.3 is dead, state DOWN
3	*Feb 9 23:28:52.943: OSPF-1 ADJ Fa0/0: Neighbor change event
4	*Feb 9 23:28:52.943: OSPF-1 ADJ Fa0/0: DR/BDR election
5	*Feb 9 23:28:52.943: OSPF-1 ADJ Fa0/0: Elect BDR 2.2.2.2
6	*Feb 9 23:28:52.943: OSPF-1 ADJ Fa0/0: Elect DR 2.2.2.2
7	*Feb 9 23:28:52.943: OSPF-1 ADJ Fa0/0: DR: 2.2.2.2 (Id) BDR: 2.2.2.2 (Id)
8	*Feb 9 23:28:52.943: OSPF-1 ADJ Fa0/0: Remember old DR 3.3.3.3 (id)
9	*Feb 9 23:29:00.867: OSPF-1 ADJ Fa0/0: Neighbor change event
10	*Feb 9 23:29:00.867: OSPF-1 ADJ Fa0/0: DR/BDR election
11	*Feb 9 23:29:00.867: OSPF-1 ADJ Fa0/0: Elect BDR 1.1.1.1
12	*Feb 9 23:29:00.867: OSPF-1 ADJ Fa0/0: Elect DR 2.2.2.2
13	*Feb 9 23:29:00.867: OSPF-1 ADJ Fa0/0: Elect BDR 1.1.1.1
14	*Feb 9 23:29:00.867: OSPF-1 ADJ Fa0/0: Elect DR 2.2.2.2
15	*Feb 9 23:29:00.867: OSPF-1 ADJ Fa0/0: DR: 2.2.2.2 (Id) BDR: 1.1.1.1 (Id)
16	*Feb 9 23:29:00.867: OSPF-1 ADJ Fa0/0: Neighbor change event
17	*Feb 9 23:29:00.867: OSPF-1 ADJ Fa0/0: DR/BDR election
18	*Feb 9 23:29:00.867: OSPF-1 ADJ Fa0/0: Elect BDR 1.1.1.1
19	*Feb 9 23:29:00.867: OSPF-1 ADJ Fa0/0: Elect DR 2.2.2.2
20	*Feb 9 23:29:00.867: OSPF-1 ADJ Fa0/0: DR: 2.2.2.2 (Id) BDR: 1.1.1.1 (Id)

FIGURE 10.26: Election of DR and BDR when DR fails - Debug output messages of R1.

triggered. The debug output of R1 (see Figure 10.26) shows this sequence of events in messages 1 to 3. The `NeighborChange` event prompts the reelection of the DR and BDR, which is signaled in messages 4 to 8 of the debug output. Interestingly, in this first execution, the election algorithm at R1 determines that R2 is both the DR and the BDR, and this is announced by R1 in packet 34.

The reason for this outcome is the following. In step 2 of the algorithm, R1 determines that R2 is the BDR because R2 previously declared itself as BDR and no one else did. Then, in step 3, the algorithm determines that R2 is the DR, since R3 failed and no router declared itself as DR. Finally, because the role of R1 did not change as a result of previous steps, the algorithm stops.

Failure detection at R2 We do not show the debug output of router R2, but it is easy to see that in the first pass of the algorithm over steps 2 and 3, router R2 concludes the same as R1 (i.e. that R2 is both DR and BDR). However, since the role of R2 (which is the router doing the computation) changed from BDR to DR due to condition 4, the algorithm is forced to perform a second pass over steps 2 and 3. In step 2, R2 is excluded from the candidate list since it is now the DR, and R1 is promoted to BDR. In step 3, the algorithm confirms that R2 is the DR and then stops. This is why router R2 announces the correct DR and BDR in the first HELLO packet it sends following the DR failure (packet 36).

R1 corrects its computation of BDR When R1 receives this HELLO from R2, the `NeighborChange` event is again triggered since R2 is newly declaring itself as DR. Thus R1 runs again the election algorithm. In step 2, it declares itself as BDR since R2 is excluded from this step and, in step 3, R2 is confirmed as DR. Since the role of the BDR changed, the algorithm does a second pass on steps 2 and 3, and then stops. This corresponds to messages 9 to 15 of the debug output.

Finally, since the role of the router changed to BDR, the `NeighborChange` event is again triggered and the election algorithm is run one last time, arriving at the same conclusion. This corresponds to messages 16 to 20 of the debug output. At this point, the network converged again.

LS UPDATE and LS ACKNOWLEDGMENT packets It is also interesting to trace the OSPF packets exchanged to update the LSDB. When R2 is elected the new DR, the identifier of the shared link changes. Thus, R1 and R2 flood new Router-LSA instances (packets 29 and 30) with the new link identifier. In OSPFv2, the link identifier is carried in the `Link ID` field of type 2 link descriptions, and this field changes to 222.222.10.2 (it was previously 222.222.10.3). Note that the LS UPDATE sent by R1 (packet 29) is sent to the `AllDRouters` multicast address (224.0.0.6), because when R1 sends this packet it still believes it is a non-DR router (neither the DR nor the BDR).

Router R2 also has to flood a new Network-LSA, since it is the new DR. This is done in packet 31. This LS UPDATE also carries the Router-LSA of R1. This is because R2 is the new DR, and it is the role of the DR to retransmit (using the `AllSPFRouters` address) the LSAs it receives from non-DR routers (received on the `AllDRouters` address). Finally, the LS ACKNOWLEDGMENT packets (packets 32 and 33) acknowledge the reception of the various LSAs.

ADDITIONAL EXPERIMENT - Connect a fourth router to the shared link of Figure 10.16, say router R4 (with RID 4.4.4.4). Start the four routers and let the network converge. At this point, router R4 should become the DR and router R3 the BDR. Now change the `Router Priority` of router R1 to 2, using the command **ip ospf priority** at interface f0/0. Then, provoke the failure of the DR, i.e. of router R4. When the network converges, use the **show ip ospf neighbor** command to verify who the DR and BDR are. You should observe that the new DR is R3, because it was previously the BDR, and the new BDR is R1, because it is the router that, among the remaining ones (R1 and R2), has the highest `Router Priority`.

10.2.2 IS-IS

We will address three different cases: (i) cold start, (ii) change of DIS, and (iii) DIS failure.

MAC addresses and their ordering In the Wireshark captures of this section we added, in the `Source` column, information on the router each MAC address belongs to. Specifically, address `ca:03:73:ec:00:00` (named r3) is from R3, `ca:02:d2:74:00:00` (named r2) is from R2, and `ca:01:c0:14:00:00` (named r1) is from R1. The ordering of these addresses is defined by the second octet (since the first octet is the same in all addresses): R3 has the highest MAC address and R2 has the second highest MAC address. This information is relevant since the MAC address is used as the tiebreaker in the DIS election, whenever the interface priorities are the same.

We have also removed the `Destination` column from the Wireshark captures since the destination is the same for all packets: the `AllL1ISs` multicast address (`01:80:c2:00:00:14`).

10.2.2.1 Cold start

This experiment analyses the IS-IS behavior when routers are all switched on at the same time at a shared link.

Experimental procedure and outcome The procedure is similar to the OSPF cold start experiment described in Section 10.2.1.1, except that now an **isis** filter must be configured in Wireshark. Since all interfaces have the same priority, router R3 will become the DIS at the shared link since its MAC address is the highest.

Wireshark capture The Wireshark capture is shown in Figure 10.27. No packets were removed from the original capture. We start by describing the main parts of the Wireshark capture:

- **Packets 8, 13, 20, 22, 24, 28, 30, 32** - Initial HELLO packets, before DIS election.

- **Packets 29, 33, 34, 38** - Nonpseudonode-LSPs and the Pseudonode-LSP sent after the DIS election.

- **Packets 36, 37, 39, 40, 42, 46 to 48** - HELLO packets advertising the DIS.

- **Packet 41** - First CSNP.

Initial HELLO packets In the initial HELLO packets, sent before the DIS election, the `LAN ID` field includes the SID of the sending router and a tag generated by the router (which is "01" in all cases). This tag will become the Pseudonode ID of the shared link if the sending router becomes the DIS. In this phase, the HELLOs sent by R1 have a `LAN ID` of R1.01, the ones sent by R2 have a `LAN ID` of R2.01, and the ones sent by R3 have a `LAN ID` of R3.01.

The first two HELLOs (packets 8 and 13) have no IS Neighbors TLV since, at this point, no neighbor is known by the sending routers. Packet 20 is an immediate HELLO sent by R3 in response to packet 13. In this packet, R3

No.	Time	Source	Protocol	Info
8	1.14539	ca:03:73:ec:00:00 (r3)	ISIS HELLO	L1, LAN ID: R3.01, HT: 30s
13	1.64568	ca:01:c0:14:00:00 (r1)	ISIS HELLO	L1, LAN ID: R1.01, HT: 30s
20	2.11645	ca:03:73:ec:00:00 (r3)	ISIS HELLO	L1, LAN ID: R3.01, HT: 30s, ISN: r1
22	2.26739	ca:02:d2:74:00:00 (r2)	ISIS HELLO	L1, LAN ID: R2.01, HT: 30s, ISN: r3
24	2.60835	ca:01:c0:14:00:00 (r1)	ISIS HELLO	L1, LAN ID: R1.01, HT: 30s, ISN: r2, r3
28	3.12115	ca:03:73:ec:00:00 (r3)	ISIS HELLO	L1, LAN ID: R3.01, HT: 30s, ISN: r1, r2
29	3.2415	ca:03:73:ec:00:00 (r3)	ISIS LSP	L1, LSP-ID: R3.00-00, SN=2, RL: 1199s, ISN: R3.01
30	3.27108	ca:02:d2:74:00:00 (r2)	ISIS HELLO	L1, LAN ID: R2.01, HT: 30s, ISN: r1, r3
32	3.6231	ca:01:c0:14:00:00 (r1)	ISIS HELLO	L1, LAN ID: R1.01, HT: 30s, ISN: r2, r3
33	3.64265	ca:03:73:ec:00:00 (r3)	ISIS LSP	L1, LSP-ID: R3.01-00, SN=1, RL: 1199s, ISN: R1.00, R2.00, R3.00
34	3.81344	ca:01:c0:14:00:00 (r1)	ISIS LSP	L1, LSP-ID: R1.00-00, SN=2, RL: 1199s, ISN: R3.01
36	4.11517	ca:03:73:ec:00:00 (r3)	ISIS HELLO	L1, LAN ID: R3.01, HT: 10s, ISN: r1, r2
37	4.25608	ca:02:d2:74:00:00 (r2)	ISIS HELLO	L1, LAN ID: R3.01, HT: 30s, ISN: r1, r3
38	4.48774	ca:02:d2:74:00:00 (r2)	ISIS LSP	L1, LSP-ID: R2.00-00, SN=2, RL: 1199s, ISN: R3.01
39	4.60708	ca:01:c0:14:00:00 (r1)	ISIS HELLO	L1, LAN ID: R3.01, HT: 30s, ISN: r2, r3
40	7.25655	ca:03:73:ec:00:00 (r3)	ISIS HELLO	L1, LAN ID: R3.01, HT: 10s, ISN: r1, r2
41	9.68542	ca:03:73:ec:00:00 (r3)	ISIS CSNP	L1, TLV Entries: R1.00-00, R2.00-00, R3.00-00, R3.01-00
42	10.5492	ca:03:73:ec:00:00 (r3)	ISIS HELLO	L1, LAN ID: R3.01, HT: 10s, ISN: r1, r2
46	12.8979	ca:01:c0:14:00:00 (r1)	ISIS HELLO	L1, LAN ID: R3.01, HT: 30s, ISN: r2, r3
47	13.6624	ca:03:73:ec:00:00 (r3)	ISIS HELLO	L1, LAN ID: R3.01, HT: 10s, ISN: r1, r2
48	13.7557	ca:02:d2:74:00:00 (r2)	ISIS HELLO	L1, LAN ID: R3.01, HT: 30s, ISN: r1, r3

ISN = IS-IS Neighbor, HT = Holding Timer, RL = Remaining Lifetime, SN = Sequence Number, SN=A means SN=0x0000000A

FIGURE 10.27: DIS election in cold start scenario - Wireshark capture.

already lists R1 as neighbor in an IS Neighbors TLV of type 6. In this type of TLV the neighbors are identified by their MAC addresses. When R1 receives packet 20, it becomes adjacent to R3, since it sees itself listed in the packet sent by R3.

Packet 22, sent by R2, is also an immediate HELLO, but in response to packets 8 or 20. When R2 scheduled this packet for transmission, it still didn't recognize R1 as neighbor, since only R3 is listed. This is the last HELLO that doesn't list all neighbors of the sending router. When R3 receives packet 22, it becomes adjacent to R2, since R3 sees itself listed in the HELLO sent by R2.

When relationships become bidirectional As in the case of the OSPF cold start experiment, it is interesting to determine when neighborhood re-

Relationships	Adjacent
R1 → R2	when packet 30 received
R1 → R3	when packet 20 received
R2 → R1	when packet 24 received
R2 → R3	when packet 28 received
R3 → R1	when packet 24 received
R3 → R2	when packet 22 received

FIGURE 10.28: DIS election in cold start scenario - When neighborhood relationships become bidirectional.

lationships become bidirectional; this is shown in Figure 10.28. Note that all relationships have done it by packet 30.

When the DIS gets elected The first packet transmitted after the DIS election is packet 29, transmitted at second 3.2. This packet is a Nonpseudonode-LSP where R3 describes itself and its interface with the shared link (pseudonode). The shared link identifier (R3.01) is included in the IS Neighbors TLV of type 2 of the Nonpseudonode-LSP. Note that this LSP could only have been generated after the DIS election because the interface with the shared link is described through the shared link identifier, and this identifier is based on the SID of the DIS.

The IS-IS specification [1] says that there must be a waiting period of $2 \times$ iSISHelloTimer (20 seconds default) before the DIS is elected. Judging from the LSP transmission time, the waiting period of this implementation is approximately 3 seconds, much less than what is suggested by the specification.

The remaining LSPs are the Nonpseudonode-LSPs sent by R1 and R2 (packets 34 and 38) and the Pseudonode-LSP sent by R3 (packet 33). The last LSP is sent around second 4.5. When the routers receive all LSPs, their LSDBs become completely synchronized. In IS-IS, this process is faster than in OSPF.

HELLO packets after DIS election Packet 36 is the first HELLO where the new DIS is recognized; the packet is sent by the DIS itself. What tells us that R3 recognized itself as DIS is the Holding Timer value, which is now 10 seconds. All subsequent HELLO packets indicate that the DIS is R3 in the LAN ID field (which has value R3.01). Packets 37 and 39 are immediate HELLOs sent in reaction to packet 36. The remaining HELLO packets are periodic HELLOs. Note that the frequency of the HELLOs sent by the DIS is higher (and the Holding Timer is smaller) than the one of remaining packets.

First CSNP Packet 41 is the first CSNP. It contains the LSDB summary, with three Nonpseudonode-LSPs describing each router and its interfaces (R1.00-00, R2.00-00, and R3.00-00), and the Pseudonode-LSP describing the shared link (R3.01-00). The CSNP is transmitted near second 10, which is in agreement with the periodicity of CSNP transmissions. Recall that the time displayed in Wireshark is relative to the first captured packet, and that the first packet is sent sometime after the router has been switched on. The IS-IS specification [1] does not indicate a specific behavior for the first CSNP. From this experiment, we guess that this packet is transmitted after the expiration of the first `completeSNPInterval` (which has a default value of 10 seconds).

10.2.2.2   Change of DIS

This experiment analyzes the behavior of IS-IS when the DIS is changed through a configuration action. The DIS is initially router R3 (due to its higher interface MAC address), and we will change the priority of interface f0/0 of router R1, so that it becomes the DIS.

Experimental procedure The procedure is the following: (i) start all routers at the same time and wait until the network converges, (ii) install a Wireshark probe at the shared link with an **isis** filter configured, (iii) enter the command **isis priority 100** at the f0/0 interface of router R1. The experiment can be stopped when the network converges again, i.e. when the new DIS is elected and the LSDBs are updated.

Wireshark capture The Wireshark capture is shown in Figure 10.29. No packets were removed from the original capture. We start by describing the main parts of the Wireshark capture:

- **Packets 21, 23, 24, 28, 29** - HELLO packets sent before changing the interface priority.

- **Packet 22** - CSNP sent by R3 before changing the interface priority.

- **Packet 30, 32, 36, 39, 43, 45 to 48** - HELLO packets sent after changing the interface priority.

- **Packet 31** - New Pseudonode-LSP sent by R1.

- **Packets 33 and 34** - Deletion of old Pseudonode-LSP.

- **Packets 35, 37, 38** - New Nonpseudonode-LSPs instances sent by R1, R2, and R3.

- **Packet 40** - CSNP sent by R1 after changing the interface priority.

Initial HELLO packets The initial HELLO packets show that the interfaces agree that router R3 is the DIS (since the `LAN ID` value is R3.01). Moreover, the initial CSNP (packet 22) indicates that the Pseudonode-LSP describing the shared link is originated by R3 (since the LSP ID is R3.01-00). The `Priority` value indicated in all these HELLO packets is 64, i.e. the default value of Cisco IOS. The periodicity and `Holding Timer` differs in the HELLO packets sent by the DIS and by the other routers: the DIS uses a periodicity of 3 seconds and an `Holding Timer` of 10 seconds; the remaining routers use a periodicity of 10 seconds and a `Holding Timer` of 30 seconds.

No.	Time	Source	Protocol	Info
21	16.554217	ca:03:73:ec:00:00 (r3)	ISIS HELLO	L1, HT: 10s, Pr: 64, LAN ID: R3.01
22	17.508757	ca:03:73:ec:00:00 (r3)	ISIS CSNP	L1, TLV Entries: R1.00-00, R2.00-00, R3.00-00, R3.01-00
23	19.482002	ca:01:c0:14:00:00 (r1)	ISIS HELLO	L1, HT: 30s, Pr: 64, LAN ID: R3.01
24	19.59576	ca:03:73:ec:00:00 (r3)	ISIS HELLO	L1, HT: 10s, Pr: 64, LAN ID: R3.01
28	22.758713	ca:03:73:ec:00:00 (r3)	ISIS HELLO	L1, HT: 10s, Pr: 64, LAN ID: R3.01
29	24.504715	ca:02:d2:74:00:00 (r2)	ISIS HELLO	L1, HT: 30s, Pr: 64, LAN ID: R3.01
30	25.463263	ca:01:c0:14:00:00 (r1)	ISIS HELLO	L1, HT: 10s, Pr: 100, LAN ID: R1.01
31	25.465273	ca:01:c0:14:00:00 (r1)	ISIS LSP	L1, LSP-ID: R1.01-00, SN=1, RL=1199s, ISN: R1.00, R2.00, R3.00
32	25.474294	ca:03:73:ec:00:00 (r3)	ISIS HELLO	L1, HT: 30s, Pr: 64, LAN ID: R1.01
33	25.475296	ca:01:c0:14:00:00 (r1)	ISIS LSP	L1, LSP-ID: R3.01-00, SN=1, RL=0s
34	25.50337	ca:02:d2:74:00:00 (r2)	ISIS LSP	L1, LSP-ID: R3.01-00, SN=1, RL=0s
35	25.512394	ca:01:c0:14:00:00 (r1)	ISIS LSP	L1, LSP-ID: R1.00-00, SN=3, RL=1199s, ISN: R1.01
36	25.51841	ca:02:d2:74:00:00 (r2)	ISIS HELLO	L1, HT: 30s, Pr: 64, LAN ID: R1.01
37	25.534531	ca:03:73:ec:00:00 (r3)	ISIS LSP	L1, LSP-ID:R3.00-00, SN=3, RL=1199s, ISN: R1.01
38	25.584586	ca:02:d2:74:00:00 (r2)	ISIS LSP	L1, LSP-ID: R2.00-00, SN=3, RL=1199s, ISN: R1.01
39	28.104285	ca:01:c0:14:00:00 (r1)	ISIS HELLO	L1, HT: 10s, Pr: 100, LAN ID: R1.01
40	28.970589	ca:01:c0:14:00:00 (r1)	ISIS CSNP	L1, TLV Entries: R1.00-00, R1.01-00, R2.00-00, R3.00-00, R3.01-00
43	30.890776	ca:01:c0:14:00:00 (r1)	ISIS HELLO	L1, HT: 10s, Pr: 100, LAN ID: R1.01
45	33.300098	ca:03:73:ec:00:00 (r3)	ISIS HELLO	L1, HT: 30s, Pr: 64, LAN ID: R1.01
46	33.524026	ca:01:c0:14:00:00 (r1)	ISIS HELLO	L1, HT: 10s, Pr: 100, LAN ID: R1.01
47	33.66206	ca:02:d2:74:00:00 (r2)	ISIS HELLO	L1, HT: 30s, Pr: 64, LAN ID: R1.01
48	36.379286	ca:01:c0:14:00:00 (r1)	ISIS HELLO	L1, HT: 10s, Pr: 100, LAN ID: R1.01

ISN = IS-IS Neighbor, HT = Holding Timer, RL = Remaining Lifetime, Pr = Priority, SN = Sequence Number, SN=A means SN=0x0000000A

FIGURE 10.29: DIS reelection after changing the router priority - Wireshark capture.

The interface priority is changed When the new priority is configured at router R1, the router immediately sends a HELLO packet (packet 30). In this packet, R1 indicates a `Priority` of 100, and advertises itself as being the DIS, since the `LAN ID` is R1.01 and the `Holding Timer` is 10 seconds. From then on, all HELLO packets advertise this DIS. Note that this HELLO was sent less than 10 seconds apart from the previous HELLO of router R1 (packet 23): packet 23 was sent at second 19.4 and packet 30 at second 25.4. This shows that packet 30 is an immediate HELLO triggered by the change in the router priority.

LSPs sent after the new DIS election Immediately after the HELLO packet, R1 sends the new Pseudonode-LSP that describes the link, with LSP ID equal to R1.01-00 (packet 31). It also sends an indication to delete the old Pseudonode-LSP, originated by the previous DIS, with LSP ID equal to R3.01-00 (packet 33). The indication to delete is easily recognized by having a `Remaining Lifetime` of 0. This behavior is markedly different from that of OSPF: in IS-IS, it is the new DIS that must delete the old Pseudonode-LSP; in OSPF, LSAs can only be deleted by their originating routers. Note that the delete indication just includes the LSP header and not its full contents.

Packet 34 deserves some attention since it seems to correspond to a non-standard behavior. This packet is again a delete indication of the old Pseudonode-LSP, transmitted in reaction to packet 33. When packet 33 arrives at R2, R2 replaces the stored LSP instance by the incoming one, since the incoming instance is considered fresher; recall that instances with a zero `Remaining Lifetime` are considered fresher than those with the same SN and non-zero `Remaining Lifetime`. According to Section 7.3.16.4 of the IS-IS specification [1], this instance should now be flooded, i.e. it should be transmitted on all interfaces except the one where the packet was received. However, according to what we observe in this experiment, the packet is also transmitted through the receiving interface. This does not hurt the correctness of the protocol, but seems unnecessary.

Note the old DIS (i.e. R3) did not transmit an indication to delete its (old) Pseudonode-LSP. This was by chance: if you run the experiment several times you may see this indication, possibly even before the indication sent by the new DIS. In fact, when R3 learns that it must resign from being the DIS, it sets the `Remaining Lifetime` of its Pseudonode-LSP to zero and schedules the flooding of this LSP through all its interfaces. However, if in the meantime, it receives the same LSP instance at some interface, it cancels its transmission through that interface. This is what happened in this case: the reception at R3 of the delete indication sent by R1 canceled the transmission of the same indication already scheduled at R3.

The remaining LSPs (packets 35, 37, and 38) are the new instances of the three Nonpseudonode-LSPs. These instances had to be created and flooded by their originating routers since the shared link identifier changed to R1.01 (previously it was R3.01). Note that the IS Neighbors TLV of these LSPs includes R1.01 as neighbor. Note also that each of these LSPs is seen only once at the link. This is because, unlike the case of packet 34, these LSPs follow the usual flooding procedure and, therefore, are not retransmitted at receiving interfaces.

CSNP sent after the new DIS election The CSNP sent after the DIS change (packet 40) already lists the new Pseudonode-LSP, which has LSP ID equal to R1.01-00. It may seem strange that it also lists the old Pseudonode-LSP (with LSP ID R3.01-00). However, expired LSPs must be kept at the LSDB for an additional minute (`ZeroAgeLifetime*`) before being purged (see Section 6.5.7). You may keep checking the contents of the CSNPs to confirm that the reference to the old Pseudonode-LSP disappears after 1 minute.

Change in the periodicity of HELLO transmissions Notice that the way HELLO packets are transmitted changed with the election of the new DIS. The HELLO packets sent by R1 now have the same behavior as the HELLO packets sent by R3 before the priority change: they are sent every 3 seconds, and have an `Holding Timer` of 10 seconds.

Convergence time We highlight that the network reaction to the DIS change was very fast: it took only a bit more than 100 ms from the first packet signaling the DIS change (packet 30) until the last LSP required for the LSDB synchronization (packet 38).

ADDITIONAL EXPERIMENTS - Repeat this experiment several times, but now just concentrated in the delete indications. We suggest that you use an **isis.lsp** filter at Wireshark, and keep alternating the commands **isis priority 100** and **no isis priority 100** at interface f0/0 of router R1. The filter keeps just the LSPs, which are only transmitted in the transient period following the priority changes. You can easily recognize the delete indications by looking at the `Remaining Lifetime` value of each packet. You will observe that delete indications may be sent by R1 and R2 (as above), by R1 and R3, or by R3 and R2.

10.2.2.3 DIS failure

This experiment analyzes the behavior of IS-IS when the DIS fails. The DIS is initially router R3 (because of its higher interface MAC address). After network convergence, we switch off the DIS to simulate its failure, and the new DIS will then become router R2.

Experimental procedure The procedure is similar to the OSPF DR failure experiment (see Section 10.2.1.5), except that the debug feature is not used this time. The procedure is the following: (i) start all routers at the same time and wait until the network converges, (ii) install a Wireshark probe at the shared link, and (iii) switch off router R3 (to simulate its failure). The experiment can be stopped when the network converges again, i.e. when the new DIS is elected and the LSDBs are updated.

Wireshark capture The Wireshark capture is shown in Figure 10.30. No packets were removed from the original capture. We start by describing the main parts of the Wireshark capture:

• **Packet 53 -** Last HELLO sent by R3, before its failure.

• **Packets 54, 58 -** HELLO packets sent before detection of the DIS failure.

• **Packets 59, 60, 66, 68 -** HELLO packets sent after the new DIS election.

• **Packet 61 -** New Pseudonode-LSP sent after the new DIS election.

• **Packets 63 and 64 -** Deletion of old Pseudonode-LSP.

No.	Time	Source	Protocol	Info
53	52.709	ca:03:73:ec:00:00 (r3)	ISIS HELLO	L1, LAN ID: R3.01, HT: 10s, ISN: r1, r2
54	53.512	ca:02:d2:74:00:00 (r2)	ISIS HELLO	L1, LAN ID: R3.01, HT: 30s, ISN: r1, r3
58	58.077	ca:01:c0:14:00:00 (r1)	ISIS HELLO	L1, LAN ID: R3.01, HT: 30s, ISN: r2, r3
59	62.711	ca:01:c0:14:00:00 (r1)	ISIS HELLO	L1, LAN ID: R3.01, HT: 30s, ISN: r2
60	62.712	ca:02:d2:74:00:00 (r2)	ISIS HELLO	L1, LAN ID: R2.01, HT: 10s, ISN: r1
61	62.743	ca:02:d2:74:00:00 (r2)	ISIS LSP	L1, LSP-ID: R2.01-00, SN=1, RL: 1199s, ISN:R1.00, R2.00
62	62.771	ca:01:c0:14:00:00 (r1)	ISIS LSP	L1, LSP-ID: R1.00-00, SN=3, RL: 1199s, ISN: R2.01
63	62.773	ca:02:d2:74:00:00 (r2)	ISIS LSP	L1, LSP-ID: R3.01-00, SN=1, RL: 0s
64	62.802	ca:01:c0:14:00:00 (r1)	ISIS LSP	L1, LSP-ID: R3.01-00, SN=1, RL: 0s
65	62.843	ca:02:d2:74:00:00 (r2)	ISIS LSP	L1, LSP-ID: R2.00-00, SN=3, RL: 1199s, ISN: R2.01
66	63.714	ca:01:c0:14:00:00 (r1)	ISIS HELLO	L1, LAN ID: R2.01, HT: 30s, ISN: r2
67	65.311	ca:02:d2:74:00:00 (r2)	ISIS CSNP	L1, TLV Entries: ISN:R1.00-00, R2.00-00, R2.01-00, R3.00-00, R3.01-00
68	65.331	ca:02:d2:74:00:00 (r2)	ISIS HELLO	L1, LAN ID: R2.01, HT: 10s, ISN: r1

ISN = IS-IS Neighbor, HT = Holding Timer, RL = Remaining Lifetime, SN = Sequence Number, SN=A means SN=0x0000000A

FIGURE 10.30: Reaction to DIS failure - Wireshark capture.

- **Packets 62, 65** - New Nonpseudonode-LSP instances sent by R1 and R2.

- **Packet 67** - CSNP sent by R2 after the new DIS election.

Initial HELLO packets The last HELLO sent by R3 (packet 53) is seen at second 52.7. In the two subsequent HELLOs (packets 54 and 58) R1 and R2 still recognize R3 as neighbor.

Failure detection at R1 Packet 59 (sent at second 62.7) is the immediate HELLO sent by R1 after it detected the failure. We know that because R3 is no longer listed as neighbor, and the packet was sent less than 10 seconds apart from the last periodic HELLO of R1 (which was packet 58, sent at second 58.1). Since the `Holding Timer` of the DIS is 10 seconds, we guess that the DIS failed at approximately second 52.7 (just after sending the last HELLO). Note that, in this HELLO, the `LAN ID` value is still R3.01, despite the fact that R3 is no longer recognized as neighbor. This is because R2, the new DIS, still did not provide the new shared link identifier: R1 knows that R2 is the new DIS, and knows its SID, but still doesn't know the Pseudonode ID that R2 assigned to the shared link.

R2 recognizes itself as DIS Router R2 recognizes itself as DIS, for the first time, in packet 60; it no longer lists R3 as neighbor, and advertises a `Holding Timer` of 10 seconds and a `LAN ID` of R2.01. From then on, all HELLO packets advertise the same `LAN ID`.

LSPs sent after the new DIS election The behavior regarding the LSP transmissions is the same as the DIS change experiment (Section 10.2.2.2). Following the election of the new DIS, R2 floods a new Pseudonode-LSP (packet 61) and a new instance of its Nonpseudonode-LSP with SN equal to 3 (packet 65). R1 also floods a new instance of its Nonpseudonode-LSP (packet 62) with SN equal to 3. Moreover, R2 takes the initiative of deleting the old Pseudonode-LSP (packet 63), and this packet is also flooded by R1 (packet 64).

CSNP sent after the new DIS election Finally, packet 67 is the first CSNP sent after the new DIS election. It lists all new LSPs, as well as the old Pseudonode-LSP, which stays in the LSDB (with a `Remaining Lifetime` of 0) for an additional minute (`ZeroAgeLifetime*`).

Convergence time The outcome of this experiment is not very different from that of the DIS change experiment (Section 10.2.2.2). The main difference is the waiting time until the DIS failure is detected (approximately 10 seconds). Once the failure is detected, the election of the new DIS and the LSDB update is very fast: it takes only a bit more than 100 ms from the first packet signaling the failure (packet 59) until the last packet required to fully update the LSDB (packet 65).

10.3   Flooding procedure

This section presents several experiments that illustrate the flooding process in OSPF and IS-IS. This issue was discussed in Section 6.4.

Experimental setup The experiments use the network topology of Figure 10.31. The network includes four routers and three links. R4 is connected to R1 and R2 via point-to-point links; R1, R2, and R3 are connected through a Fast Ethernet link. The routers must be initially configured with the basic routing configurations. The experiments use Wireshark.

Main purpose of experiments The main purpose of these experiments is to analyze the reaction to an interface cost change. Specifically, we will change the cost of interface s0/0 of router R4. This will surely trigger the origination and flooding of new LSA (OSPF) or LSP (IS-IS) instances.

10.3.1   OSPF shared links

Experimental procedure The procedure is the following: (i) start all routers at the same time and wait until the network converges; (ii) install a Wireshark probe at the shared link and start a capture with an **ospf** filter; (iii) keep alternating the cost of interface s0/0 of router R4 between 100 and 64 (the default value). Alternating the interface cost can be achieved by alternating

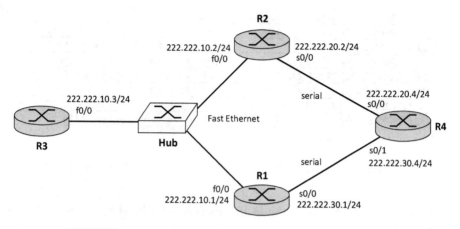

FIGURE 10.31: Flooding experiments - Network topology.

the commands **ip ospf cost 100** and **no ip ospf cost**. The first command
sets the cost of the interface to 100 and the second returns it to its default
value, i.e. 64. The objective of these changes is to observe different sequences
of packets at the shared link. Concentrate on the first LS UPDATE packet
transmitted on the shared link: sometimes this packet is transmitted by R1,
and other times by R2. You can stop the experiment when you have observed
both cases.

Since all routers are switched on at the same time, R3 will become the DR
and R2 the BDR. Check that this is true before step (ii) of the experiment,
e.g. using the command **show ip ospf neighbor**.

Experiment outcome Changing the OSPF cost of interface s0/0 of router
R4 forces the generation of a new Router-LSA instance. Specifically, the
Metric value associated with the type 1 and type 3 link descriptions of interface s0/0 is changed from 64 to 100, and vice-versa. When a new instance is
generated, the SN is incremented by one, and R4 transmits the instance on
all its interfaces, encapsulated in LS UPDATE packets. The two Router-LSA
instances are then received by R1 and R2, and retransmitted by these routers
on the shared link using again LS UPDATE packets. Either of them can be
the first to transmit an LS UPDATE packet on the link; this depends on the
relative load of the routers and links on the path from R4 to the shared link.
The behavior is different in each case, since in one case the LS UPDATE is
transmitted by the BDR and in another by a non-DR router.

It can also happen that one of the transmissions initiated by R4 is so
delayed that either R1 or R2 first receives the new LSA instance through
the shared link (i.e. at the f0/0 interface, and not at the s0/0 interface). In
this case, the router that receives the delayed LSA will not transmit it at the
shared link. We could not observe this behavior in our OSPF experiments.

No.	Time	Source	Destination	Protocol	Info
9	12.844499	222.222.10.1	224.0.0.6	OSPF	LS Update, Router-LSA, R4, SN: 2
10	12.844499	222.222.10.2	224.0.0.5	OSPF	LS Update, Router-LSA, R4, SN: 2
11	12.855357	222.222.10.3	224.0.0.5	OSPF	LS Update, Router-LSA, R4, SN: 2
15	15.369859	222.222.10.2	224.0.0.5	OSPF	LS Acknowledge, Router-LSA, R4, SN: 2
150	171.19171	222.222.10.2	224.0.0.5	OSPF	LS Update, Router-LSA, R4, SN: 11
151	171.19171	222.222.10.1	224.0.0.6	OSPF	LS Update, Router-LSA, R4, SN: 11
153	173.7053	222.222.10.3	224.0.0.5	OSPF	LS Acknowledge, Router-LSA, R4, SN: 11

SN=A means SN=0x8000000A

FIGURE 10.32: OSPF flooding on shared links - Wireshark capture.

However, we did see it in the IS-IS experiments related to shared links (see Section 10.3.3).

Wireshark capture The Wireshark capture is shown in Figure 10.32. In relation to the original capture, we only kept two groups of update-acknowledgment interactions, one relative to an LS UPDATE sent by R1 and another to an LS UPDATE sent by R2.

R1 transmits first on the shared link In the first update-acknowledgment interaction (packets 9 to 15), it is R1 that transmits the first LS UPDATE on the shared link (packet 9). Since R1 is a non-DR router, the packet is sent to the DR using the `AllDRouters` multicast address (224.0.0.6). The Router-LSA is then retransmitted by the DR to all other routers encapsulated on another LS UPDATE packet (packet 11), but now using the `AllSPFRouters` multicast address (224.0.0.5). In the meantime, R2 also transmits the Router-LSA received from R4 encapsulated in an LS UPDATE packet (packet 10); since R2 is the BDR it transmits the packet using the `AllSPFRouters` address. The final packet of this group is the LS ACKNOWLEDGMENT sent by R2 (packet 15). Note that R1 doesn't need to acknowledge the Router-LSA sent by the DR, since the LSA is a duplicate (it is already in R1's LSDB) and is an implied acknowledgment to the LSA previously sent by R1 (packet 9).

R2 transmits first on the shared link In the second update-acknowledgment group (packets 150 to 153), the first LS UPDATE packet (packet 150) is sent by R2, which is the BDR. The DR does not retransmit LSAs sent by the DR, but it must acknowledge them, which it does at packet 153. R1 also transmits the Router-LSA sent by R4 (packet 151). However, when this LSA arrives at the DR, it is treated as a duplicate since the DR already received the same instance from R2; therefore, the LSA is no longer flooded. R2 does not acknowledge the LSA sent by R1 since the LSA is a duplicate (it is already in R2's LSDB) and is an implied acknowledgment to the LSA previously sent by R2 (packet 150).

No.	Time	Source	Destination	Protocol	Info
17	38.914683	222.222.20.4	224.0.0.5	OSPF	LS Update, Router-LSA, R4, SN: 2
20	41.426992	222.222.20.2	224.0.0.5	OSPF	LS Acknowledge, Router-LSA, R4, SN: 2
25	50.785842	222.222.20.4	224.0.0.5	OSPF	LS Update, Router-LSA, R4, SN: 3
28	53.272794	222.222.20.2	224.0.0.5	OSPF	LS Acknowledge, Router-LSA, R4, SN: 3

SN=A means SN=0x8000000A

FIGURE 10.33: OSPF flooding on point-to-point links - Wireshark capture.

10.3.2 OSPF point-to-point links

Experimental procedure In this experiment, we repeat the procedure of OSPF shared link experiment, but now placing the Wireshark probe at the point-to-point link between R4 and R2. Moreover, we just do two interface cost changes.

Wireshark capture The Wireshark capture is shown in Figure 10.33. In relation to the original capture, we only kept the first two LS UPDATEs and the corresponding LS ACKNOWLEDGMENTs.

Experiment outcome The flooding process is similar to the one of shared links. When the interface cost is changed at interface s0/0 of R4, R4 generates a new Router-LSA instance and transmits it at both interfaces, encapsulated in LS UPDATE packets. Packets 17 and 25 are two LS UPDATE packets transmitted from R4 to R2 in reaction to an interface cost change. When these packets are received at R2, R2 replies confirming its correct reception, using LS ACKNOWLEDGMENT packets (packets 20 and 28). To correlate the updates with acknowledgments, the LS ACKNOWLEDGMENT packets carry the header of the LSA being acknowledged. Packet 28 is shown in Figure 10.34. Notice that the packet only carries the LSA header, indicating that the Router-LSA instance being acknowledged was originated by router R4 (`Advertising Router` field is 4.4.4.4) and had an SN is 0×80000003. Notice also that the instance is really fresh (its `LS Age` is just 1 second).

10.3.3 IS-IS shared links

Experimental procedure In this experiment we want to make sure that R3 is the DIS. To do that, initially enter the command **isis priority 100** at interface f0/0 of router R3. Besides that, the procedure of this experiment is similar to the one of Section 10.3.1: (i) start all routers at the same time and wait until the network converges; (ii) install a Wireshark probe at the shared link and start a capture with an **isis** filter; (iii) keep alternating the cost of interface s0/0 of router R4, between the 30 and 10 (the default value). Alternating the interface costs can be achieved by alternating the commands **isis metric 30** and **no ip ospf cost**. The first command sets the cost of the interface to 30 and the second returns it to the default value, i.e. 10.

```
> Frame 28: 68 bytes on wire (544 bits), 68 bytes captured (544 bi
> Cisco HDLC
> Internet Protocol Version 4, Src: 222.222.20.2, Dst: 224.0.0.5
v Open Shortest Path First
    > OSPF Header
    v LSA-type 1 (Router-LSA), len 72
        .000 0000 0000 0001 = LS Age (seconds): 1
        0... .... .... .... = Do Not Age Flag: 0
      > Options: 0x22 ((DC) Demand Circuits, (E) External Routing)
        LS Type: Router-LSA (1)
        Link State ID: 4.4.4.4
        Advertising Router: 4.4.4.4
        Sequence Number: 0x80000003
        Checksum: 0x1460
        Length: 72
```

FIGURE 10.34: OSPF flooding on point-to-point links - LSA ACKNOWL-
EDGMENT sent by router R2, confirming the reception of a Router-LSA
from R4.

Experiment outcome R4 generates a new Nonpseudonode-LSP instance
whenever the cost of an interface changes. In this experiment, what will be
changed is the `Default Metric` value related to the interface with R2 (identified as R2.00-00). This field appears both in the IS Neighbors TLV and in the
IP Internal Reachability Information TLV of the Nonpseudonode-LSP. Once
generated, this LSP instance is transmitted on interfaces s0/0 and s0/1 of R4.
When R1 and R2 receive this LSP instance, they may (or may not) retransmit
it on the shared link.

Depending on the relative loads of the routers and links in the path taken
by the Nonpseudonode-LSPs from R4 to the shared link, at least one router
will retransmit the LSP on the shared link, but both routers may do it. The
specific outcome is unpredictable, and that is why we keep alternating the
interface costs in this experiment. The experiment can be stopped when both
patterns have been observed (one LSP or two LSPs).

Wireshark capture The Wireshark capture is shown in Figure 10.35. In
relation to the original capture, we only kept the CSNP and LSP packets of
the time period we want to analyze.

The CSNPs First notice that the CSNPs are transmitted regularly, with
time spacings between 8 and 10 seconds. In fact, the periodicity of CSNP
transmissions is not affected by the arrival of LSPs; unlike HELLO packets
there is no such thing as an immediate CSNP.

The LSPs In the capture, there are two groups of LSPs: the first group with a
single LSP (packet 37), and the second group with two LSPs (packets 62 and
63). These two patterns were observed in successive interface cost changes:
the first group when the cost was first changed to 30, and the second group
when the cost was changed back to the default value. That these patterns
were observed in successive cost changes was mere luck; the patterns occur

randomly and, as will be explained later, the second pattern is more probable than the first one.

Before the first cost change Packet 37 is the CSNP transmitted immediately before the first cost change. The CSNP advertises four Nonpseudonode-LSPs, each representing one router (R1.00-00, R2.00-00, R3.00-00, and R4.00-00), and one Pseudonode-LSP, representing the shared link and originated by router R3 (R3.01-00). Note that the SN of the Nonpseudonode-LSPs from R4 (R4.00-00) is 3 (i.e. 0×00000003).

Reaction to the first cost change Packet 41 is the LSP advertising the new cost of interface s0/0 of router R4. It is transmitted on the shared link by router R2. Notice that the SN increased to 4. You can also analyze the contents of the packet to check that, in the IS Neighbors TLV, the `Default Metric` value related to the interface with R2 changed to 30.

Before the second cost change Packets 46 and 52 are two CSNPs sent after the first LSP transmission and before the next interface cost change. Notice that the SN of R4.00-00 is now 4. These CSNPs acknowledge to R2 that the LSP it previously sent (packet 41) was received by R3. That is how acknowledgments are performed in IS-IS shared links (see Section 6.4.2).

Reaction to the second cost change When the second interface cost change is made (now back to the default value), R4 originates another Nonpseudonode-LSP instance, now with SN equal to 5. This time the LSP is transmitted at the shared link by both R1 and R2 (packets 62 and 63).

The CSNP that follows the previous LSP transmissions at the shared link (Packet 66) already advertises the Nonpseudonode-LSP of R4 with an SN of 5. As in the case of packet 41, this CSNP acknowledges that the LSPs previously transmitted by R1 and R2 were correctly received.

Explanation for having one or two LSPs transmitted at the shared link Figure 10.36 illustrates the flooding process at the shared link when two LSPs or one LSP are transmitted at the shared link. In the first case (Figure 10.36.a), the two LSPs arrive at approximately the same time at R2 and R1. Then R2 and R1 replace the stored instance (with SN=3) by the incoming one (with SN=4), and both routers transmit the new LSP instance on the shared link. When each router receives (at its f0/0 interface) the LSP transmitted by the other on the shared link, both routers discard the incoming packet since the stored instance has equal freshness.

In the second case (Figure 10.36.b), there is a large delay in the reception of the LSP transmitted through the s0/1 interface of R4. In the meantime, R2 receives the LSP, replaces the stored instance by the incoming one, and transmits the new LSP instance on the shared link. When R1 receives this LSP (at its f0/0 interface), it replaces the stored instance by the incoming one, just like R2 did. When the LSP transmitted by the s0/1 interface of R4 finally arrives at R4, it is discarded since there is already a stored instance with equal freshness. In this case, it may also happen that R1 transmits the

No.	Time	Source	Protocol	Info
37	35.37964	c2:03:3f:34:00:00 (r3)	ISIS CSNP	LSP Entries: R1.00-00 (SN=3), R2.00-00 (SN=3), R3.00-00 (SN=2), R3.01-00 (SN=1), R4.00-00 (SN=3)
41	40.36491	c2:02:17:24:00:00 (r2)	ISIS LSP	LSP-ID: R4.00-00 (SN=4, RL: 1197s)
46	43.30774	c2:03:3f:34:00:00 (r3)	ISIS CSNP	LSP Entries: R1.00-00 (SN=3) , R2.00-00 (SN=3), R3.00-00 (SN=2), R3.01-00 (SN=1), R4.00-00 (SN=4)
52	51.53764	c2:03:3f:34:00:00 (r3)	ISIS CSNP	LSP Entries: R1.00-00 (SN=3), R2.00-00 (SN=3), R3.00-00 (SN=2), R3.01-00 (SN=1), R4.00-00 (SN=4)
62	59.0381	c2:02:17:24:00:00 (r2)	ISIS LSP	LSP-ID: R4.00-00 (SN=5, RL: 1197s)
63	59.0386	c2:01:42:28:00:00 (r1)	ISIS LSP	LSP-ID: R4.00-00 (SN=5, RL: 1197s)
66	60.5396	c2:03:3f:34:00:00 (r3)	ISIS CSNP	LSP Entries: R1.00-00 (SN=3), R2.00-00 (SN=3), R3.00-00 (SN=2), R3.01-00 (SN=1), R4.00-00 (SN=5)

RL = Remaining Lifetime, SN=A means SN=0x0000000A

FIGURE 10.35: IS-IS flooding on shared links - Wireshark capture.

new LSP instance through its s0/0 interface, in which case R4 would receive a self-originated LSP instance.

ADDITIONAL EXPERIMENT - Repeat this experiment but making sure that, initially, the DIS is either R1 or R2. You just have to increase the priority associated with the interface f0/0 of one of these routers to a value higher than 100. The objective is to ensure that, at least sometimes, the first Nonpseudonode-LSP transmitted on the shared link is transmitted by the DIS. You will notice a similar behavior. Sometimes both routers transmit the Nonpseudonode-LSP instance generated by R4 when the interface cost is changed; other times only one LSP is transmitted and, in this case, sometimes it is transmitted by R1 and other times by R2. The CSNPs keep being transmitted periodically at the shared link and are updated whenever a new Nonpseudonode-LSP instance of R4 is received.

10.3.4 IS-IS point-to-point links

Experimental procedure In this experiment, we repeat the procedure of the IS-IS shared link experiment, but now placing the Wireshark probe at the point-to-point link between R4 and R2. Moreover, we just do one interface cost change.

Wireshark capture The Wireshark capture is shown in Figure 10.37. In relation to the original capture, we only kept the LSP and PSNP packets. The Wireshark capture does not recognize the source and destination addresses

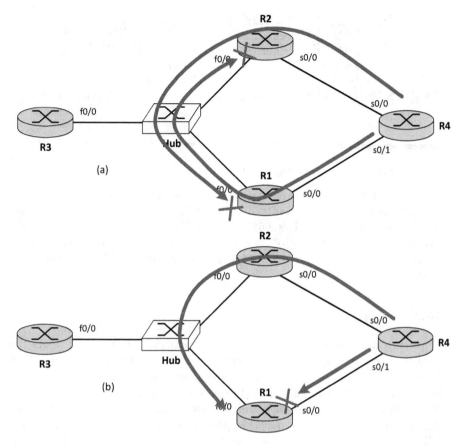

FIGURE 10.36: IS-IS flooding on shared links - Reaction to the origination of new LSP by R4 when (a) two LSPs are retransmitted at the shared link and (b) only one LSP is retransmitted.

(after all, this is a point-to-point link); however, this can be easily inferred from other informations contained in the packets.

Experiment outcome The change in interface cost results just in the exchange of two packets. When the cost is changed at router R4, R4 generates a new Nonpseudonode-LSP instance, and floods it through all its interfaces. The LSP transmitted through interface s0/0 (towards R2) is packet 5. Notice that the SN of this LSP is 3. When R2 receives this packet, it acknowledges its reception through a PSNP, which identifies the acknowledged LSP through the LSP Entries TLV (packet 6). Notice that this TLV has an LSP ID of R4.00-00 and an SN of 3, which indicates it is indeed acknowledging the previous LSP.

No.	Time	Source	Protocol	Info
5	3.878912	N/A (R4)	ISIS LSP	LSP-ID: R4.00-00 (SN=3, RL: 1199s)
6	4.857307	N/A (R2)	ISIS PSNP	LSP Entries: R4.00-00 (SN=3)

RL = Remaining Lifetime, SN=A means SN=0x0000000A

FIGURE 10.37: IS-IS flooding on point-to-point links - Wireshark capture.

10.4   Deletion procedure

It is instructive to think about events that trigger the deletion of LSAs and
LSPs. Considering the network of Figure 10.31, one such situation is when
the designated router (DR or DIS) is either R1 or R2 and is replaced at the
shared link without failing. In this case, the designated router still has the
opportunity to delete its previously originated Network-LSA or Pseudonode-
LSP.

10.4.1   OSPF

Experimental procedure In this experiment, the configuration of R1 has
to be first modified such that R1 becomes initially the DR at the shared link.
To do this, you have to enter the command **ip ospf priority 2** at interface
f0/0. You can either modify directly the startup-config file, or (i) switch on
the router, (ii) make and save the configuration, and then (iii) switch off
the router again. Having done this preliminary configuration, perform the
following steps: (i) switch on all routers at the same time and wait until the
network converges; (ii) confirm that R1 is indeed the DR at the shared link
(e.g. using the command **show ip ospf neighbor** at R3) and take note of the
SN of the Network-LSA stored at the LSDB (using the command **show ip
ospf database**); (iii) install a Wireshark probe at the shared link and start a
capture with an **ospf** filter; (iv) shutdown interface f0/0 of R1 (entering the
command **shutdown** at interface f0/0).

Experiment outcome Initially, R1 is the DR at the shared link, and R3 is
the BDR (since its RID is higher than the one of R2). When the interface
f0/0 of R1 is shutdown, the router immediately understands it ceased being
the DR and floods a new instance of its Router-LSA and an indication to
delete the Network-LSA representing the shared link. These LSAs are first
transmitted to router R4 and then retransmitted from R4 to R2, which injects
them at the shared link. Many LS UPDATE and LS ACKNOWLEDGMENT
packets appear at the shared link, but we just want to concentrate in the
indication to delete the Network-LSA. We show it in Figure 10.38. Notice
that the LS Age field equals 3600 seconds, which is the distinguishing feature
of a delete indication. It is also true that the Sequence Number carried in the

```
> Frame 44: 98 bytes on wire (784 bits), 98 bytes captured (784 bits)
> Ethernet II, Src: c2:02:27:64:00:00 (c2:02:27:64:00:00), Dst: IPv4m
> Internet Protocol Version 4, Src: 222.222.10.2, Dst: 224.0.0.6
˅ Open Shortest Path First
    > OSPF Header
    ˅ LS Update Packet
        Number of LSAs: 1
        ˅ LSA-type 2 (Network-LSA), len 36
            .000 1110 0001 0000 = LS Age (seconds): 3600
            0... .... .... .... = Do Not Age Flag: 0
          > Options: 0x22 ((DC) Demand Circuits, (E) External Routing)
            LS Type: Network-LSA (2)
            Link State ID: 222.222.10.1
            Advertising Router: 1.1.1.1
            Sequence Number: 0x80000002
            Checksum: 0x98b7
            Length: 36
            Netmask: 255.255.255.0
            Attached Router: 1.1.1.1
            Attached Router: 2.2.2.2
            Attached Router: 3.3.3.3
```

FIGURE 10.38: OSPF deletion procedure - Indication to delete the Network-LSA originated by R1.

packet (0×80000002) is the same as the stored instance. Notice also that the Network-LSA still lists three routers as being attached to the shared link; the new Network-LSA, originated by R3 (the new DR), will list only two routers (R2 and R3).

ADDITIONAL EXPERIMENT - An experiment that is very easy to perform and triggers the transmission of delete indications is the clearance of the ospf process at a router without switching it off. Again in the network of Figure 10.31, (i) start all routers at the same time, (ii) install a Wireshark probe at the shared link with an **ospf** filter, and (iii) enter the command **clear ip ospf process** in the privileged EXEC level.

10.4.2 IS-IS

Experimental procedure This experiment is easier to perform in IS-IS than in OSPF, since the DIS is easier to change than the DR. The procedure is the following: (i) switch on all routers at the same time and wait until the network converges; (ii) verify which router is the DIS and take note of the SN of the Pseudonode-LSP stored at the LSDB (using the command **show isis database**); (iii) install a Wireshark probe at the shared link and start a capture with an **isis** filter; (iv) decrease the priority of interface f0/0 of the DIS. The easiest way to learn who the DIS is, is to check in the LSDB which router is advertising the Pseudonode-LSP, i.e. the LSP with non-zero Pseudonode

```
> Frame 65: 60 bytes on wire (480 bits), 60 bytes captured (480 bits) on interface 0
> IEEE 802.3 Ethernet
> Logical-Link Control
> ISO 10589 ISIS InTRA Domain Routeing Information Exchange Protocol
v ISO 10589 ISIS Link State Protocol Data Unit
     PDU length: 27
     Remaining lifetime: 0
     LSP-ID: 0000.0000.0003.01-00
     Sequence number: 0x00000001
     [Checksum: [missing]]
     [Checksum Status: Not present]
   > Type block(0x01): Partition Repair:0, Attached bits:0, Overload bit:0, IS type:1
```

FIGURE 10.39: IS-IS deletion procedure - Indication to delete the Pseudonode-LSP originated by R3.

ID (using the command **show isis database**). The default priority value is 64, and you will have to change it to a lower value using the command **isis priority** at interface f0/0 of the DIS.

Experiment outcome In our case, the DIS was initially R3 and we changed the priority of its interface f0/0 to 63. When this was done, the router immediately transmitted an indication to delete the Pseudonode-LSP it owned. This LSP is shown in Figure 10.39. Notice that the `Remaining Lifetime` equals zero, which is a distinguishing feature of delete indications. The LSP ID identifies the LSP being deleted: R3-01.00 indicates that this LSP is originated by R3 and that the LSP is a Pseudonode-LSP, since the Pseudonode ID is 1 (non-zero). Notice also that, unlike OSPF, the delete indication does not carry the full contents of the LSP being deleted; it only carries the LSP ID, `Sequence Number` and `Remaining Lifetime`.

10.5 Initial LSDB synchronization

This section presents several experiments that illustrate the initial LSDB synchronization process of OSPF and IS-IS. This issue was discussed in Section 6.6.

Experimental setup The experiments use the network topology of Figure 10.40, with three routers and two links; the links are a shared link between R1 and R2 and a point-to-point link between R2 and R3. They are performed with the IPv4 versions of both protocols since the behavior of the IPv6 versions is similar. The routers must be initially configured with the basic routing configurations. Both Wireshark and the debug feature of Cisco IOS are used in these experiments.

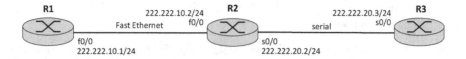

FIGURE 10.40: Experiments related to the initial LSDB synchronization process - Network topology.

10.5.1 OSPF

The objective of this experiment is to analyze the various phases of the LSDB synchronization process of OSPF, described in Section 6.6.1. We concentrate on the synchronization over the shared link (between R1 and R2). The experimental procedure is designed such that when a router is switched on, it finds at the neighbor self-originated LSAs with higher freshness.

Experimental procedure The procedure is the following: (i) switch on all routers at the same time and wait until the network converges; (ii) switch off router R1 and wait until the network converges again; (iii) install a Wireshark probe at the shared link and start a capture with an **ospf** filter; (iv) run the command **debug ip ospf adj** at router R2; (v) switch on router R1 again.

Experimental outcome When R1 is switched on for the first time, the SN of its Router-LSA is incremented several times during the convergence process because of the interface state change (the interface is elected BDR) and of the neighbor state change (when its relationship with R2 reaches the Full state). Actually, after network convergence the Router-LSA will have an SN of 3. This ensures that when R1 is switched on again it will find at the neighbor an instance of its self-originated Router-LSA with higher SN.

After R1 being switched off, the shared link turns from a transit shared link to a stub shared link. This prompts the deletion of the Network-LSA previously originated by R2 and the update of the Router-LSA of R2. The Router-LSA of R2 has to be updated since interfaces with transit and stub shared links are described differently (with different link description types).

When R1 is switched on again the shared link turns again to a transit shared link, and R2 originates a new Network-LSA and a new instance of its Router-LSA.

Note that the outcome of this experiment is very close to the example of Section 6.6.1.

Wireshark capture and debug output of R2 Figure 10.41 shows the Wireshark capture and Figure 10.42 shows the debug output of router R2. In relation to the initial Wireshark capture, we removed all HELLO packets; in relation to the original debug output, we removed all messages related to the designated router election, except the messages related to the first election. We start by describing the main parts of the Wireshark capture:

- **Packets 27 to 32, 34** - DB DESCRIPTION packets that exchange the LSDB summaries.

- **Packet 33** - LS REQUEST sent by R1 asking for the complete LSAs owned by R2.

- **Packet 35** - LS UPDATE sent by R2 in response LS REQUEST (packet 33).

- **Packets 36, 37, 42** - LS UPDATE disseminating the new Network-LSA, and the new Router-LSA instances of R1 and R2.

- **Packets 40, 44** - LS ACKNOWLEDGMENTs confirming the reception of the LSAs sent in previous LS UPDATEs.

First DB DESCRIPTION packets Packets 27 and 28 are the first DB DE-SCRIPTION packets sent by R1 and R2. R1 chooses a ddSN of 5384 and R2 chooses a ddSN of 1397, and they both assume being the master (MS-bit set). According to the debug output, R2 received the first DB DESCRIPTION from R1 (message 1) before reaching the 2-Way state with this neighbor (message 2). This means that R1 reached this state before R2, since DB DESCRIP-TION packets can only be sent to a neighbor after verification of bidirectional communication with that neighbor. These two packets carry no LSA headers.

R2 runs the designated router election algorithm When R2 reaches the 2-Way state with R1, it immediately runs the designated router election algorithm (messages 3 to 7 of the debug output), concluding that it keeps being the DR and that R1 is the new BDR. It then sends its first DB DESCRIPTION packet which, as referred above, is packet 28; this event is also signaled in messages 8 and 9 of the debug output.

R1 abandons the ExStart state In the next DB DESCRIPTION sent by R1 (packet 29), R1 already understood it is not the master, since the MS-bit is cleared and the ddSN is the same as in packet 28 sent by R2 (i.e. 1397). The I-bit is also cleared since R1 understood it is the slave and, therefore, abandoned the ExStart state. In this packet, R1 advertises its newly created Router-LSA with SN=1 (actually 0×80000001).

R2 abandons the ExStart state When packet 29 is received at R2 (message 10 of the debug output), R2 also changes to the Exchange state (message 11), since it received a DB DESCRIPTION with both the I-bit and the MS-bit cleared, and its RID is higher than the neighbor's one.

R2 describes its LSAs R2 then sends a DB DESCRIPTION packet (packet 30 and message 12 of the debug output), where it describes all its LSAs, i.e. the Router-LSA of R1 (with SN=3), the Router-LSA of R2 (with SN=4), and the Router-LSA of R3 (with SN=1). Because this is a new DB DESCRIPTION packet sent by the master, the ddSN is incremented by 1 in relation to the previous one; it is now 1398.

R1 and R2 abandon the Exchange phase At this point, R1 and R2 have

No.	Time	Source	Destination	Protocol	Info
27	67.94105	222.222.10.1	222.222.10.2	OSPF	DB Description, ddSN=5384, I=1, M=1, MS=1
28	67.95162	222.222.10.2	222.222.10.1	OSPF	DB Description, ddSN=1397, I=1, M=1, MS=1
29	67.96282	222.222.10.1	222.222.10.2	OSPF	DB Description, ddSN=1397, I=0, M=1, MS=0, R1 Router-LSA SN=1
30	67.97296	222.222.10.2	222.222.10.1	OSPF	DB Description, ddSN=1398, I=0, M=1, MS=1, R1 Router-LSA SN=3, R2 Router-LSA SN=4, R3 Router-LSA SN=1
31	67.98381	222.222.10.1	222.222.10.2	OSPF	DB Description, ddSN=1398, I=0, M=0, MS=0
32	67.99386	222.222.10.2	222.222.10.1	OSPF	DB Description, ddSN=1399, I=0, M=0, MS=1
33	68.00447	222.222.10.1	222.222.10.2	OSPF	LS Request, R1 Router-LSA, R2 Router-LSA, R3 Router-LSA
34	68.01449	222.222.10.1	222.222.10.2	OSPF	DB Description, ddSN=1399, I=0, M=0, MS=0
35	68.01532	222.222.10.2	222.222.10.1	OSPF	LS Update, R1 Router-LSA SN=3, R2 Router-LSA SN=4, R3 Router-LSA SN=1
36	68.46187	222.222.10.2	224.0.0.5	OSPF	LS Update, R2 Router-LSA SN=5
37	68.52375	222.222.10.2	224.0.0.5	OSPF	LS Update, R2 Network-LSA SN=1
40	70.53188	222.222.10.1	224.0.0.5	OSPF	LS Acknowledge, R1 Router-LSA SN=3, R2 Router-LSA SN=4, R3 Router-LSA SN=1, R2 Router-LSA SN=5, R2 Network-LSA SN=1
42	73.02315	222.222.10.1	224.0.0.5	OSPF	LS Update, R1 Router-LSA SN=4
44	75.51349	222.222.10.2	224.0.0.5	OSPF	LS Acknowledge, R1 Router-LSA SN=4

SN=A means SN=0x8000000A

FIGURE 10.41: Initial LSDB synchronization of OSPF - Wireshark capture.

completed the exchange of their LSDB summaries. However, the reception of this information must still be acknowledged before both routers can complete the database description phase and abandon the Exchange state. Packet 31 is the first DB DESCRIPTION packet sent by R1 where the M-bit is cleared, indicating that R1 has its Database summary list empty; the list is empty since the reception of packet 29 (sent by R1) was acknowledged by packet 30 (sent by R2). The reception of this packet at R2 is signaled in message 13 of the debug output.

Packet 31 (sent by R1) is also an acknowledgment that packet 30 (sent by R2) was received correctly. Thus, the Database summary list of R2 is

1	*Mar 1 00:04:49.487: OSPF: Rcv DBD from 1.1.1.1 on FastEthernet0/0 seq 0x1508 opt 0x52 flag 0x7 len 32 mtu 1500 state INIT
2	*Mar 1 00:04:49.491: OSPF: 2 Way Communication to 1.1.1.1 on FastEthernet0/0, state 2WAY
3	*Mar 1 00:04:49.491: OSPF: Neighbor change Event on interface FastEthernet0/0
4	*Mar 1 00:04:49.491: OSPF: DR/BDR election on FastEthernet0/0
5	*Mar 1 00:04:49.491: OSPF: Elect BDR 1.1.1.1
6	*Mar 1 00:04:49.491: OSPF: Elect DR 2.2.2.2
7	*Mar 1 00:04:49.491: DR: 2.2.2.2 (Id) BDR: 1.1.1.1 (Id)
8	*Mar 1 00:04:49.491: OSPF: Send DBD to 1.1.1.1 on FastEthernet0/0 seq 0x575 opt 0x52 flag 0x7 len 32
9	*Mar 1 00:04:49.491: OSPF: First DBD and we are not SLAVE
10	*Mar 1 00:04:49.511: OSPF: Rcv DBD from 1.1.1.1 on FastEthernet0/0 seq 0x575 opt 0x52 flag 0x2 len 52 mtu 1500 state EXSTART
11	*Mar 1 00:04:49.511: OSPF: NBR Negotiation Done. We are the MASTER
12	*Mar 1 00:04:49.511: OSPF: Send DBD to 1.1.1.1 on FastEthernet0/0 seq 0x576 opt 0x52 flag 0x3 len 92
13	*Mar 1 00:04:49.527: OSPF: Rcv DBD from 1.1.1.1 on FastEthernet0/0 seq 0x576 opt 0x52 flag 0x0 len 32 mtu 1500 state EXCHANGE
14	*Mar 1 00:04:49.531: OSPF: Send DBD to 1.1.1.1 on FastEthernet0/0 seq 0x577 opt 0x52 flag 0x1 len 32
15	*Mar 1 00:04:49.547: OSPF: Rcv LS REQ from 1.1.1.1 on FastEthernet0/0 length 60 LSA count 3
16	*Mar 1 00:04:49.547: OSPF: Send UPD to 222.222.10.1 on FastEthernet0/0 length 148 LSA count 3
17	*Mar 1 00:04:49.555: OSPF: Rcv DBD from 1.1.1.1 on FastEthernet0/0 seq 0x577 opt 0x52 flag 0x0 len 32 mtu 1500 state EXCHANGE
18	*Mar 1 00:04:49.559: OSPF: Exchange Done with 1.1.1.1 on FastEthernet0/0
19	*Mar 1 00:04:49.559: OSPF: Synchronized with 1.1.1.1 on FastEthernet0/0, state FULL
20	*Mar 1 00:04:49.991: OSPF: Build router LSA for area 0, router ID 2.2.2.2, seq 0x80000005
21	*Mar 1 00:04:50.063: OSPF: Build network LSA for FastEthernet0/0, router ID 2.2.2.2
22	*Mar 1 00:04:50.063: OSPF: Build network LSA for FastEthernet0/0, router ID 2.2.2.2
33	*Mar 1 00:04:54.547: OSPF: Rcv LS UPD from 1.1.1.1 on FastEthernet0/0 length 64 LSA count 1

FIGURE 10.42: Initial LSDB synchronization of OSPF - Debug output messages of R2.

emptied, and R2 sends a DB DESCRIPTION packet with the M-bit cleared (packet 32 and message 14 of the debug output). Router R1 has to reply to this message (since it is the slave), which it does in packet 34.

R2 goes to the Full state When R2 receives packet 34, it abandons the

Exchange state and changes directly to the **Full** state (messages 17 to 19 of the debug output). R2 doesn't go through the **Loading** phase since it has no interest in the LSAs that R1 owns: R1 advertised one Router-LSA, but R2 has an instance of that LSA with higher SN.

R2 floods a new Router-LSA instance Since R2 became fully adjacent to R1, it now has to generate a new Router-LSA instance, with an SN incremented by one in relation to the previous one (message 20 of the debug output). This is required since the shared link changed from being a stub shared link to a transit shared link. Thus, the corresponding link description must be changed from type 3 (link to stub network) to type 2 (link to transit network). The new Router-LSA instance is advertised in the LS UPDATE packet (packet 36) sent to the **AllSPFRouters** multicast address (224.0.0.5).

R2 floods a new Network-LSA Likewise, since R2 is the DR and became adjacent to a new router (in this case, the adjacency is the first one on the shared link), it has to originate a new Network-LSA (messages 21 and 22 of the debug output). As in the case of the new Router-LSA, the Network-LSA is advertised in an LS UPDATE packet (packet 37) sent to the **AllSPFRouters** multicast address (224.0.0.5).

R1 requests several LSAs When R1 receives packet 32, it changes from the **Exchange** state to the **Loading** state. Based on the information provided by R2 in packet 30, R1 learns that it is missing several (complete) LSAs that R2 owns and requests them in an LS REQUEST packet (packet 33). Note that R1 also requests the complete contents of its self-originated Router-LSA. However, this specific request is unnecessary, since R1 already knows (from packet 30) that the SN of the LSA is 3, and that it will have to flood a new instance of this LSA with an SN incremented by one in relation to the neighbor's one.

R2 sends the complete LSAs When R2 receives this request (message 15 of the debug output), it replies with an LS UPDATE unicasted to R1, including the complete contents of the Router-LSAs that R2 owns (packet 35 and message 16 of the debug output).

R1 floods a new instance of the self-originated Router-LSA R1 will send a final LS UPDATE, since it noted that R2 had an instance of its self-originated Router-LSA with an SN higher than its own. Thus, it creates a new instance of this Router-LSA with the SN incremented by one in relation to the neighbors one, i.e. with an SN of 4. The reception of this message at R2 is signaled by message 33 of the debug output.

Acknowledging the reception of LSAs Finally, packets 40 and 44 are LS ACKNOWLEDGMENTs that confirm the reception of the LSAs sent in previous LS UPDATE packets. Notice that the LS ACKNOWLEDGMENT sent by R1 acknowledges LSAs that were transmitted in different LS UPDATE packets. This gives an example of delayed acknowledgments.

ADDITIONAL EXPERIMENT - Repeat this experiment but now with the Wireshark probe place at the point-to-point link; moreover, start the Wireshark probe before switching off R1. In this way, you will be able to observe the reaction of R2 when R1 is switched off, and when it is switched on again. When R1 is switched off nothing will happen for approximately 40 seconds, which is the time R2 takes to detect the failure of R1. Then R2 sends a new instance of its Router-LSA and an indication to delete the Network-LSA of the transit shared link in LS UPDATE packets. You will notice that the Network-LSA is a delete indication because the value of the LS Age field is 3600 seconds. R2 sends these LSA instances because when router R1 fails, the shared link changes from being a transit link to a stub link. You will also observe an LS ACKNOWLEDGMENT packet sent by R3 acknowledging the reception of the two LSAs sent by R2.

Later when R1 is switched on again, R2 sends to R3 a new instance of its Router-LSA, a new Network-LSA representing the transit shared link (since the shared link became again a transit link) and the new Router-LSA of R1, and R3 sends the corresponding acknowledgments.

10.5.2   IS-IS

The initial LSDB synchronization of IS-IS was described in Section 6.6.2 and is analyzed in this section.

10.5.2.1   Shared links

The objective of this experiment is to analyze the various phases of the IS-IS synchronization process on shared links.

Experimental setup and procedure The experimental setup and procedure are similar to those of Section 10.5.1, except that an **isis** filter must be configured in Wireshark; moreover the debug feature is not used in this experiment. Note that the DIS of the shared link is R2 because its interface has a higher MAC address.

Wireshark capture The Wireshark capture is shown in Figure 10.43. In relation to the original capture, we removed the HELLO packets and the CSNP packets outside the time window of the experiment. We start by describing the main parts of the Wireshark capture:

- **Packet 8** - Last CSNP sent by R2 before R1 being switched on.

- **Packet 23, 24, 27** - Initial LSPs sent by R1 and R2.

- **Packet 28** - Nonpseudonode-LSP of R1 sent by R2 because it received an instance of this LSP with an SN lower than the stored one.

- **Packet 29** - Nonpseudonode-LSP of R1 sent by R1 because it received an instance of this LSP with an SN higher than the stored one.

- **Packet 35** - First CSNP received by R1.

No.	Time	Source	Protocol	Info
8	16.4663	c2:02:2f:c4:00:00 (r2)	ISIS CSNP	LSP Entries: R1.00-00 (SN=6) , R2.00-00 (SN=7), R2.01-00 (SN=5), R3.00-00 (SN=6)
23	29.3411	c2:02:2f:c4:00:00 (r2)	ISIS LSP	LSP-ID: R2.00-00, SN=9, RL: 1199s
24	29.3512	c2:02:2f:c4:00:00 (r2)	ISIS LSP	LSP-ID: R2.01-00, SN=6, RL: 1199s
27	30.5106	c2:01:0d:fc:00:00 (r1)	ISIS LSP	LSP-ID: R1.00-00, SN=2, RL: 1199s
28	30.5216	c2:02:2f:c4:00:00 (r2)	ISIS LSP	LSP-ID: R1.00-00, SN=6, RL: 540s
29	30.532	c2:01:0d:fc:00:00 (r1)	ISIS LSP	LSP-ID: R1.00-00, SN=7, RL: 1199s
35	38.0717	c2:02:2f:c4:00:00 (r2)	ISIS CSNP	LSP Entries: R1.00-00 (SN=7) , R2.00-00 (SN=9), R2.01-00, (SN=6), R3.00-00 (SN=6)
36	39.0971	c2:01:0d:fc:00:00 (r1)	ISIS PSNP	LSP Entries: R3.00-00
37	39.1296	c2:02:2f:c4:00:00 (r2)	ISIS LSP	LSP-ID: R3.00-00, SN=6, RL: 559s

RL = Remaining Lifetime, SN=A means SN=0x0000000A

FIGURE 10.43: Initial LSDB synchronization on IS-IS shared links - Wireshark capture.

- **Packet 36 and 37** - PSNP sent by R1 to request missing LSP and response sent by R2.

Last CSNP before switching on R1 Packet 8 is the last CSNP transmitted before R1 being switched on. Note that the Nonpseudonode-LSP of R1 is still in the LSDB of R2, despite R1 being switched off. This is because the age of this LSP is still counting down to zero. Note also that the Pseudonode-LSP that describes the shared link (R2.01-00) is also still in the LSDB. When R1 was switched off, the shared link changed from a transit shared link to a stub shared link and, therefore, the Pseudonode-LSP describing the shared link should have been removed. However, in this implementation, it stays in the LSDB for 20 minutes (`MaxAge*`).

Initial LSPs sent by R1 and R2 When R1 is switched on, R1 and R2 become adjacent and send to each other their self-originated LSPs. In packet 23, R2 transmits on the link its Nonpseudonode-LSP (R2.00-00 with SN=9) and, in packet 24, transmits the Pseudonode-LSP describing the transit shared link (R2.01-00 with SN=6), since R2 is the link DIS. Note that the SNs of these LSPs are higher than the ones advertised by the previous CSNP (packet 8); this is because the shared link changed from a stub shared link to a transit shared link. In packet 27, R1 advertises its newly created Nonpseudonode-LSP (R1.00-00 with SN=2).

Reaction to outdated self-originated LSPs When R2 receives the Nonpseudonode-LSP of R1 (packet 27), it verifies that its stored instance has higher SN (the stored instance has SN=6 and the one sent by R1 has SN=2). Thus, R2 floods on the link its stored instance of R1.00-00 (packet

28). When R1 receives this LSP, it verifies that it is a self-originated LSP with higher SN, and floods a new instance of the LSP with the SN incremented by one in relation to the SN received from R2 (packet 29).

CSNP from R2 and request of missing information by R1 Packet 35 is a CSNP sent by R2 (the link DIS), where it advertises all LSPs contained in its LSDB. This is the first CSNP received by R1. When R1 examines this packet, it concludes that it is still missing the Nonpseudonode-LSP of R3 (R3.00-00). Thus, it transmits a PSNP requesting this LSP (packet 36), and R2 sends it immediately (packet 37).

10.5.2.2   Point-to-point links

The objective of this experiment is to analyze the various phases of the IS-IS LSDB synchronization process on point-to-point links.

Experimental setup and procedure The experimental setup and procedure are similar to those of Section 10.5.2.1, except that the Wireshark probe now has to be placed at the point-to-point link, and the router that is switched on and off is router R3.

Moreover, we want to make sure that the Nonpseudonode-LSP instance of R3 that remains in the LSDB of R1 and R2 after R3 being switched off has higher SN than the first Nonpseudonode-LSP instance flooded by R3 when it is switched on again. To enforce this, we must increase the SN of the Nonpseudonode-LSP during the period when R3 is switched on. One easy way to accomplish this is to change the interface cost (e.g using the command `isis metric` 30 at interface f0/0 of R3).

The procedure is then the following: (i) switch on all routers at the same time and wait until the network converges; (ii) change the cost of interface f0/0 of router R3 (at least one time); (iii) switch off router R3 and wait until the network converges again; (iii) install a Wireshark probe at the point-to-point link and start a capture with an **isis** filter; (v) switch on router R3 again.

Wireshark capture The Wireshark capture is shown in Figure 10.44. In relation to the initial Wireshark capture, we removed all HELLO packets. We start by describing the main parts of the Wireshark capture:

- **Packets 24 and 25** - Nonpseudonode-LSPs of R2 and R3, sent initially.

- **Packet 26 an 27** - CSNPs describing the LSDB summaries of R2 and R3.

- **Packet 29 to 31** Complete LSPs sent by R2 and R3.

- **Packets 32 and 33** - PSNPs acknowledging the reception of previously sent LSPs.

Initial Nonpseudonode-LSPs Routers R2 and R3 send initially to each other their self-originated Nonpseudonode-LSPs. R2 sends the LSP with LSP ID R2.00-00 and SN=9 (packet 25), and R3 sends the the LSP with LSP ID R3.00-00 and SN=2 (packet 24).

No.	Time	Source	Protocol	Info
24	101.1079	N/A (R3)	ISIS LSP	LSP-ID: R3.00-00, SN=2, RL: 1200s
25	101.1129	N/A (R2)	ISIS LSP	LSP-ID: R2.00-00, SN=9, RL: 1200s
26	101.5561	N/A (R3)	ISIS CSNP	LSP Entries: R2.00-00 (SN=9), R3.00-00 (SN=2)
27	101.5571	N/A (R2)	ISIS CSNP	LSP Entries: R1.00-00 (SN=3), R2.00-00 (SN=9), R2.01-00 (SN=2), R3.00-00 (SN=6)
29	102.0548	N/A (R3)	ISIS LSP	LSP-ID: R3.00-00, SN=7, RL: 1199s
30	102.0748	N/A (R2)	ISIS LSP	LSP-ID: R1.00-00, SN=3, RL: 910s
31	102.0958	N/A (R2)	ISIS LSP	LSP-ID: R2.01-00, SN=2, RL: 861s
32	102.1189	N/A (R3)	ISIS PSNP	LSP Entries: R1.00-00 (SN=3), R2.00-00 (SN=9), R2.01-00 (SN=2)
33	103.0672	N/A (R2)	ISIS PSNP	LSP Entries: R3.00-00 (SN=7)

RL = Remaining Lifetime, SN=A means SN=0x0000000A

FIGURE 10.44: Initial LSDB synchronization on IS-IS point-to-point links (1st experiment) - Wireshark capture.

Exchange of LSDB summaries R2 and R3 then exchange CSNPs with the summaries of their LSDBs. At this point, R3 has only its own Nonpseudonode-LSP (R3.00-00 with SN=2) and the Nonpseudonode-LSP received from R2 on packet 25 (R2.00-00 with SN=9). R2 advertises all LSPs describing the network, including a Nonpseudonode-LSP from R3 with a higher SN than the one just generated by R3 (packet 24), but that is now outdated.

R3 floods new instance of self-originated Nonpseudonode-LSP From the CSNP sent by R2 (packet 27) R3 understands that R2 has an instance of its self-originated Nonpseudonode-LSP with a higher SN than its own. Thus, it floods, on packet 29, a new instance with the SN incremented by one in relation to the instance held by R2: it increases the SN from 6 to 7.

R2 floods LSPs that R3 is missing Based on the information provided by R3, R2 understands that R3 is still missing R1.00-00 and R2.01.00, and sends these LSPs on packets 30 and 31, respectively.

Acknowledging the LSPs Packets 32 and 33 are the PSNPs acknowledging the LSPs received by each router. In packet 32, R3 acknowledges the reception of R1.00 with SN=3, R2.00 with SN=9, and R2.01 with SN=2. In packet 33, R2 acknowledges the reception of R3.00-00 with SN=7. Note that although R2 has received two instances of R3.00-00, one with SN=2 (packet 24) and another with SN=7 (packet 29), it only acknowledges the last one.

A slight variation of the previous experiment Suppose now that we want to ensure that, except for the Nonpseudonode-LSPs of R2 and R3, the other LSP instances are the same when R2 and R3 synchronize for the second time. To achieve this, we repeat the procedure of the previous experiment but instead of switching off and on router R3, we just switch off and on its interface s0/0 (using the commands **shutdown** and **no shutdown**).

Wireshark capture The Wireshark capture is shown in Figure 10.45. As in

No.	Time	Source	Protocol	Info
22	41.66376	N/A (R3)	ISIS LSP	LSP-ID: R3.00-00, SN=5, RL: 1200s
23	41.672	N/A (R2)	ISIS LSP	LSP-ID: R2.00-00, SN=6, RL: 1200s
24	42.13392	N/A (R2)	ISIS CSNP	LSP Entries: R1.00-00 (SN=2), R2.00-00 (SN=6), R2.01-00 (SN=1), R3.00-00 (SN=5)
25	42.13453	N/A (R3)	ISIS CSNP	LSP Entries: R1.00-00 (SN=2), R2.00-00 (SN=6), R2.01-00 (SN=1), R3.00-00 (SN=5)
27	42.67679	N/A (R2)	ISIS PSNP	LSP Entries: R3.00-00 (SN=5)
28	42.67679	N/A (R3)	ISIS PSNP	LSP Entries: R2.00-00 (SN=6)

RL = Remaining Lifetime, SN=A means SN=0x0000000A

FIGURE 10.45: Initial LSDB synchronization on IS-IS point-to-point links (2nd experiment) - Wireshark capture.

the previous experiment, routers R2 and R3 first exchange their self-originated Nonpseudonode-LSPs (packets 22 and 23). These LSPs are the only ones that change when interface s0/0 of R3 is switched off and on. The routers then exchange their LSDB summaries (packets 24 and 25). The summaries indicate that the LSDB is the same in both routers. Therefore, no router needs to request further LSPs to its neighbor. Finally, the two last PSNPs acknowledge the two Nonpseudonode-LSPs sent initially: packet 27 acknowledges the LSP sent in packet 22, and packet 28 acknowledges the LSP sent in packet 23.

10.5.2.3 Dealing with outdated self-originated LSPs

In this section, we compare the outcome of two experiments, designed to analyze the role of the checksum during the initial LSDB synchronization process, when a previous incarnation of a self-originated LSP has the same SN as the newly created one.

We use a procedure similar to Section 10.5.2.2. Upon network convergence, R3 is switched off and switched on again, such that when it is switched on it finds a previous incarnation of its self-originated Nonpseudonode-LSP (R3.00-00) at R2. We perform two experiments where we want to ensure that: (i) the SNs of the old and new instances are the same in both experiments and, (ii) in one experiment the checksums of the old and new instances are the same, but in the other experiment the checksums are different.

Experimental procedure (experiment 1) The first experiment leads to LSP instances with equal SNs and checksums. The procedure is the following: (i) switch on all routers at the same time and wait until the network converges; (ii) switch off router R3 and wait until the network converges again; (iii) install a Wireshark probe at the point-to-point link and start a capture with an **isis** filter; (iv) switch on router R3 again.

Experimental procedure (experiment 2) The second experiment is more complex and involves changing the startup-config while router R3 is switched off. The goal is to ensure that when R3 is switched on for the second time, it

No.	Time	Source	Protocol	Info
19	41.11884	N/A (R2)	ISIS LSP	LSP-ID: R2.00-00 (SN=6, RL: 1200s, Chk: 0x9ccc)
20	41.11935	N/A (R3)	ISIS LSP	LSP-ID: R3.00-00 (SN=2, RL: 1200s, Chk: 0x1450)
21	41.55336	N/A (R3)	ISIS CSNP	LSP Entries: R2.00-00 (SN=6, RL: 1198, Chk: 0x9ccc), R3.00-00 (SN=2, RL: 1199, Chk: 0x1450)
22	41.5549	N/A (R2)	ISIS CSNP	LSP Entries: R1.00-00 (SN=2, RL: 912, Chk: 0x2e4f), R2.00-00 (SN=6, RL: 1199, Chk: 0x9ccc), R2.01-00 (SN=1, RL: 914, Chk: 0x7cd6), R3.00-00 (SN=2, RL: 912, Chk: 0x1450)
25	42.08239	N/A (R2)	ISIS LSP	LSP-ID: R1.00-00, (SN=2, RL: 912s, Chk: 0x2e4f)
26	42.11363	N/A (R2)	ISIS LSP	LSP-ID: R2.01-00 (SN=1, RL: 914s, Chk: 0x7cd6)
27	42.11511	N/A (R3)	ISIS PSNP	LSP Entries: R1.00-00 (SN=2, RL: 910, Chk: 0x2e4f), R2.00-00 (SN=6, RL: 1198, Chk: 0x9ccc), R2.01-00 (SN=1, RL: 912, Chk: 0x7cd6)
28	42.14623	N/A (R2)	ISIS PSNP	LSP Entries: R3.00-00 (SN=2, RL: 911, Chk: 0x1450)

RL = Remaining Lifetime, Chk = Checksum, SN=A means SN=0x0000000A

FIGURE 10.46: Role of checksum in the IS-IS initial LSDB synchronization process (experiment 1) - Wireshark capture.

generates a Nonpseudonode-LSP with the same SN of the old LSP but with a different checksum. One way to achieve this is to modify the IP address assigned to interface s0/0. The procedure is then the following: (i) switch on all routers at the same time and wait until the network converges; (ii) switch off router R3 and wait until the network converges again; (iii) in the startup-config file of R3, modify the IP address of interface s0/0 from 222.222.10.3 to 222.222.10.4; (iv) install a Wireshark probe at the point-to-point link and start a capture with an **isis** filter; (v) switch on router R3 again.

Wireshark captures The Wireshark captures of the two experiments are shown in Figures 10.46 and 10.47. In relation to the original captures, we removed all HELLO packets. In the figures, we included information relative to the **Remaining Lifetime** and **Checksum** of the various LSP Entries TLVs, since this information is relevant for the interpretation of the experiments.

Outcome of experiment 1 We start by discussing experiment 1. The first two packets (packets 19 and 20) are the self-originated Nonpseudonode-LSPs of R2 and R3. The next two packets (packets 21 and 22) are the CSNPs that advertise the LSDB summaries of R2 and R3. R3 advertises the LSP it just created (R3.00-00) and the one just received from R2 in packet 19 (R2.00-00); R2 advertises four LSPs (R1.00-00, R2.00-00, R2.01-00, and R3-00.00). Based on the information provided in the CSNP of packet 21 (sent by R3), R2 learns that R3 is still missing R1.00-00 and R2.01-00 and sends to R3 these LSPs in packets 25 and 26. Packets 27 and 28 are the PSNPs acknowledging the reception of the previous LSPs.

Analyzing in detail R3.00-00 Let us now concentrate on R3.00-00. In

No.	Time	Source	Protocol	Info
18	41.068508	N/A (R3)	ISIS LSP	LSP-ID: R3.00-00 (SN=2, Lifetime=1200s, Checksum=0x283b)
19	41.072484	N/A (R2)	ISIS LSP	LSP-ID: R2.00-00 (SN=6, Lifetime=1200s, Checksum=0x9ccc)
20	41.518508	N/A (R2)	ISIS CSNP	LSP Entries: R1.00-00 (SN=2, Lifetime=822, Checksum=0x2e4f), R2.00-00 (SN=6, Lifetime=1199, Checksum=0x9ccc), R2.01-00 (SN=1, Lifetime=824, Checksum=0x7cd6), R3.00-00 (SN=2, Lifetime=0, Checksum=0xe415)
21	41.52051	N/A (R3)	ISIS CSNP	LSP Entries: R2.00-00 (SN=6, Lifetime=1198, Checksum=0x9ccc), R3.00-00 (SN=2, Lifetime=1199, Checksum=0x263c)
23	42.033861	N/A (R3)	ISIS LSP	LSP-ID: R3.00-00 (SN=3, Lifetime=1199s, Checksum=0x263c)
24	42.034864	N/A (R2)	ISIS LSP	LSP-ID: R1.00-00 (SN=2, Lifetime=821s, Checksum=0x2e4f)
25	42.068989	N/A (R2)	ISIS LSP	LSP-ID: R2.01-00 (SN=1, Lifetime=824s, Checksum=0x7cd6)
26	42.070959	N/A (R3)	ISIS PSNP	LSP Entries: R1.00-00 (SN=2, Lifetime=819, Checksum=0x2e4f), R2.00-00 (SN=6, Lifetime=1198, Checksum=0x9ccc), R2.01-00 (SN=1, Lifetime=822, Checksum=0x7cd6)
27	42.522566	N/A (R2)	ISIS PSNP	LSP Entries: R3.00-00 (SN=2, Lifetime=1197, Checksum=0x263c)

SN=A means SN=0x0000000A

FIGURE 10.47: Role of checksum in the IS-IS initial LSDB synchronization process (experiment 2) - Wireshark capture.

packet 20, R3 sends the instance of R3.00-00 it just created, with SN=2, `Remaining Lifetime` = 1200s, and `Checksum` = 0×1450; the `Remaining Lifetime` of this instance is the highest possible. Then, in the CSNP sent by R2 (packet 22), R2 advertises an instance of R3.00-00 with the same SN and `Checksum`, but a significantly lower `Remaining Lifetime` (922 seconds). This was the instance left at R2 when R3 was switched off. Notice that, in the CSNP of packet 22, the `Remaining Lifetime` of R2.00-00 is only 1 second older than the one of packet 19, but the `Remaining Lifetime` of R3.00-00 is 188 seconds older than the one of packet 20. Moreover, in the subsequent packets, R3 does not flood an instance of R3 with the SN incremented by one. In fact, upon receiving packet 20, R2 understands, based on the SN and the `Checksum` of R3.00-00, that the contents of the stored and incoming instances of R3.00-00 is exactly the same (they have the same freshness), and decides not to overwrite the stored instance. Moreover, when R3 receives the CSNP of R2 (packet 22), it learns, based again on the SN and the `Checksum`, that the instance of R3.00 stored at R2 is fully updated (the stored and incoming instances are equally fresh), and decides that there is no need to flood an additional instance.

Outcome of experiment 2 The outcome of experiment 2 is not the same. The starting point of experiment 2 is the same as experiment 1. Thus, R3 creates initially an instance of R3.00-00 with SN=2 and `Checksum` = 0×1450, and this is the instance that is left at R2 and R1 when R3 is switched off. When R3 is switched on again it creates immediately an instance of R3.00-00 with SN=2 (as in experiment 1) but with a different `Checksum`, because of the change made in the configuration of R3 while it was switched off; packet 18 reveals that the Checksum is now 0×283b. When R2 receives packet 18, it understands, based on the Checksum, that the stored and incoming instances of R3.00 are not the same: the Checksum of the stored instance is 0×1450 and the one of the incoming instance is 0×283b. As shown in the CSNP of packet 20, R2 then decides to expire this instance by setting its Remaining Lifetime to 0; it also modifies its Checksum to 0×e415. When R3 receives packet 20, it learns that the instance of R3.00 stored at R2 expired, and floods a new instance of that LSP with SN=3 (packet 23); because the SN is different than the one of packet 18, the Checksum is also different (it changed to 0×263c).

11

Experiments on Hierarchical Networks

Both OSPF and IS-IS networks can be structured in a two-level hierarchy of areas, where the upper-level can only have one area and the lower-level areas attach directly to the upper one and cannot attach to each other. This issue was addressed in Chapter 7. In this chapter we present several experiments that illustrate the various features of OSPF and IS-IS hierarchical networks. Section 11.1 explains the router configurations and the LSDB structure. Section 11.2 shows how the restrictions imposed on the shortest path selection process can lead to non-optimal routing. The experiments are carried out only with the IPv4 versions of the protocols, since the IPv6 versions have similar behavior.

11.1 LSDB structure

We analyze here the LSDB structure of hierarchical OSPF and IS-IS networks, containing area-internal, area-external, and domain-external addressing information. The experiments will be based on the network topology of Figure 11.1. The network includes two ASes connected through BGP (AS 100 and AS 200), and AS 200 includes two areas running hierarchical OSPF or IS-IS. The lower-level area is area 1 in both OSPF and IS-IS; the backbone is area 0 in OSPF and area 2 in IS-IS. Recall that in OSPF the backbone is necessarily area 0. Router R1 is internal to area 1; routers R2 and R3 are the ABRs between area 1 and the backbone; router R4 is internal to the backbone and, at the same time, is the DBR that communicates with router R5 (the DBR of area 100). AS 100 includes the prefix 150.150.0.0/16 that is to be injected in AS 200. The prefixes of AS 200 belong to the range 222.222.0.0/16. The OSPF interface costs are all 10 (the default value), except the ones of interface f0/0 of R1 (with cost 15) and interface f0/1 of R3 (with cost 30).

11.1.1 Router configurations

The configurations required for the support of hierarchical routing in OSPF and IS-IS are not very different from the basic routing configurations (see Section 8.4). Figure 11.2 shows the complete configurations of router R2.

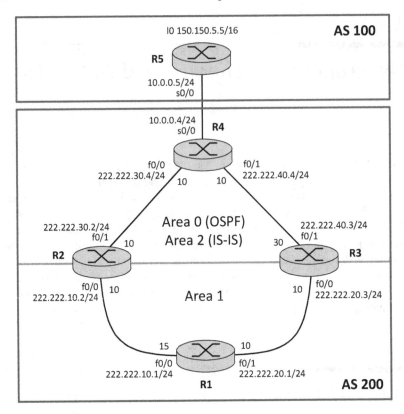

FIGURE 11.1: Experiments related to the LSDB structure of hierarchical networks - Network topology.

OSPF The router configurations of OSPF are very similar to the single-area ones. The only modifications are in the ABRs where, in the **network** command of the **router ospf** mode, each subnet must be declared in the area it belongs to. For example, in the configuration of R2 (Figure 11.2.a) subnet 222.222.10.0/24 is declared in area 1 and subnet 222.222.30.0/24 is declared in area 0.

IS-IS The IS-IS router configurations are slightly more complex than the OSPF ones. The area-internal routers must be configured as L1 or L2 routers, depending on the type of area they belong to. Declaring a router as L1 or as L2 in the **router isis** mode forces all router interfaces to become L1-only or L2-only, which is precisely what we desire. In our case, R1 must be configured as an L1 router (through the command **is-type level-1**) and R2 as an L2 router (through the command **is-type level-2-only**). The ABRs must be configured as L1/L2 routers. Cisco IOS assumes this configuration per default, so no specific configuration is needed in the **router isis** mode. However, an L1/L2 router has its interfaces configured as L1/L2 interfaces per default, and

```
!R2
configure terminal
interface f0/0
ip address 222.222.10.2 255.255.255.0
no shutdown
interface f0/1
ip address 222.222.30.2 255.255.255.0
(a) no shutdown
exit
router ospf 1
router-id 2.2.2.2
network 222.222.10.0 0.0.0.255 area 1
network 222.222.30.0 0.0.0.255 area 0
end
write
```

```
!R2
configure terminal
interface f0/0
ip address 222.222.10.2 255.255.255.0
ip router isis
isis circuit-type level-1
no shutdown
interface f0/1
ip address 222.222.30.2 255.255.255.0
ip router isis
(b) isis circuit-type level-2-only
no shutdown
exit
router isis
net 49.0001.0000.0000.0002.00
metric-style wide
redistribute isis ip level-2 into level-1 distribute-list 100
exit
access-list 100 permit ip any any
end
write
```

FIGURE 11.2: Router configurations for hierarchical routing in (a) OSPF and (b) IS-IS (router R2).

we need to configure the interfaces connected to L1 areas as L1-only interfaces, and the ones connected to the L2 area as L2-only interfaces. This is shown in Figure 11.2.b for the case of router R2: interface f0/0 is configured as L1-only through the command **isis circuit-type level-1**, and the f0/1 interface is configured as L2-only using the command **isis circuit-type level-2-only**.

Besides the above configurations, L1/L2 routers must be explicitly told to redistribute the prefixes of the L2 LSDB into the L1 LSDB. This is done through the command **redistribute isis ip level-2 into level-1 distribute-**

Router Link States (Area 1)					
Link ID	ADV Router	Age	Seq#	Checksum	Link count
1.1.1.1	1.1.1.1	451	0x80000003	0x004483	2
2.2.2.2	2.2.2.2	452	0x80000003	0x00176C	1
3.3.3.3	3.3.3.3	452	0x80000003	0x00BFA5	1

Net Link States (Area 1)				
Link ID	ADV Router	Age	Seq#	Checksum
222.222.10.2	2.2.2.2	452	0x80000001	0x0093C8
222.222.20.3	3.3.3.3	452	0x80000001	0x001F2A

Summary Net Link States (Area 1)				
Link ID	ADV Router	Age	Seq#	Checksum
222.222.30.0	2.2.2.2	483	0x80000002	0x00A0AD
222.222.30.0	3.3.3.3	452	0x80000001	0x00B17B
222.222.40.0	2.2.2.2	452	0x80000001	0x0098A2
222.222.40.0	3.3.3.3	493	0x80000001	0x00DE4E

Summary ASB Link States (Area 1)				
Link ID	ADV Router	Age	Seq#	Checksum
4.4.4.4	2.2.2.2	452	0x80000001	0x00EA2F
4.4.4.4	3.3.3.3	454	0x80000001	0x00956C

Type-5 AS External Link States					
Link ID	ADV Router	Age	Seq#	Checksum	Tag
150.150.0.0	4.4.4.4	441	0x80000001	0x000B07	100

FIGURE 11.3: LSDB structure of hierarchical OSPF - LSDB of area 1.

list 100 entered in the **router isis** mode (see Figure 11.2.b). This further requires the configuration of an access list to define which prefixes will be redistributed. With the command **access-list 100 permit ip any any** we allow the redistribution of all IPv4 prefixes.

11.1.2 OSPF LSDB structure

The LSDB structure of OSPF hierarchical networks was addressed in Section 7.3.1. The experiment of this section is based on the network topology of Figure 11.1. Figures 11.3 and 11.4 show the summaries of the LSDB of area 1 and area 0. These LSDBs can both be obtained at the ABRs using the command **show ip ospf database**.

Topological and area-internal addressing information Let us first

```
                    Router Link States (Area 0)

Link ID        ADV Router    Age    Seq#          Checksum      Link count
2.2.2.2        2.2.2.2       353    0x80000003    0x00E474      1
3.3.3.3        3.3.3.3       353    0x80000002    0x00EE3A      1
4.4.4.4        4.4.4.4       352    0x80000003    0x00D286      2

                    Net Link States (Area 0)

Link ID        ADV Router    Age    Seq#          Checksum
222.222.30.4   4.4.4.4       352    0x80000001 0x00DC55
222.222.40.4   4.4.4.4       352    0x80000001 0x00A083

                    Summary Net Link States (Area 0)

Link ID        ADV Router    Age    Seq#          Checksum
222.222.10.0   2.2.2.2       379    0x80000002 0x007DE4
222.222.10.0   3.3.3.3       348    0x80000001 0x00F758
222.222.20.0   2.2.2.2       348    0x80000001 0x0075D9
222.222.20.0   3.3.3.3       379    0x80000002 0x00F063

                    Type-5 AS External Link States

Link ID        ADV Router    Age    Seq#          Checksum      Tag
150.150.0.0    4.4.4.4       333    0x80000001 0x000B07        100
```

FIGURE 11.4: LSDB structure of hierarchical OSPF - LSDB of area 0.

concentrate on the topological and area-internal addressing information which, in OSPFv2, is provided by the Router-LSAs and Network-LSAs. The LSDB of each area only describes the topological elements of that area. The Network-LSAs of area 1 only describe the subnets 222.222.10.0/24 and 222.222.20.0/24, and the Network-LSAs of area 0 only describe 222.222.30.0/24 and 222.222.40.0/24. R4 has no Router-LSA in area 1, and R1 has no Router-LSA in area 0. Moreover, the Router-LSAs originated in one area only describe the interfaces that belong to that area. For example, R2 originates one Router-LSA in area 1 describing its interface f0/0, and originates another Router-LSA in area 0 describing its interface f0/1. Figure 11.5.a shows the complete contents of the Router-LSA originated by R2 in area 0. It has only one link description characterizing the interface with the shared link with prefix 222.222.10.0/24 (identified as Link connected to: a Transit Network).

Area-external addressing information The area-external prefixes are advertised using Network-Summary-LSAs. They can be seen in the Summary Net Link States part of the display (Figures 11.3 and 11.4). There are four LSAs of this type in each LSDB, two of them originated by R2 and the other two

(a)
```
Routing Bit Set on this LSA
LS age: 669
Options: (No TOS-capability, DC)
LS Type: Router Links
Link State ID: 2.2.2.2
Advertising Router: 2.2.2.2
LS Seq Number: 80000003
Checksum: 0x176C
Length: 36
Area Border Router
Number of Links: 1

  Link connected to: a Transit Network
  (Link ID) Designated Router address: 222.222.10.2
  (Link Data) Router Interface address: 222.222.10.2
    Number of TOS metrics: 0
     TOS 0 Metrics: 10
```

(b)
```
LS age: 928
Options: (No TOS-capability, DC, Upward)
LS Type: Summary Links(Network)
Link State ID: 222.222.30.0 (summary Network Number)
Advertising Router: 3.3.3.3
LS Seq Number: 80000001
Checksum: 0xB17B
Length: 28
Network Mask: /24
     TOS: 0  Metric: 40
```

(c)
```
LS age: 818
Options: (No TOS-capability, DC, Upward)
LS Type: Summary Links(AS Boundary Router)
Link State ID: 4.4.4.4 (AS Boundary Router address)
Advertising Router: 3.3.3.3
LS Seq Number: 80000001
Checksum: 0x956C
Length: 28
Network Mask: /0
     TOS: 0  Metric: 30
```

FIGURE 11.5: LSDB structure of hierarchical OSPF; (a) Router-LSA originated by R2 on area 1, (b) Network-Summary-LSA originated by R3 on area 1, and (c) ASBR-Summary-LSA originated by R3 on area 1.

by R3. The LSAs of area 1 describe the prefixes of area 0, and vice-versa. The LSAs also include the cost of the shortest intra-area path from the originating ABR to the advertised prefix. Figure 11.5.b shows the Network-Summary-LSA originated by R3 on area 0 to advertise prefix 222.222.30.0/24. The prefix is described in the Link State ID and Network Mask fields, and the path cost is included in the Metric field. In this case, the path cost is the sum of the

```
C      222.222.20.0/24 is directly connected, FastEthernet0/1
C      222.222.10.0/24 is directly connected, FastEthernet0/0
O IA   222.222.40.0/24 [110/35] via 222.222.10.2, 00:09:01, FastEthernet0/0
O IA   222.222.30.0/24 [110/25] via 222.222.10.2, 00:09:01, FastEthernet0/0
O E2   150.150.0.0/16 [110/1] via 222.222.10.2, 00:08:50, FastEthernet0/0
```

FIGURE 11.6: LSDB structure of hierarchical OSPF - Forwarding table of router R1.

interface costs of the f0/1 interface of R3 (which is 30) and the f0/0 interface of R4 (which is 10).

Domain-external routing information The advertisement of a domain-external prefix in OSPF hierarchical networks requires two LSA types, one to advertise the prefix and another to advertise the ASBR that injected the prefix into the domain. The prefix is advertised through AS-External-LSAs, which are flooded with domain flooding scope. This means that the LSAs are not modified by the ABRs and reach all routers as injected by the originating ASBR. This can be confirmed by the LSDBs of the two areas. The AS-External-LSA can be seen in the `Type-5 AS External Link States` part of the display. Notice that both LSDBs include the same AS-External-LSA, originated by R4 and advertising prefix 150.150.0.0/16.

As explained in Section 7.3.1, this LSA gives no indication on how to route to the ASBR outside the ASBR's area. This information is provided by the ASBR-Summary-LSA, which is disseminated as a distance vector within the ABR overlay. Thus, both R2 and R3 inject an LSA of this type in area 1. These LSAs can be seen in the `Summary ASB Link States` part of the display (Figure 11.3). Figure 11.5.c shows the ASBR-Summary-LSA originated by router R3. The ASBR identifier is included in the `Link State ID` field, and the path cost from the originating ABR to the ASBR (which is this case is 30) is included in the `Metric` field.

Forwarding table of R1 Finally, Figure 11.6 shows the forwarding table of router R1. The entries relative to area-external prefixes are identified through the keyword "IA" (for inter-area). Note that router R1 selects R2 as the outgoing ABR to reach all external destinations, since the next hop is 222.222.10.2 for all corresponding entries. The shortest path cost is 15+10=25 to 222.222.30.0/24 and to the ASBR, and is 15+10+10=35 to 222.222.40.0/24. The path cost to the same destinations via router R3 is 10+30=40 to 222.222.40.0/24 and to the ASBR, and is 10+30+10=50 to 222.222.30.0/24. The entry relative to the domain-external prefix (150.150.0.0/16) is identified by the keyword "E2", as in the case of single-area networks.

LSPID	LSP Seq Num	LSP Checksum	LSP Holdtime	ATT/P/OL
R1.00-00	0x00000004	0xCE58	1001	0/0/0
Area Address: 49.0001				
NLPID: 0xCC				
Hostname: R1				
IP Address: 222.222.20.1				
Metric: 15 IP 222.222.10.0/24				
Metric: 10 IP 222.222.20.0/24				
Metric: 10 IS-Extended R3.01				
Metric: 15 IS-Extended R2.01				
R2.00-00	* 0x00000005	0xA129	999	1/0/0
Area Address: 49.0001				
NLPID: 0xCC				
Hostname: R2				
IP Address: 222.222.10.2				
Metric: 10 IP 222.222.10.0/24				
Metric: 10 IS-Extended R2.01				
Metric: 10 IP-Interarea 150.150.0.0/16				
Metric: 10 IP-Interarea 222.222.30.0/24				
Metric: 20 IP-Interarea 222.222.40.0/24				
R2.01-00	* 0x00000001	0x66DA	801	0/0/0
Metric: 0 IS-Extended R2.00				
Metric: 0 IS-Extended R1.00				
R3.00-00	0x00000005	0xBEB7	996	1/0/0
Area Address: 49.0001				
NLPID: 0xCC				
Hostname: R3				
IP Address: 222.222.20.3				
Metric: 10 IP 222.222.20.0/24				
Metric: 10 IS-Extended R3.01				
Metric: 30 IP-Interarea 150.150.0.0/16				
Metric: 40 IP-Interarea 222.222.30.0/24				
Metric: 30 IP-Interarea 222.222.40.0/24				
R3.01-00	0x00000001	0x67D7	984	0/0/0
Metric: 0 IS-Extended R3.00				
Metric: 0 IS-Extended R1.00				

FIGURE 11.7: LSDB structure of hierarchical IS-IS - L1 LSDB.

11.1.3   IS-IS LSDB structure

The LSDB structure of IS-IS hierarchical networks was addressed in Section 7.3.2. The experiment of this section is based on the network topology of Figure 11.1. Figures 11.7 and 11.8 show the L1 and L2 LSDB of router R2. It can be obtained through the command **show isis database detail**.

The L1 LSDB includes the LSPs that describe the lower-level area, which are the Nonpseudonode-LSPs R1.00-00, R2.00-00, and R3.00-00, and the Pseudonode-LSPs R2.01-00 and R3.01-00. Likewise, the L2 LSDB includes

LSPID	LSP Seq Num	LSP Checksum	LSP Holdtime	ATT/P/OL
R2.00-00	*** 0x00000003**	**0x287A**	**985**	**0/0/0**
Area Address: 49.0001				
NLPID: 0xCC				
Hostname: R2				
IP Address: 222.222.30.2				
Metric: 10	IP 222.222.30.0/24			
Metric: 10	IS-Extended R4.01			
Metric: 10	IP 222.222.10.0/24			
Metric: 20	IP 222.222.20.0/24			
R3.00-00	**0x00000003**	**0x500D**	**990**	**0/0/0**
Area Address: 49.0001				
NLPID: 0xCC				
Hostname: R3				
IP Address: 222.222.40.3				
Metric: 30	IP 222.222.40.0/24			
Metric: 30	IS-Extended R4.02			
Metric: 25	IP 222.222.10.0/24			
Metric: 10	IP 222.222.20.0/24			
R4.00-00	**0x00000003**	**0xF5AB**	**844**	**0/0/0**
Area Address: 49.0002				
NLPID: 0xCC				
Hostname: R4				
IP Address: 222.222.40.4				
Metric: 10	IS-Extended R4.02			
Metric: 10	IS-Extended R4.01			
Metric: 0	IP 150.150.0.0/16			
Metric: 10	IP 222.222.30.0/24			
Metric: 10	IP 222.222.40.0/24			
R4.01-00	**0x00000001**	**0x9430**	**769**	**0/0/0**
Metric: 0	IS-Extended R4.00			
Metric: 0	IS-Extended R2.00			
R4.02-00	**0x00000001**	**0xA121**	**759**	**0/0/0**
Metric: 0	IS-Extended R4.00			
Metric: 0	IS-Extended R3.00			

FIGURE 11.8: LSDB structure of hierarchical IS-IS - L2 LSDB.

the LSPs that describe the backbone, i.e. the Nonpseudonode-LSPs R2.00-00, R3.00-00, and R4.00-00, and the Pseudonode-LSPs R4.01-00 and R4.02-00.

Topological information We first concentrate on the topological information, which is provided by the Extended IS Reachability TLVs (identified in the display through the keyword "IS-Extended" that follows the `Metric` value). The LSP IDs of the Pseudonode-LSPs reveal that the DIS in each shared link is always the router with higher SID, i.e. R2 in the R2-R1 link (LSP ID is R2.01-00), R3 in the R3-R1 link (LSP ID is R3.01-00), R4 in the R4-R2 link (LSP ID is R4.01-00), and R4 in the R4-R3 link (LSP ID is R4.02-00).

The topological information only describes the area it refers to. Note that the L1 LSDB does not include the descriptions of router R4 (R4.00-00) and of links R4-R2 (R4.01-00) and R4-R3 (R4.02-00). Likewise, the L2 LSDB does not include the descriptions of router R1 (R1.00-00) and of links R2-R1 (R2.01-00) and R3-R1 (R3.01-00). R3 and R2, since they are L1/L2 routers, have Nonpseudonode-LSPs in both the L1 and L2 LSDB. However, in the L1 LSDB only the interfaces with lower-level area are represented, and in the L2 LSDB only the ones with the backbone are represented. In the L1 LSDB, R2.00-00 describes only the interface with R2.01-00 and R3.00-00 describes only the interface with R3.01-00. Likewise, in the L2 LSDB, R2.00-00 describes only the interface with R4.01-00 and R3.00-00 describes only the interface with R4.02-00. Thus, as in the case of OSPF, the topological information is kept inside areas.

The ATT bit is used to indicate, inside a lower-level area, whether a router is an L1/L2 router. Note that this bit is set in the Nonpseudonode-LSPs originated by R2 (R2.00-00) and R3 (R3.00-00) on area 1 (Figure 11.7).

Addressing information The addressing information is identified through the keywords "IP" and "IP-Interarea" that follow the Metric value. Although this is not clear from the display, all prefixes are advertised through Extended IP Reachability TLVs (type 135), whether they are area-internal, area-external, or domain-external; you can check this using Wireshark. The keyword "IP-Interarea" is assigned to prefixes redistributed from the L2 LSDB into the L1 LSDB. These prefixes are the domain-external prefix 150.150.0.0/16, and the area-external prefixes 222.222.30.0/24 and 222.222.40.0/24. The TLVs that advertise them have the Up/Down bit set to 1. The Metric value is the path cost from the originating L1/L2 router to the destination. For example, R3 advertises a cost of 40 to 222.222.30.0/24, which is the cost of path R3 → R4 → 222.222.30.0/24 (cost of f0/1 interface of R3 + cost of f0/0 interface of R4). R2 advertises a path cost of 10 to the same subnet, since the path only includes the interface f0/1 of R2, which has a cost of 10.

The prefixes redistributed in the opposite direction, i.e. from the L1 LSDB into the L2 LSDB, are indistinguishable from the area-internal prefixes. For example, in the L2 LSDB the Nonpseudonode-LSP of R3 (Figure 11.8), advertises in the same way the area-external prefixes 222.222.10.0/24 and 222.222.20.0/24, with path costs 25 and 10, respectively, and the area-internal prefix 222.222.40.0/24, with path cost 30 (this path cost coincides with the cost of the f0/1 interface of R3). Note also that, as already pointed out for the case of single-area networks, a prefix assigned to a shared link is advertised by all routers attached to it. For example, the prefix 222.222.40.0/24 is advertised in the Nonpseudonode-LSP of R3 (with cost 30) and in the Nonpseudonode-LSP of R4 (with cost 10).

In the L1 LSDB (Figure 11.7), the area-internal and the external prefixes are distinguished by the Up/Down bit. The TLVs advertising area-internal prefixes (i.e. 222.222.10.0/24 and 222.222.20.0/24) have the Up/Down bit cleared.

Forwarding table of router R1 The forwarding table of router R1 is shown

```
C     222.222.20.0/24 is directly connected, FastEthernet0/1
C     222.222.10.0/24 is directly connected, FastEthernet0/0
i ia  222.222.40.0/24 [115/35] via 222.222.10.2, FastEthernet0/0
i ia  222.222.30.0/24 [115/25] via 222.222.10.2, FastEthernet0/0
i ia  150.150.0.0/16 [115/25] via 222.222.10.2, FastEthernet0/0
i*L1  0.0.0.0/0 [115/10] via 222.222.20.3, FastEthernet0/1
```

FIGURE 11.9: LSDB structure of hierarchical IS-IS - Forwarding table of router R1.

in Figure 11.9. The routes towards the various destinations are the same as OSPF (see Section 11.1.2). Router R2 is again the ABR used to reach all external destinations. The keyword "ia" indicates that the corresponding entry is an inter-area path. Note that there is no distinction between the area-external and domain-external prefixes. The last entry is a default route (0.0.0.0/0), which IS-IS always includes in the forwarding table of the internal routers of L1 areas when the area includes L1/L2 routers. The entry points to the closest L1/L2 router (in the shortest path sense), which in this case is R3.

11.2 Restrictions on shortest path selection

In this section, we illustrate the consequences of restricting the shortest path selection process. The restrictions are similar in OSPF and IS-IS, and we will show them using OSPFv2 as an example. As discussed in Section 7.4, the distance vectors originated by the ABRs cannot advertise inside an area (i) routes to area-internal destinations and (ii) routes to area-external destinations that cross that area. Moreover, (iii) when building the forwarding tables routers prefer the intra-area routes over the inter-area ones. These restrictions may sometimes lead to non-optimal path selection.

Experimental setup The experiment is based on the network topology of Figure 11.10. The network has two areas. Routers R1 and R2 are internal to area 1, and routers R3 and R4 are ABRs. Subnets 222.222.10.0/24, 222.222.20.0/24, and 222.222.30.0/24 belong to area 1, and subnet 222.222.40.0/24 belong to area 0. All interfaces have cost 10, except interface s0/0 of router R2, which must be configured with a cost of 100.

Experimental outcome The forwarding table of R2 (Figure 11.11.a) shows that R2 is not using the shortest path to 222.222.20.0/24. The selected path is via R1 (next hop is 222.222.10.1) with a cost of 110, but the shortest path is via R4 with a cost of 30. In fact, the ABRs (R3 and R4) are not allowed to advertise 222.222.20.0/24 in area 1 using Network-Summary-LSAs, even if they have information about it. Figure 11.11.b shows the complete contents

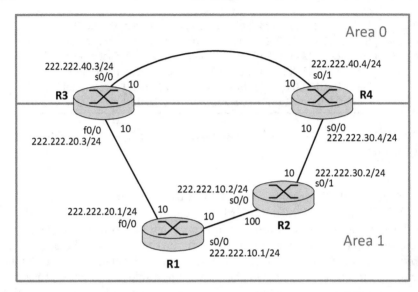

FIGURE 11.10: Experiments related to the restrictions on shortest path selection - Network topology.

of the Network-Summary-LSAs of area 0 referring to 222.222.20.0/24. These
LSAs are stored at both R3 and R4. The first LSA is originated by R3 and
indicates that R3 provides a cost of 10 to 222.222.20.0/24. The second LSA
indicates that R4 provides a cost of 120 to the same prefix. Thus, R4 knows
that it can provide a cost of 10+10=20 to 222.222.20.0/24, but is not allowed
to inject this information in area 1. This can be confirmed in Figure 11.11.c,
which shows summary information about the Network-Summary-LSAs of area
1 (obtained through the command **show ip ospf database summary**). It
can be seen that only the area-external prefix (222.222.40.0/24) is advertised
through these LSAs.

Notice also the decision taken by R4. As already mentioned, R4 has access
to the two routes from itself to 222.222.20.0/24. On one side, it is directly
attached to area 1 and knows, from the LSDB of this area, that the cost of
the intra-area path from itself to 222.222.20.0/24 is 120. On the other side,
R4 received a Network-Summary-LSA from R3 advertising a cost of 10 to
222.222.20.0/24 and, from this information, it determines that there is an
inter-area path to the prefix with cost 20. However, according to the forwarding table of Figure 11.11.d, R4 selects the path with the highest cost (via R2).
This is because, when building its forwarding table, the router considers first
the intra-area routes and, only if there is none, considers the inter-area routes.

Consequences of shutting down interface s0/0 of R2 Now, let us shutdown interface s0/0 of R2. We can do that by entering the command **shutdown** in the s0/0 interface. Figure 11.12 shows the resulting forwarding tables

```
     ┌─────────────────────────────────────────────────────────────────────┐
     │ O    222.222.20.0/24 [110/110] via 222.222.10.1, 00:02:46, Serial0/0  │
 (a) │ C    222.222.10.0/24 is directly connected, Serial0/0                 │
     │ O IA 222.222.40.0/24 [110/20] via 222.222.30.4, 00:02:46, Serial0/1   │
     │ C    222.222.30.0/24 is directly connected, Serial0/1                 │
     └─────────────────────────────────────────────────────────────────────┘
```

```
     ┌─────────────────────────────────────────────────────────────────────┐
     │ LS age: 308                                                           │
     │ Options: (No TOS-capability, DC, Upward)                              │
     │ LS Type: Summary Links(Network)                                       │
     │ Link State ID: 222.222.20.0 (summary Network Number)                  │
     │ Advertising Router: 3.3.3.3                                           │
     │ LS Seq Number: 80000001                                               │
     │ Checksum: 0xF262                                                      │
     │ Length: 28                                                            │
     │ Network Mask: /24                                                     │
     │     TOS: 0  Metric: 10                                                │
 (b) │                                                                       │
     │ LS age: 300                                                           │
     │ Options: (No TOS-capability, DC, Upward)                              │
     │ LS Type: Summary Links(Network)                                       │
     │ Link State ID: 222.222.20.0 (summary Network Number)                  │
     │ Advertising Router: 4.4.4.4                                           │
     │ LS Seq Number: 80000001                                               │
     │ Checksum: 0x25BD                                                      │
     │ Length: 28                                                            │
     │ Network Mask: /24                                                     │
     │     TOS: 0  Metric: 120                                               │
     └─────────────────────────────────────────────────────────────────────┘
```

	Link ID	ADV Router	Age	Seq#	Checksum
(c)	222.222.40.0	3.3.3.3	1266	0x80000001	0x00162B
	222.222.40.0	4.4.4.4	1256	0x80000001	0x00F745

```
     ┌─────────────────────────────────────────────────────────────────────┐
     │ O  222.222.20.0/24 [110/120] via 222.222.30.2, 00:03:21, Serial0/0    │
 (d) │ O  222.222.10.0/24 [110/110] via 222.222.30.2, 00:03:21, Serial0/0    │
     │ C  222.222.40.0/24 is directly connected, Serial0/1                   │
     │ C  222.222.30.0/24 is directly connected, Serial0/0                   │
     └─────────────────────────────────────────────────────────────────────┘
```

FIGURE 11.11: Restrictions on shortest path selection - (a) Forwarding table of R2, (b) Network-Summary-LSAs of area 0 relative to 222.222.20.0/24, (c) Summary information about the Network-Summary-LSAs of area 1, (d) Forwarding table of R4.

of routers R2 and R4, and the complete contents of the Network-Summary-LSA relative to 222.222.20.0/24 included in the LSDB of R2. When the interface is shutdown, area 1 becomes partitioned, and router R4 ceases to receive routing information relative to 222.222.20.0/24 via R2. R4 has only access to the inter-area path (via R3) and, as shown in Figure 11.12.b, installs this path in its forwarding table. Moreover, from the perspective of R4, 222.222.20.0/24 no longer belongs to a directly attached area. Thus, R4 is allowed to advertise the prefix to R2 (through its interface s0/0). The corresponding Network-Summary-LSA is shown in Figure 11.12.c. It is based

```
       O IA  222.222.20.0/24 [110/30] via 222.222.30.4, 00:07:29, Serial0/1
(a)    O IA  222.222.40.0/24 [110/20] via 222.222.30.4, 00:07:29, Serial0/1
       C     222.222.30.0/24 is directly connected, Serial0/1
```

```
       O IA  222.222.20.0/24 [110/20] via 222.222.40.3, 00:09:28, Serial0/1
(b)    C     222.222.40.0/24 is directly connected, Serial0/1
       C     222.222.30.0/24 is directly connected, Serial0/0
```

```
       Routing Bit Set on this LSA
       LS age: 489
       Options: (No TOS-capability, DC, Upward)
       LS Type: Summary Links(Network)
       Link State ID: 222.222.20.0 (summary Network Number)
(c)    Advertising Router: 4.4.4.4
       LS Seq Number: 80000001
       Checksum: 0x390E
       Length: 28
       Network Mask: /24
         TOS: 0  Metric: 20
```

FIGURE 11.12: Restrictions on shortest path selection, after shutting down interface s0/0 of R2 - (a) Forwarding table of R2, (b) Forwarding table of R4, (c) Network-Summary-LSA of R2 relative to 222.222.20.0/24.

on this LSA that R2 installs in its forwarding table an inter-area path to 222.222.20.0/24 with cost 30, as shown in Figure 11.12.a. Thus, the shortest path from R2 to 222.222.20.0/24 got finally installed when the intra-area path broke!

Glossary

ABR (Area Border Router): OSPF router located in the frontier between areas; called L1/L2 in IS-IS (page 20).

ABR overlay: logical network of ABRs used for the exchange of inter-area routing information (page 116).

ACK protection: protecting the transmission of a packet by forcing the receiver to acknowledge the correct reception of the packet (page 36).

address block: group of contiguous addresses characterized by a common prefix (page 8).

ACK (DELETE): generic designation of control message that acknowledges the reception of NR delete indications; see Figure 6.1 for correspondence with OSPF and IS-IS (page 101).

ACK (UPDATE): generic designation of control message that acknowledges the reception of NR instances; see Figure 6.1 for correspondence with OSPF and IS-IS (page 101).

addressing information: information on address prefixes, i.e. IPv4 or IPv6 prefixes (page 43).

adjacent neighbors: neighbors whose connection fulfilled all conditions to become part of the network map (page 188).

aep-NR (area-external-prefix-NR): generic designation of NR that describes an address prefix external to an area; see Figure 5.2 for correspondence with OSPF and IS-IS (page 122).

aip-NR (area-internal-prefix-NR): generic designation of NR that describes an address prefix internal to an area; see Figure 5.2 for correspondence with OSPF and IS-IS (page 70).

AR (Advertising Router): router that originates an LSA/LSP, and starts its flooding; same as originating router (page 148).

Area Addresses TLV: IS-IS TLV that lists all Area IDs assigned to the originating router (page 168).

Area ID (AID): identifier of an area (page 151).

area flooding scope: refers to the flooding of an NR/LSA/LSP only inside an area (page 120).

architectural constants: protocol parameters that are fixed by the specification and cannot be changed by the network manager (page 145).

AS (Autonomous System): administrative domain of the Internet, usually under the responsibility of a single administration (e.g. an ISP) and having its own routing policy; may include one or more routing domains (page 19).

ASBR (Autonomous System Border Router): router located in the frontier between Autonomous Systems (page 19).

ASBR-Summary-LSA: OSPFv2 LSA that describes an ASBR external to an area; equivalent to dbr-NR (page 242).

AS-External-LSA: OSPFv2 and OSPFv3 LSA that describes an address prefix external to a routing domain; equivalent to dep-NR (page 175).

backbone: upper-level area of hierarchical networks (page 143).

BDR (Backup Designated Router): router that replaces the DR if the DR fails, in OSPF (page 152).

cold start: starting a distributed algorithm by switching on all its elements simultaneously (page 35).

connectivity graph: graph of the connectivity expressed by the network map (page 58).

converged network: a network in a stable state (page 250).

CSNP (COMPLETE SEQUENCE NUMBER PDU): IS-IS control packet that transmits the summaries of all LSPs contained in an LSDB to a neighboring router; equivalent to LSDB SUMMARY (page 181).

DB DESCRIPTION: OSPF control packet that transmits the summaries of all LSAs contained in an LSDB to a neighboring router; equivalent to LSDB SUMMARY (page 181).

DBR (Domain Border Router): router located in the frontier between routing domains (page 19).

dbr-NR (domain-border-router-NR): generic designation of NR that describes a domain border router external to an area; see Figure 5.2 for correspondence with OSPF and IS-IS (page 138).

DELETE: generic designation of control message that transports NR delete indications to neighboring routers; see Figure 6.1 for correspondence with OSPF and IS-IS (page 101).

dep-NR (domain-external-prefix-NR): generic designation of NR that describes an address prefix external to a routing domain; see Figure 5.2 for correspondence with OSPF and IS-IS (page 82).

DIS (Designated Intermediate System): router elected on a shared link to represent the link (IS-IS designation) (page 152).

domain flooding scope: refers to the flooding of an NR/LSA/LSP through the whole routing domain (page 120).

DR (Designated Router): router elected on a shared link to represent the link (generic and OSPF designation) (page 56).

DVR (Distance Vector Routing): routing approach where routers compute their paths based on information provided by their neighbors and on their knowledge of the state of the links with those neighbors (page 129).

Dynamic Hostname TLV: IS-IS TLV that associates the SID with a name encoded in ASCII (page 169).

Extended IP Reachability TLV: IPv4 IS-IS TLV with extended metrics that represents address prefixes internal to an area (equivalent to aip-NR), address prefixes external to an area (equivalent to aep-NR), or address prefixes external to a routing domain (equivalent to dep-NR) (page 163).

Extended IS Reachability TLV: IS-IS (IPv4 and IPv6) TLV with extended metrics that describes the neighbors of a router or transit shared link (pseudonode); equivalent to router-NR when included in a Nonpseudonode-LSP, and to a slink-NR when included in a Pseudonode-LSP (page 163).

flooding scope: zone of the network where a control message is disseminated (page 99).

forwarding table (layer-3): table that indicates at a router how to forward packets to the next router interface or host interface (page 5).

fully adjacent neighbors: neighbors that have synchronized their LSDBs (page 188).

HELLO: generic (page 45), OSPF (page 180), and IS-IS (page 180) designation of control message transmitted periodically to neighboring routers, to establish and maintain neighborhood relationships; IS-IS has different types of HELLO packets: POINT-TO-POINT IS-IS HELLO (page 180), LAN IS-IS HELLO (page 180), L1 HELLO (page 189), and L2 HELLO (page 189); see Figure 6.1 for correspondence with OSPF and IS-IS.

hierarchical network: network structured in areas, where areas are organized in levels with a hierarchical relationship between them (page 143).

hosts: end devices which are the sources and sinks of information (e.g. laptops, tablets, servers, sensors, actuators) (page 3).

ingoing interface: part of an interface that receives packets (from link to equipment) (page 3).

Inter-Area-Prefix-LSA: OSPFv3 LSA that describes an address prefix external to an area; equivalent to aep-NR (page 241).

Intra-Area-Prefix-LSA: OSPFv3 LSA that describes an address prefix internal to an area; equivalent to aip-NR (page 171).

inter-area path: a path that crosses more than one area (page 134).

Inter-Area-Router-LSA: OSPFv3 LSA that describes an ASBR external to an area; equivalent to dbr-NR (page 242).

inter-area routing protocol: routing protocol that runs on area border routers to determine paths between areas (page 20).

inter-AS routing protocol: routing protocol that runs on ASBRs to determine paths between ASes (BGP is the de facto standard for this type of protocol) (page 19).

inter-domain routing protocol: routing protocol that runs on domain border routers to determine paths between routing domains (page 19).

interface acceptance rules: rules that define if a control packet can be accepted by a receiving interface (page 184).

intra-area path: a path that crosses only network elements of one area (page 134).

intra-domain routing protocol: routing protocol that runs inside a routing domain to determine paths within the domain (e.g. OSPF and IS-IS) (page 19).

IOS (Internetwork Operating System): operating system of Cisco routers (page 253).

IP Interface Address TLV: IS-IS TLV that lists the IPv4 addresses assigned to the IS-IS interfaces of the originating router (page 169).

IPv6 Interface Address TLV: IS-IS TLV that lists the IPv6 addresses assigned to the IS-IS interfaces of the originating router (page 169).

IP External Reachability Information TLV: IPv4 IS-IS TLV that describes an address prefix external to a routing domain; equivalent to dep-NR (page 175).

IP Internal Reachability Information TLV: IPv4 IS-IS TLV that represents address prefixes internal to an area (equivalent to aip-NR) or address prefixes external to an area (equivalent to aep-NR) (page 163).

IPv6 Reachability TLV: IPv6 IS-IS TLV that represents address prefixes internal to an area (equivalent to aip-NR), address prefixes external to an area (equivalent to aep-NR), or address prefixes external to a routing domain (equivalent to dep-NR) (page 163).

IS Neighbors TLV (type 2): IS-IS (IPv4 and IPv6) TLV that describes the neighbors of a router or transit shared link (pseudonode); equivalent to router-NR when included in a Nonpseudonode-LSP, and to a slink-NR when included in a Pseudonode-LSP (page 163).

IS Neighbors TLV (type 6): IS-IS TLV that describes neighboring interfaces on shared links, using their MAC addresses (page 186).

L1/L2 router: IS-IS router that supports L1-only, L2-only, and L1/L2 interfaces, and is located in the border between areas; called ABR in OSPF (page 145).

L1-only interface: interface of IS-IS router that only transmits/receives L1 control packets (page 144).

L2-only interface: interface of IS-IS router that only transmits/receives L2 control packets (page 144).

L1/L2 interface: interface of IS-IS router that transmits/receives both L1 and L2 control packets (page 144).

L1 router: IS-IS router that supports only L1-only interfaces (page 145).

L2 subdomain: IS-IS upper-level area, also called backbone (page 143).

L2 router: IS-IS router that supports only L2-only interfaces (page 145).

LAN ID: field of IS-IS LAN HELLO packets that identifies the DIS, using its SID and Pseudonode ID (page 201).

la-NR (link-address-NR): generic designation of NR that describes addresses provided by the next-hop router for communications within the link; see Figure 5.2 for correspondence with OSPF and IS-IS (page 77).

layer-2 link: logical link between layer-3 devices (hosts or routers); can correspond to a point-to-point link or a relatively complex layer-2 network (page 13).

layer-2 switch: switching equipment that forwards packets according to layer-2 addresses (e.g. MAC addresses) (page 14).

link information: information on the addresses used to transport packets between neighboring routers (page 43).

link flooding scope: refers to the flooding of an NR/LSA/LSP only within a link (page 77).

Link-LSA: OSPFv3 LSA that describes addresses provided by the next-hop router for communications within the link; equivalent to la-NR (page 177).

LINK STATE PDU (LSP packet): IS-IS control packet that transmits an LSP instance to a neighboring router; equivalent to UPDATE and not to be confused with LSP record (page 181).

LSA (Link State Advertisement): element of an OSPF LSDB, which contains elementary routing information and is indivisible from a flooding perspective; equivalent to NR (page 148).

LS ACKNOWLEDGMENT: OSPF control packet that acknowledges the reception of LSA instances; equivalent to ACK (UPDATE) (page 181).

LSA ID: identifier of LSA (page 155).

LSDB (Link State Database): database that stores the routing information (page 43).

LSDB SUMMARY: generic designation of control message that transmits the summaries of all NRs contained in an LSDB to neighboring routers; see Figure 6.1 for correspondence with OSPF and IS-IS (page 108).

LSDB SUMMARY REQUEST: generic designation of control message that requests NRs from neighboring routers; see Figure 6.1 for correspondence with OSPF and IS-IS (page 108).

LSP (Link State PDU): element of the IS-IS LSDB, which is a container of TLVs, and is indivisible from a flooding perspective; not to be confused with LSP packet (page 148).

LSP Entries TLV: IS-IS TLV that advertises LSP summaries in PSNPs and CSNPs control packets (page 183).

LSP ID: identifier of LSP, consisting of SID, Pseudonode ID, and LSP Number (page 156).

LSP Number: field that uniquely identifies an LSP fragment, part of the LSP ID (page 156).

LSR (Link State Routing): routing approach where routers exchange information to build and maintain a map of the complete network and on the available address prefixes (page 21).

LS REQUEST: OSPF control packet that requests LSAs from neighboring routers; equivalent to PARTIAL LSDB REQUEST (page 181).

LS UPDATE: OSPF control packet that transmits LSA instances to neighboring routers; equivalent to UPDATE (page 181).

NET (Network Entity Title): complete identifier of an IS-IS router, which includes its area identifier (page 151).

network map: part of the LSDB that represents the network topology (page 43).

Network-LSA: OSPFv2 and OSPFv3 LSA that describes a transit shared link; in OSPFv2 also describes prefix assigned to a transit shared link; equivalent to slink-NR and, in OSPFv2, also equivalent to aip-NR (page 160).

Network-Summary-LSA: OSPFv2 LSA that describes an address prefix external to an area; equivalent to aep-NR (page 240).

Node ID: identifier of a neighbor in IS-IS, comprising the SID and the Pseudonode ID (page 164).

Nonpseudonode-LSP: LSP that describes a router, its links, and the address prefixes assigned to the router and its links (page 148).

NR (Network Record): generic designation of a piece of the LSDB; equivalent to LSA in OSPF and TLV in IS-IS, from a routing information perspective (page 50).

originating router: router responsible for creating, updating, deleting, and disseminating an NR/LSA/LSP (page 50).

outgoing interface: part of an interface that transmits packets (from equipment to link) (page 3).

PARTIAL LSDB REQUEST: generic designation of control message that requests NR instances from a neighboring router; see Figure 6.1 for correspondence with OSPF and IS-IS (page 108).

prefix: higher-order (leftmost) bits of an address (page 8).

Protocols Supported TLV: IS-IS TLV that identifies the network layer protocols supported by the originating router (page 168).

point-to-point link: link that connects two, and only two, layer-3 devices (page 31).

Point-to-Point Three-Way Adjacency TLV: IS-IS TLV that advertises the state of a point-to-point link interface, as being Down, Initializing or Up (page 192).

Pseudonode ID: local tag assigned by the DIS to differentiate among transit shared links, which is part of the shared link identifier and of the LSP ID (page 152).

Pseudonode-LSP: LSP that describes a transit shared link, topologically (page 148).

PSNP (PARTIAL SEQUENCE NUMBER PDU): IS-IS control packet used in the acknowledgment (point-to-point links) and the request of LSP instances (shared links); equivalent to ACK (UPDATE) and PARTIAL LSDB REQUEST (page 181).

RID (Router ID): identifier of OSPF router (page 149).

router: switching equipment that forwards packets according to layer-3 addresses (e.g. IPv4 and IPv6 addresses) (page 3).

Router-LSA: OSPFv2 and OSPFv3 LSA that describes a router and its attached links; in OSPFv2 it also describes prefixes assigned to the router and its links; equivalent to router-NR and, in OSPFv2, also equivalent to aip-NR (page 157).

router-NR: generic designation of NR that describes a router and its links; see Figure 5.2 for correspondence with OSPF and IS-IS (page 64).

routing domain: a domain belonging to a single administration and running a single routing protocol (page 19).

routing information: information required to build the router forwarding tables (page 43).

routing protocol: distributed algorithm that runs on switching equipment and cooperates through the exchange of control messages to determine paths within a network (page 7).

separation principle: separation between the various types of routing information present at the LSDB—a good design practice (page 68).

shared link: link that abstracts a layer-2 network and can potentially connect many layer-3 devices (page 31).

SID (System ID): identifier of IS-IS router of specific area (page 151).

slink-NR (shared-link-NR): generic designation of NR that describes a shared link; see Figure 5.2 for correspondence with OSPF and IS-IS (page 64).

SN (Sequence Number): integer number that is incremented sequentially to express the freshness of NRs/LSAs/LSPs (page 96).

stub shared link: shared link with only one attached router (page 32).

subnet: logical network gathering a block of contiguous IP addresses sharing a common prefix and assigned to interfaces that can communicate among themselves without the intervention of a router (page 15).

Stop-and-Wait (SW) protocol: error control protocol where the reception of a message must be acknowledged, and transmitting the next message is only possible after receiving the acknowledgment of the previous one (page 36).

switching equipment: equipment that transfers packets from ingoing to outgoing interfaces according to forwarding tables (e.g. routers and layer-2 switches) (page 3).

TLV (Type-Length-Value): variable length record that includes information on its type and length besides the actual contents (page 156).

topological elements: the elements that define the network topology, i.e. the routers and the links (page 51).

topological identifiers: identifiers of routers and links (page 68).

topological information: information on the network topology, i.e. on the routers and links between routers (page 43).

transit shared link: shared link with more than one attached router (page 32).

UPDATE: generic designation of control message that transports NR instances (full contents) to neighboring routers; see Figure 6.1 for correspondence with OSPF and IS-IS (page 101).

References

[1] Information technology Telecommunications and information exchange between systems Intermediate System intra-domain routeing information exchange protocol for use in conjunction with the protocol for providing the connectionless-mode network service (ISO 8473). ISO/IEC 10589, November 2002.

[2] IP Routing: ISIS Configuration Guide, Cisco IOS XE Release 3SE (Catalyst 3650 Switches). Cisco Systems, 2013.

[3] P. Albitz and C. Liu. *DNS and BIND*. O'Reilly Media, 5th edition, 2009.

[4] R. Albrightson, J. Garcia-Luna-Aceves, and J. Boyle. EIGRP-A fast routing protocol based on distance vectors. In *Interop 94*, 1994.

[5] D. Allan and N. Bragg. *802.1aq Shortest Path Bridging Design and Evolution: The Architect's Perspective*. Wiley, 2012.

[6] D. Bertsekas and R. Gallager. *Data Networks*. Prentice-Hall, 2nd edition, 1992.

[7] R. Callon. Use of OSI IS-IS for Routing in TCP/IP and Dual Environments. RFC 1195, December 1990.

[8] R. Coltun, D. Ferguson, J. Moy, and A. Lindem. OSPF for IPv6. RFC 5340, July 2008.

[9] D. Comer. *Internetworking with TCP/IP: Principles, Protocols, and Architectures*. Prentice Hall, 4th edition, 2000.

[10] J. Day. *Patterns in Network Architecture: A Return to Fundamentals*. Prentice Hall, 2008.

[11] J. Doyle. *OSPF and IS-IS: Choosing an IGP for Large-Scale Networks*. Addison Wesley, 2006.

[12] B. Edgeworth, A. Foss, and R. Garza Rios. *IP Routing on Cisco IOS, IOS XE, and IOS XR: An Essential Guide to Understanding and Implementing IP Routing Protocols*. Cisco Press, 2005.

[13] J. Garcia-Lunes-Aceves. Loop-free routing using diffusing computations. *IEEE/ACM Transactions on Networking*, 1(1):130–141, February 1993.

[14] R. Graziani. *IPv6 Fundamentals: A Straighforward Approach to Understanding IPv6.* Cisco Press, 2013.

[15] H. Gredler and W. Goralski. *The Complete IS-IS Routing Protocol.* Springer, 2005.

[16] A. Gurtov. *Host Identity Protocol (HIP): Towards the Secure Mobile Internet.* Wiley, 2008.

[17] S. Hagen. *IPv6 Essentials.* O'Reilly, 2nd edition, 2006.

[18] S. Halabi. *Internet Routing Architectures.* Cisco Press, 2nd edition, 2001.

[19] J. Harrison, J. Berger, and M. Bartlett. IPv6 Traffic Engineering in IS-IS. RFC 6119, February 2011.

[20] C. Hopps. Routing IPv6 with IS-IS. RFC 5308, October 2008.

[21] K. Ishiguro, V. Manral, A. Davey, and A. Lindem. Traffic Engineering Extensions to OSPF Version 3, September 2008.

[22] A. Johnston. *SIP: Understanding the Session Initiation Protocol.* Artech House, 3rd edition, 2009.

[23] D. Katz, K. Kompella, and D. Yeung. Traffic Engineering (TE) Extensions to OSPF Version 2. RFC 3630, September 2003.

[24] D. Katz and R. Saluja. Three-Way Handshake for Intermediate System to Intermediate System (IS-IS) Point-to-Point Adjacencies. RFC 3373, September 2002.

[25] D. Katz, R. Saluja, and D. Eastlake 3rd. Three-Way Handshake for IS-IS Point-to-Point Adjacencies. RFC 5303, October 2008.

[26] Z. Kou, F. Yang, and H. Ma. Update to OSPF Hello procedure. Internet draft, December 2006.

[27] J. Kurose and K. Ross. *Computer Networking: A Top-Down Approach.* Addison Wesley, 6th edition, 2013.

[28] T. Li, T. Przygienda, and H. Smit. Domain-wide Prefix Distribution with Two-Level IS-IS. RFC 2966, February 2000.

[29] T. Li and H. Smit. IS-IS Extensions for Traffic Engineering. RFC 5305, October 2008.

[30] G. Malkin. RIP Version 2. RFC 2453, November 1998.

[31] A. Martey. *IS-IS Network Design Solutions.* Cisco Press, 2002.

[32] J. McQuillan, I. Richer, and E. Rosen. The New Routing Algorithm for the ARPANET. *IEEE Transactions on Communications*, 28(5), May 1980.

[33] D. Medhi and K. Ramasamy. *Network Routing Algorithms, Protocols, and Architectures*. Morgan Kaufmann, 2007.

[34] J. Moy. Multicast Extensions to OSPF. RFC 1584, March 1994.

[35] J. Moy. *Anatomy of an Internet Routing Protocol*. Addison Wesley, 1998.

[36] J. Moy. OSPF Version 2. RFC 2328, April 1998.

[37] J. Moy. *OSPF Complete Implementation*. Addison Wesley, 2001.

[38] C. Murthy and B. Manoj. *Ad Hoc Wireless Networks Architectures and Protocols*. Prentice-Hall, 2004.

[39] T. Narten, E. Nordmark, and W. Simpson. Neighbor Discovery for IP Version 6 (IPv6). RFC 2461, December 1998.

[40] R. Perlman. Fault-tolerant broadcast of routing information. *Computer Networks*, 7:395–405, 1983.

[41] R. Perlman. *Interconnections - Bridges, Routers, Switches, and Internetworking Protocols*. Addison Wesley, 2000.

[42] L. Peterson and B. Davie. *Computer Networks: a Systems Approach*. Morgan Kaufmann, 5th edition, 2012.

[43] D. Plummer. An Ethernet Address Resolution Protocol. RFC 826, RFC Editor, November 1982.

[44] T. Przygienda, N. Shen, and N. Sheth. M-ISIS: Multi Topology (MT) Routing in Intermediate System to Intermediate Systems (IS-ISs). RFC 5120, February 2008.

[45] P. Psenak, S. Mirtorabi, A. Roy, L. Nguyen, and P. Pillay-Esnault. Multi-Topology (MT) Routing in OSPF. RFC 4915, June 2007.

[46] Y. Rekhter, T. Li, and S. Hares. A Border Gateway Protocol 4 (BGP-4). RFC 4271, January 2006.

[47] J. Saltzer. On the Naming and Binding of Network Destinations. RFC 1498, August 1993.

[48] N. Shen and H. Smit. Dynamic Hostname Exchange Mechanism for IS-IS. RFC 2763, February 2000.

[49] J. Shoch. Internetwork naming, addressing, and routing. In *IEEE Proceedings COMPCON*, pages 72–79, 1978.

[50] T. Thomas. *OSPF Network Design Solutions*. Cisco Press, 2nd edition, 2003.

[51] R. Valadas. OSPF extension for the support of multi-area networks with arbitrary topologies. arXiv 1704.08916, April 2017.

[52] I. van Beijnum. *BGP*. O'Reilly, 2002.

Index